# Lecture Notes in Bioinformatics 5688

Edited by S. Istrail, P. Pevzner, and M. Waterman

Editorial Board: A. Apostolico  S. Brunak  M. Gelfand
T. Lengauer  S. Miyano  G. Myers  M.-F. Sagot  D. Sankoff
R. Shamir  T. Speed  M. Vingron  W. Wong

Subseries of Lecture Notes in Computer Science

T0142093

Pierpaolo Degano   Roberto Gorrieri (Eds.)

# Computational Methods in Systems Biology

7th International Conference, CMSB 2009
Bologna, Italy, August 31- September 1, 2009
Proceedings

 Springer

Volume Editors

Pierpaolo Degano
Università di Pisa, Dipartimento di Informatica
Largo Bruno Pontecorvo, 3, 56127, Pisa, Italy
E-mail: degano@di.unipi.it

Roberto Gorrieri
Università degli Studi di Bologna
Dipartimento di Scienze dell'Informazione
Mura Anteo Zamboni 7, 40127 Bologna, Italy
E-mail: gorrieri@cs.unibo.it

Library of Congress Control Number: 2009932359

CR Subject Classification (1998): J.3, I.6, F.1.1, F.4.1, G.2.2

LNCS Sublibrary: SL 8 – Bioinformatics

ISSN      0302-9743
ISBN-10   3-642-03844-1 Springer Berlin Heidelberg New York
ISBN-13   978-3-642-03844-0 Springer Berlin Heidelberg New York

Typesetting: Camera-ready by author, data conversion by Scientific Publishing Services, Chennai, India
Printed on acid-free paper      SPIN: 12735143      06/3180      5 4 3 2 1 0

# Preface

This volume contains the proceedings of the 7th Conference on Computational Methods in Systems Biology (CMSB 2009), held in Bologna, from August 31 to September 1, 2009.

The first CMSB was held in Trento in 2003, bringing together life scientists, computer scientists, engineers and physicists. The goal was to promote the convergence of different disciplines aiming at a new understanding and description of biological systems, firmly ground in formal models, supported by computational languages and tools, and offering new methods of analysis. The conference then moved to Paris in 2004, Edinburgh in 2005, Trento in 2006, Edinburgh in 2007 and Rostock/Warnemünde in 2008.

This year the conference attracted about 45 submissions form 18 countries, mainly from Europe and North America, but also from Asia and Australia. We wish to thank all authors for their interest in CMSB 2009. After careful discussions, the Programme Committee eventually selected 18 papers for presentation at the conference. Each of them was accurately refereed by at least three reviewers, who delivered detailed and insightful comments and suggestions. The Conference Chairmen warmly thank all the members of the Programme Committee and all their sub-referees for the excellent support they gave, as well as for the friendly and constructive discussions. We also would like to thank the authors for having revised their papers to address the comments and suggestions by the referees.

This year we also had a poster session, hosting ten short presentations, each refereed by at least two members of the Programme Committee. We would like to thank their authors for the bright presentations of their work-in-progress. The abstracts of these posters are collected in a separate Technical Report, available at http://compass2.di.unipi.it/TR/Files/TR-09-09.pdf.gz.

The conference programme was enriched by the outstanding invited talks of Rita Casadio, John K. Heath and Corrado Priami, whom we warmly thank. Their contributions are also included here.

The conference this year was jointly organized with CMSB 2009, emphasizing the close connections and similarities between concurrent, artificial systems, and biological, natural systems. The joint invited talk by Corrado Priami further illustrated this correspondence.

We would like to thank all the people who contributed to the organization of CMSB 2009, and the generous support from the Alma Mater Studiorum – Università degli Studi di Bologna and from Microsoft Research Cambridge. We are also grateful to Andrei Voronkov, who allowed us to use the wonderful free conference software system EasyChair, which we used for the electronic submission of papers, the refereeing process and the Programme Committee work.

Pierpaolo Degano
Roberto Gorrieri

# Organization

## Steering Committee

| | |
|---|---|
| Finn Drabløs | Norwegian University of Science and Technology, Trondheim, Norway |
| François Fages | INRIA Rocquencourt, France |
| David Harel | Weizmann Institute of Science, Israel |
| Monika Heiner | TU Cottbus, Germany |
| Michael Hörnquist | Linköping University, Sweden |
| Satoru Miyano | University of Tokyo, Japan |
| Gordon Plotkin | University of Edinburgh, UK |
| Corrado Priami | MSR - University of Trento Centre for Comp. and Systems Biology, Italy |
| Adelinde M. Uhrmacher | University of Rostock, Germany |

## Programme Committee

| | |
|---|---|
| Pierpaolo Degano | Università di Pisa, Italy (Co-chair) |
| Finn Drabløs | Norwegian University of Science and Technology, Norway |
| François Fages | INRIA Rocquencourt, France |
| Stephen Gilmore | University of Edinburgh, UK |
| Roberto Gorrieri | Università di Bologna, Italy (Co-chair) |
| Monika Heiner | TU Cottbus, Germany |
| Adaoha Elizabeth C. Ihekwaba | CoSBI, Italy |
| Marta Kwiatkowska | University of Oxford, UK |
| Pietro Liò | Computing Lab Cambridge, UK |
| Satoru Miyano | University of Tokyo, Japan |
| Mark van Rossum | University of Edinburgh, UK |
| Grzegorz Rozenberg | Leiden University, The Netherlands |
| Carolyn Talcott | Stanford Research Institute, USA |
| Adelinde M. Uhrmacher | University of Rostock, Germany |

## Organizing Committee

| | |
|---|---|
| Cinzia Di Giusto | Università di Bologna, Italy |
| Roberto Gorrieri | Università di Bologna, Italy |
| Cristian Versari | Università di Bologna, Italy (Chair) |
| Antonio Vitale | Università di Bologna, Italy |
| Gianluigi Zavattaro | Università di Bologna, Italy |

## Additional Reviewers

Marco Aldinucci
Grégory Batt
Chiara Bodei
Luca Bortolussi
Andrea Bracciali
Mario Bravetti
Matteo Cavaliere
Steven Eker
Vashti Galpin
Cinzia Di Giusto
Maria Luisa Guerriero
Russ Harmer
Jane Hillston
Espen Højsgaard
Mathias John

Roberto Larcher
Chen Li
Fei Liu
Laurence Loewe
Vittorio Maniezzo
Roberto Marangoni
Radu Mardare
Carsten Maus
Richard Mayr
Emanuela Merelli
Paolo Milazzo
Ivan Mura
Masao Nagasaki
Gethin Norman
Dave Parker

Carla Piazza
Alberto Policriti
Davide Prandi
Christian Rohr
Martin Schwarick
Teppei Shimamura
Heike Siebert
Joris Slegers
Sylvain Soliman
Ashish Tiwari
Cristian Versari
Antonio Vitale

# Table of Contents

# Prediction of Protein-Protein Interacting Sites: How to Bridge Molecular Events to Large Scale Protein Interaction Networks

Lisa Bartoli[1], Pier Luigi Martelli[1], Ivan Rossi[2], Piero Fariselli[1], and Rita Casadio[1]

[1] Biocomputing Group, University of Bologna, CIRB/Department of Biology,
Via San Giacomo 9/2, Bologna, Italy
[2] BioDec s.r.l., Casalecchio di Reno, Bologna, Italy
{casadio,lisa,gigi,ivan,piero}@biocomp.unibo.it
http://www.biocomp.it

**Abstract.** Most of the cellular functions are the result of the concerted action of protein complexes forming pathways and networks. For this reason, efforts were devoted to the study of protein-protein interactions. Large-scale experiments on whole genomes allowed the identification of interacting protein pairs. However residues involved in the interaction are generally not known and the majority of the interactions still lack a structural characterization. A crucial step towards the deciphering of the interaction mechanism of proteins is the recognition of their interacting surfaces, particularly in those structures for which also the most recent interaction network resources do not contain information. To this purpose, we developed a neural network-based method that is able to characterize protein complexes, by predicting amino acid residues that mediate the interactions. All the Protein Data Bank (PDB) chains, both in the unbound and in the complexed form, are predicted and the results are stored in a database of interaction surfaces (http://gpcr.biocomp.unibo.it/zenpatches). Finally, we performed a survey on the different computational methods for protein-protein interaction prediction and on their training/testing sets in order to highlight the most informative properties of protein interfaces.

**Keywords:** protein-protein interaction site, neural network, patch smoothing, interacting surfaces.

## 1 Introduction

The analysis of protein-protein interaction surfaces has a long history, dating back to the seventies. Chotia and Janin [1] first analyzed a small number of structures to highlight some principles of protein-protein recognition. Much later Thornton and co-workers [2] focused on the features of patches of interacting residues within homo-dimers, showing how features as solvation potential, residue interface propensity, hydrophobicity, planarity, protrusion and accessible surface area were good candidates for the description of protein interfaces. With the increasing number of proteins known with atomic resolution in the Protein Data Bank (PDB, http://www.rcsb.org/pdb/ home/home.do), an increasing number of efforts were devoted to the issue of

P. Degano and R. Gorrieri (Eds.): CMSB 2009, LNBI 5688, pp. 1–17, 2009.
© Springer-Verlag Berlin Heidelberg 2009

extracting basic features of interacting protein complexes with the aim of predicting possible protein-protein interactions [for recent reviews see 3-5]. This became particularly urgent when it was recognized that protein-protein interactions were at the basis of a new paradigm of interpreting metabolic pathways. Indeed, interactions between proteins are the core of almost all the biochemical processes that mediate cellular function [6]. Large-scale experiments on whole genomes allowed the identification of many interacting protein pairs but the residues involved in these interactions are generally not known and the vast majority of the interactions remain to be structurally characterized. The structural characterization of protein-protein complexes remains an expensive and time consuming process and it is particularly problematic for transient complexes (see below).

Molecular recognition in protein complexes can be predicted with docking techniques, based on the notions of surface complementarity and electrostatics rules, and fitting together the surfaces of two or more known structures. Significant advances have been achieved in this field: however, the molecular forces involved in the complex association process are not completely understood and the methods are also influenced by the conformational changes that often take place upon protein-protein binding [7].

A possible alternative for characterizing protein complexes is a thorough study of protein complexes known with atomic resolution and the implementation of methods capable of inferring from known examples the general rules for predicting the propensity of the amino acid residues to mediate protein-protein interactions. It is a common believe that residues present at the protein interfaces should be easier to predict provided that its distinguishing features are known [8]. With this underlying assumption many studies have attempted to characterise the residues in protein-protein interfaces and from this to highlight all those rules that may help in the development of predictive methods. In the following we will describe features of interacting protein complexes as derived from structural data bases, the-state of the art methods that have been differently trained on these features and our approach. We will also describe the results of some experiments trying to emphasize the role of data bases in getting the evaluation scores of the predictive methods.

## 2   General Features of Interacting Protein Complexes

Earlier works were restricted to a limited subset of oligomeric proteins in the PDB while more recent works have been able to extend the analysis to a larger dataset of protein structures that became available in the most recent PDB releases. When in vitro studies are also available, one possibility is to take into consideration the strength of the interaction, whether the interaction is transient or not and whether or not the complexes are formed by copies of the same chain. These later studies suggest that the composition of interacting residues in the interfaces is different in each subset: for example homo-dimers tend to have more hydrophobic residues in their interfaces than hetero-dimers. It has also been observed that small interfaces tend to have a higher content of charged and polar residues if compared to larger ones. A further distinction can be made between obligate, non-obligate, transient and permanent complexes. Obligate interactions are those where the monomers do not exist as stable structures in vivo while in non-obligate interactions the monomers can exist independently of the complex. The distinction between transient and permanent interactions was originally based on the lifetime of the association. A permanent interaction is quite stable and for this reason, in general, it exists only in its complexed form, while a transient interaction associates and dissociates in vivo. Structural and functional obligate interactions usually are routinely permanent while non-obligate interactions can be transient or permanent. The interfaces of obligate complexes have clusters of hydrophobic residues while the transient interfaces have a significant number of polar residues that contribute to electrostatic interactions [9]. Together this findings suggest that transient interfaces are in general smaller with respect to obligate ones. Recently, it has been observed that in obligate complexes, monomers share more molecular functions (following the GO annotation) than in transient complexes. Moreover, residues in the obligate interfaces tend to evolve slower than residues in the transient ones. This indicates that obligate complexes residues tend to co-evolve with the interacting partner while in transient complexes a higher rate of substitution leaves no marked signature of correlated mutations [for more references see 3-6].

Different contact patterns between obligate and non-obligate interfaces have also been detailed. It was found that contacts among obligate complexes are mainly non polar and that protein chains belonging to obligate complexes show a higher number of contacts per interface than non-obligate complex chains. Unfortunately most inter-actions do not readily fall into a definite class and for this reason classifying com-plexes by these definitions is not so simple [6]. Furthermore, secondary structure elements seem to give a major contribute to the formation of the interface in obligate complexes. The interaction between secondary structure motifs is driven by the inter-action between main chain atoms, especially in the formation of obligate interfaces. Finally, β-sheet formation across the interacting units was observed only in obligate interactions.

Peculiar  characteristics of interaction sites can be captured by position-specific sequence profiles such as those generated from PSI-BLAST multiple-sequence align-ments [8-11]. Residue conservation at the interface is observed to be slightly higher than those of general surface residues, although it is not significantly different from those in the protein interior. The discriminatory power of the evolutionary conservation has been observed to be stronger for obligate and more permanent interactions [5].

Summing up, several different chemico-physical and topological features can be associated to interacting protein surfaces. However, correlations between these fea-tures and protein-protein binding sites are so subtle that they cannot be predicted with linear models alone.

## 3  Prediction of Interacting Residues: The-State-of-the-Art Methods

In spite of the differences detailed above, protein interfaces are not endowed with features that make them simple to be computed from specific sets of rules. For this reason computational methods that attempt to identify interface residues are very valuable. The most widely adopted machine learning methods, Neural Networks (NN) and Support Vector Machines (SVM), exploit the information contained in the protein sequence and structure to predict interface residues. Two main approaches are commonly followed to address the interaction sites prediction: structure based (historically-older) and sequenced-based depending on which information is at the basis of the analysis at hand. The methods substantially differ because of the underlying training/testing dataset and also because they adopt different definitions of protein surface and contacts, namely residues belonging to the protein-protein interface. The characteristics of the results of each computational method is strongly dependent on the dataset and on the contact definition adopted [3-5]. Based on the notion that the data bases of sequences and structures recently underwent an exponential increase, in the following we will detail those methods that more recently can be listed when dealing with the state of the art of prediction of protein-protein interaction sites. By this we aim at a thorough analysis of all the method recently developed hoping to highlight pitfalls and possible developments to improve the efficacy of the approach. A further sub-distinction is made between methods that adopt or do not adopt evolutionary information [3-5]. Evolutionary based methods are those methods that essentially make use of sequence profile derived from Multiple Sequence Alignments (MSA). All the-state-of-the-art methods are summed up with their main characteristics in Table 1.

## 3.1  Sequence Based Methods

ISIS (Interaction Sites Identified from Sequence [12]) is a combination of NNs that aims at predicting interacting residues from sequence. The NN input consisted of protein residue composition, sequence profile, predicted secondary structure and solvent accessibility. ISIS first builds the sequence profile required as input by PROF (a predictor of secondary structure from the same group [13]), and secondly the profile and the PROF outputs are given as input to a second NN that classifies each residue as interacting or non-interacting. A residue is defined as belonging to an interface if the distance between any of its atoms is within 6 Å from any atom of a residue in the partner chain. ISIS was trained on a set of 1134 protein chains selected from the PDB and it reached an overall accuracy of 68% (Table 2).

## 3.2  Structure and Evolutionary Based Methods

Recently, the Relative Solvent Accessibility (RSA) of protein residues in a chain has been integrated with high resolution structural data [14] for predicting interacting surface residues. The SPPIDER classifier (Solvent accessibility-based Protein-Protein Interaction sites IDEntification and Recognition) is a 10 NNs consensus method trained with a cross validated procedure on a 435 protein chains dataset (262 from hetero-complexes and 173 from homo-complexes both derived from PDB). 19 physico-chemical features, such as the evolutionary conservation of amino acid properties (type, charge, hydrophobicity) derived from a MSA, were included in the predictor. Nevertheless the authors demonstrated that the most important feature was the difference between the predicted RSA and the real RSA (calculated in the unbound structure), averaged over a window length of 11 residues. The unbound structure was obtained from the complex considering only the chains of interest. The

**Table 1.** Recent machine learning methods for protein-protein interaction sites predictions

| Authors (Method) | Surface/ Interface definition | Dataset | Machine learning | Features |
|---|---|---|---|---|
| Porollo *et al.*,2007 (SPPIDER) | **Protein:** All-Atom **Surface:** RSA > 5% **Interface:** surface residue whose change in RSA is > 4% and whose change in exposed surface upon complex formation is > 5 Å | 435 protein chains from PDB (262 from hetero-complexes, 173 from homo-complexes). Control set of 149 chains (92 from hetero-complexes, 57 from homo-complexes) | Consensus method based on 10 NNs | 19 physico-chemical features: the most important is the WNA over 11 spatial neighbours of dSA |
| Dong Q. *et al.*, 2007 | **Protein:** All-Atom **Surface:** ASA of at least one atom is > 2 Å **Interface:** residue whose ASA is decreased > 1 Å upon complexation. | 1139 non-redundant protein chains from the PDB | SVM | Sequence profile, ASA, binary profile propensities on a 12 residues window |
| Chung J. *et al.*, 2006 | **Protein:** All-Atom **Surface:** RSA > 15% **Interface:** residue whose ASA is decreased > 1 Å upon complexation | 274 non-redundant chains of hetero-complexes taken from the PDB | SVM | Sequence profile, ASA, conservation score on a 12 residues window |
| Li J.J. *et al.*, 2007 | **Protein:** CA trace **Surface:** RSA > 16% **Interface:** CA-CA distance < 1.2 nm | 69 non-redundant chains from ref.10 | LDF | Conservation score of 11 spatial neighbors |
| Wang B. *et al.*, 2006 | **Protein:** CA trace **Surface:** RSA > 16% **Interface:** CA-CA distance < 1.2 nm | 69 non-redundant chains from ref.10 | SVM | Sequence profile, conservation score of 11 spatial neighbors |

**Table 1.** (*Continued*)

| | | | | |
|---|---|---|---|---|
| Bordner&Abagyan, 2005 | **Protein**: All-Atom **Interface**: any pairs of atoms in each molecule that are separated by < 4 Å | 632 non-redundant protein interfaces from the PDB (518 homo-dimers, 114 hetero-dimers) Testing set: 43 interfaces from ref .9 | SVM | Local surface properties combined with evolutionary conservation score |
| Li M. *et al.*, 2007 | **Protein**: All-Atom **Surface**: RSA > 15% **Interface**: All-atom distance < 5 Å | 883 chains from hetero-complexes from the PDB | CRF | Sequence profile, ASA, residue conservation and transition features of 15 spatial neighbors |
| Chen & Zhou, 2005 (Cons-PPISP) | **Protein**: All-Atom **Surface**: RSA > 10% **Interface**: All-atom distance < 5 Å | 1156 non-homologous protein chains (756 homo-dimers, 400 hetero-dimers) | Consensus between 6 NNs | Sequence profile, solvent accessibility |
| Ofran & Rost, 2007 (ISIS) | **Protein sequence** | 1134 protein chains from the PDB | NN | Amino acid composition, sequence profile, predicted secondary, structure solvent accessibility of 9 sequentially neighbors |

See text for explanations. NN: Neural Network; SVM: Support Vector Machine; CRF: Conditional Random fields; LDF: Linear Discriminant Function ASA: absolute solvent accessibility; WNA: weighted neighbour averages; dSA: difference in solvent accessibility.

observed RSA was calculated with the DSSP program, while the predicted RSA was the result of a previously developed method called SABLE [14]. An interacting residue was defined as a surface exposed residue (RSA > 5%) whose change in RSA between the isolated chain and the complex structure was greater than 4% RSA and whose change in exposed surface area was >5 Å. A filtering procedure for removing negative examples difficult to predict was adopted. An initially "non-interacting" residue was excluded from training if at least 5 of its 10 nearest neighbours belong to the positive class. When tested with a 10-fold cross validation the classifier reached an overall classification accuracy of about 74% with a correlation coefficient of 42%.

The sensitivity and specificity were 60% and 64% respectively. This scoring indexes are so far the highest reported (Table 2).

Residue interface propensities of different types of complexes (homo-permanent, homo-transient, hetero-permanent, hetero-transient) have been analyzed and combined with sequence profiles and accessible surface area (ASA) to recognize binding sites [15] (Dong et al., in Table 1). Sequence profile, ASA and binary profile propensities have been given as input to a SVM system. The training was performed over a non-redundant set comprising 1139 protein chains, chosen from PDB. The binary profile is a 20 valued vector whose elements are 1 or 0, depending on whether the amino acid frequency in a given position of the sequence profile is greater than a given threshold. A residue was defined as belonging to the surface when the ASA of at least one of its atoms was > 2 Å. An interface residue is a surface residue that decreased its ASA more than 1 Å upon complex formation. By this, the best results were obtained when the binary profile propensities were calculated for each class, as detailed above, and assigned to the corresponding class (overall accuracy is 73%, correlation coefficient is 37%); however when the propensities were assigned to a different class (for example hetero-transient propensities to hetero-permanent complexes) the performance dropped to an overall accuracy of 67% with a 22% correlation coefficient. The specificity and sensitivity were then respectively 40% and 56% (Table 2).

**Table 2.** Scoring indexes of recent machine learning methods for protein-protein interaction sites predictions

| Authors/Method | Accuracy | Sensitivity | Specificity | Correlation coefficient |
|---|---|---|---|---|
| SPPIDER | 74% | 60% | 64% | 42% |
| Dong Q. et al., 2007 | 67% | 56% | 40% | 22% |
| Chung J. et al., 2006 | - | 67% | 50% | - |
| Li J.J. et al., 2007 | - | 66% | 49% | 28% |
| Wang B. et al., 2006 | 65% | 66% | 50% | 30% |
| Bordner&Abagyan, 2005 | 67% | 67% | 22% | - |
| Li M. et al., 2007 | 71% | - | - | 30% |
| Cons-PPISP | 61% | 38% | - | - |
| ISIS | 68% | - | - | - |

See Table 1 and text for explanations. Scoring indexes are defined in section 4.2.

Chung et al. [16] combined the sequence profile, the ASA and a conservation score for training an SVM-based classifier to identify protein-protein binding sites. The residue conservation score was the result of a multiple structural alignment, built with the CE-MC algorithm, weighted on the crystallographic B factor. This procedure takes into account the structural flexibility than can affect the quality of the alignment. 274 non-redundant chains of hetero-complexes collected from the PDB were included in the dataset. Only surface residues, defined as residues whose RSA was >15%, were considered with a complement of 12 structural surface neighbours. Two residues participate into an interfacial contact if the distance between any of their heavy atoms is less than 5 Å. A predicted non-interface residue becomes an interface residue when the distance between its CB atom and the CB atoms of at least three predicted interface residues was < 6 Å. The method achieves a 50% specificity with a 67% sensitivity.

Li J.J. *et al.* [17] represented the evolutionary information through a conservation score based on a phylogenetic tree calculated from the MSA profile. The Rate4Site algorithm was used for the conservation score calculation. By introducing this score, the feature vector is 20 times reduced with respect to a sequence profile based vector. The method was tested on a 69 non-redundant chains obtained by filtering an already compiled dataset [10]. Surface residues were defined as those residues whose RSA calculated with the DSSP was greater than 16%. The protein was represented with its CA trace, choosing a 1.2 nm cut-off for the interaction distance. A Linear Discriminant Function (LDF) was tested on the dataset and reached about 49% specificity, 66% sensitivity (for the positive class) with a correlation coefficient of 28%. In a second paper from the same group, the conservation score was combined with a sequence profile as input for an SVM system [18]. The method with the same input parameters as before reaches a sensitivity of 66.3%, a specificity of 49.7%, an overall accuracy of about 65% and a correlation coefficient of 30%.

Bordner and Abagyan [19] described a SVM method trained on a combination of local surface properties and an evolutionary conservation score. A newly developed Bayesian algorithm that made use of phylogenetic trees was adopted to calculate the evolutionary score. Local surface patches of 15 residues (central residue and 14 surface neighbors obtained comparing CA distances) were generated for each surface residue. Z-scores of residue frequencies in multiple alignment and evolutionary rates for each residue of the patch were adopted as SVM input. The system was trained on a 1494 non-redundant protein interfaces dataset (518 homo-dimers, 114 hetero-dimers and 862 multimers from the PDB). A post processing procedure was then applied to remove possible false positives: residues were clustered in patches comprising those whose CA atoms were less than 6 Å apart; predicted interface patches were changed in non-interface ones if their ASA was <150 Å. For sake of comparison with other methods in Table 2 with we report the scoring indexes obtained by the authors without clustering. The method scores with an overall accuracy of 76% and a sensitivity and a specificity equal to 64% and 34%, respectively. When tested on a different set of 43 transient hetero-dimer interfaces [9] the accuracy falls to 67% and the specificity to 22%.

A different approach, based on conditional random fields (CRF), has been proposed by Li M. *et al.* [20]. Surface residue segments were collected and each residue within the segment was labeled as interface or non-interface. These segments were included as training examples. A surface residue was defined as a residue whose relative DSSP calculated ASA is at least 15%. A residue was defined as an interface residue if the distance between any of its heavy atoms and any heavy atom of a residue on the partner chain was < 5 Å. The CRF features included state features, such as profile of 15 spatially neighboring residues, ASA and residue conservation, and transition features. The database consisted of 883 chains from hetero-complexes extracted from the PDB. The method achieved an overall accuracy of 71% with a correlation coefficient of 30%.

Chen and Zhou tried to improve their first predictor [21] with a larger training dataset compiled from the PDB and containing 1156 non-homologous protein chains

(756 homo-dimers and 400 hetero-dimers). The new system [cons-PPISP, 22] was bases on a 6 NN consensus method taking as input the sequence profile of each residue contained in a 15 residue long window, the average solvent accessibility of the target residue and of its 6 nearest neighbours. Surface residues were collected fixing the relative surface exposure to solvent to at least 10%. A surface residue was defined as an interfacial contact if the distance between any of its heavy atoms and any heavy atom of a surface residue on the partner chain was less than 5 Å. When tested on a set of 100 non-homologous protein structures (42 homo-dimers and 58 hetero-dimers) the new method reached a 80% accuracy and a 51% sensitivity. To reach these performances the training set was balanced randomly removing 1/3 of the non-interface residues and removing residues belonging to interfaces other than the biggest one (for chains forming at least trimers). Tested on a set of 68 protein chains of 40 complexes collected as a docking benchmark, cons-PPISP scored with a 61% accuracy and a 38% sensitivity; when tested on 8 NMR-characterized proteins it predicted interfaces residues with an accuracy of 69% and a sensitivity of 47% (Table 2).

## 4 Details on Data Bases

From what detailed above it can be concluded that different methods are scoring differently on different data sets, when contact definition is different. It is also evident that protein-protein interaction sites prediction is still far from being optimal, as previously discussed [5]. It also evident that different methods have been trained and tested on different data sets and this may hamper their direct comparison. In this section we focus on the rules and definitions adopting for grouping sequences and structures in order to highlight the demand of common selecting criteria. To address the question of which data base is more suited to address the problem of protein-protein interaction, we analyzed a number of representative datasets described in literature to highlight differences and similarities (Table 1). The large majority of the datasets adopted for training and testing purposes are extracted from the PDB following different rules. Porollo and Meller [14] filtered the PDB data base and retained only complexes with at least one chain sharing less than 50% sequence identity with any other chain within the set. This threshold is 2fold higher than that routinely adopted by other authors. The procedure resulted in 584 protein chains.

Multiple interfaces for a single chain are taken into account, while the majority of the other methods considered only independent dimer interfaces.

Dong *et al.* [15] selected only multiple chains with a resolution greater than 4 Å and longer than 40 residues. Even if multi-chain complexes were extracted, if they contained more than two chains, each chain was selected once. Moreover, if a chain had multiple partners they chose the partner with the biggest interface. The PQS server helped to eliminate crystal packing complexes. The sequence identity threshold for avoiding redundancy was 25%. A final set of 1139 chains was obtained.

Chung *et al.* [16] retained all non-NMR structures with multiple chains and a resolution higher than 3.5 Å from the PDB (March 2004). For each complex the decrease in ASA has been calculated using the DSSP program and a pair of chains was retained if it was $\geq 450\text{Å}^2$. The pairs belonging to SCOP classes $\geq 8$ or shorter than 80 residues were discarded. All the sequences were compared with a BLAST search and clustered if the sequence identity was $\geq$30% over >90% of the sequences. Choosing a representative for each cluster the final dataset comprised a non-redundant set of hetero-complexes with <30% of sequence identity. If the sequence identity among chains of the same complex was >95% over 90% of the sequences length, the complex was identified as an homo-complex and discarded.

For chains interacting with multiple partners, only the one with the most partners was chosen as representative. A further alignment with the CE_MC algorithm choosing the homologous at the SCOP family level, retained only the chains that were aligned over 60% of their length and with a Z-score >4.0. The final dataset contained 274 non-redundant chains of hetero-complexes.

The Li J.J. *et al.* [17] dataset comprises only 69 chains with sequence identity less than 30%.

Starting from the PDB Bordner and Abagyan [19] selected all X-ray structures and used the biological unit information of the mmCIF file to build the structure of the complex. Two chains in a complex were defined as interacting if heavy atoms in each chain were closer than 4 Å with heavy atoms on the other chain. Pairs containing chains shorter than 20 residues were removed. The quality of the dataset was improved checking the consistency between PDB file and the Swiss-Prot annotation. In order to remove homo-complexes sequences were clustered and only those with less than 30% sequence identity were retained (BLAST search against nr with E-value cut-off 0.1). A further filter was applied for removing interfaces showing, according to PDB file information, mutations and immune system proteins with polymorphisms and somatic mutations. Only interfaces larger than 10 residues were retained in the last 1494 protein-protein interfaces dataset, composed by 518 homo-dimers, 114 hetero-dimers and 862 multimers. Different interfaces for the same chain are considered in an independent way.

Li M. *et al.* [20] considered the July 2005 PDB release. The resolution cut-off value chosen was 3.5 Å. Redundancy was avoided by retaining only those structures whose sequence identity was $\geq$ 30% over > 90% of the sequence, ending up with 1276 chains from hetero-complexes. In general, complexes can have more than one chain that can form more than one interface. For addressing this problem each chain was considered once and only one partner with the most interfacial residues was considered but also other cases were taken into account. Indeed minor interfaces were labeled as interface, non-interface or even excluded from the dataset and for each case a different result was reported.

Chen and Zhou [22] substantially implemented the same method of 2001 with a bigger database extracted from the January 2002 PDB release. All multiple chain structures, whose chains were longer than 40 residues and whose interface was larger than 20 residues were considered. Each chain was assigned only to the partner with the largest interface. Residues belonging to other interfaces were labeled as non interface residues or eliminated from the training set. The similarity cut-off for the PSI-BLAST all-against-all search was 30% over the aligned region that had to cover at least 90% of the two sequences. A homo-dimer was defined as a complex with highly homologous chains (95%) over a 90% aligned region. Their set comprises 798 homo-dimers and 458 hetero-dimers.

# 5  Our Approach: ISPRED

In the following we introduce the method that we implemented for the prediction of protein-protein interaction sites. The method although, described before [10], is now re-trained on new training set. The method is essentially a feed forward neural network trained with a standard back propagation algorithm.

Each protein residue is represented with its carbon alpha (CA). By this, the protein surface is represented using its CA trace and the interface residues are defined as those surface residues (residues with relative solvent accessibility > 16%) whose Euclidean distance with at least one residue of the partner chain is below 1.2 nm. Solvent exposure is separately computed for each chain, using the DSSP program. This threshold value is selected after comparison with the patches obtained using an all-atom representation of the residues.

The method exploits the evolutionary information contained in the multiple sequence alignment of protein families, encoded in the sequence profile. The profile is a L*20 matrix, where L is the protein sequence length, that for each position of the alignment contains the frequency of occurrence of each type of residue in the corresponding position of the alignment.

The system is trained using an 11 residue-long window centered on the surface residue to be predicted with 10 nearest neighbors in the surface patch. To give a rough approximation of the local surface, the residues included in the input window are close in space and not necessarily adjacent in the sequence and represent. Thus, each residue in the input window is encoded as a vector of 20 elements, whose values are the frequencies extracted fro the sequence profile.

Each protein residue is thus encoded in a 232 nodes input vector that contains the sequence profile of the residue to predict, the sequence profile of its 10 nearest spatial neighbors, the relative accessible surface area of the residue to predict and the frequency of interaction of the residues in the window (namely 11), calculated on the overall dataset.

The output of the neural network is the probability of each surface residue to be in an interaction site given the current input. Neural network predictions are uncorrelated. In order to take into account the cooperation among close surface residues, we averaged the network output over the list of neighbors. The probability for a given residue is calculated as follows:

$$P(i) = \Sigma_j \, w(i,j) \, O(R_i(j))/ \, \Sigma_j \, w(i,j)$$

where $O(R_i(j))$ is the output corresponding to the neighbor $j$ of the residue $i$ and $w(i,j)$ is a weight. We tested several weighting schemes. The uniform distribution (Uniform)

$$w^U(i,j)=1$$

the exponential decreasing with the Euclidean distance between the residue and its neighbors (Exp),

$$w^E(i,j)=exp[-d(i, R_i(j))]$$

the decreasing with the inverse of the Euclidean distance,between the residue and its neighbors (Inv),

$$w^1(i,j)=1/[\ d(i,\ R_i(j))(i-\delta(0,j)+\ \delta(0,j))]$$

With the inclusion of Kronecker delta ($\delta$), w(i,j)=1 when we refer to the current residue (j=0) otherwise it decreases with the inverse of the distance.

We selected the uniform distribution as the best performing one.

The introduction of this smoothing function reduces the number of spurious assignments and for this reason it improves the predictor performance [11].

### 5.1 ISPRED2.0: Dataset Description

Our dataset is composed of 626 protein chains belonging to 318 different protein complexes (291 dimers and 27 multimers) obtained from the PDB. Among the 116 homo-complexes, 40 structures are transient and 76 are obligate, while among the 202 hetero-complexes, 106 are classified as transient and 106 as obligate. For the selection of the transient complexes we found that 215 of the selected chains have a correspondent structure resolved both in complex and in a complex independent stable structure. Only the complexes with at least one monomer with a highly homologous ($\geq$70%) unbound form have been selected as transient complexes. We also filtered out structures whose sequences are shorter than 40 residues.

For purpose of comparison with SPPIDER, we randomly selected a benchmark subset of 50 protein chains from our initial dataset, with the constraint that they do not share more than 25% sequence similarity with any of the sequences in the Porollo dataset [14]. 18 chains belong to homo-complexes (2 transient complexes and 16 obligate) and 32 are from hetero-complexes (among with 16 transient). We also selected a subset of 417 proteins (S417) that are the complement of the intersection (S467) between the 626 proteins (S626) dataset and the complete Porollo's dataset, filtered out from the 50 proteins selected as described above. The intersection between the two sets has been computed with a BLAST alignment.

### 5.2 Scoring the Performance

The results of the methods are evaluated using the following measures. The overall accuracy (Q2) is defined as:

$$Q2= \text{Overall accuracy} = TP+TN/(TP+TN+FP+FN)$$

where:

- TP = number of true positive predictions (observed interacting residues predicted as interacting)
- TN = number of true negative predictions (observed non-interacting residues predicted as non-interacting)
- FP = number of false positive predictions (observed non-interacting residues predicted as interacting)
- FN = number of false negative predictions (observed interacting residues predicted as non-interacting)

The sensitivity (coverage) for the positive class Q[+] and for the negative one Q[-] are computed as follows

$$Q[+]= TP/(TP+FN)$$
$$Q[-] = TN/(TN+FP)$$

To measure the probability of correct predictions, we computed the specificity (namely the accuracy for each class, + and -) expressed as follows:

$$P[+] = TP/(TP+FP)$$

$$P[-] = TN/(TN+FN)$$

The correlation coefficient is:

$$C = (TP*TN-FP*FN)((TP+FP)*(TP+FN)*(TN+FN)*(TN+FP))^{1/2}$$

## 6  Results

The benchmark of the new release of ISPRED (2.0) is done considering the predictor with the highest score around: SPPIDER. The problem that we want to tackle is how good is SPIDDER performance when tested on different sets from the one used by the authors.

In the first experiments ISPRED 2.0 is directly compared with the performance of SPPIDER, as described in literature. Results are shown in Table 3, where also a second cascaded network is added to our predictor to filter out non correct assignments. Results are obtained by averaging over 10 different values for each scoring index computed by means of a cross validation procedure. SPIDDER was originally trained on 435 protein complexes (S435) and tested on 149 proteins (S149). In the last experiment of Table 3 ISPRED was tested on the SPPIDER testing set, again adopting a cross validation procedure. It is evident that a second cascaded network slightly improves our performance. Values however are still lower than that declared for SPPIDER. As discussed above (section 4)  we are now comparing results obtained with different training and testing methods.

We then used  datasets comprising 50 randomly selected proteins from the 626 of our data set and we tested SPPIDER performance. The results are reported in Table 4.

It evident that on small testing sets, different from the original one, SPPIDER performances are well below the reported values. This is so even when consider independent test sets of larger dimension  never seen before by the two methods (Table 5).

All the PDB proteins were filtered with ISPRED to predict putative contact sites. The predictions are available at our web site with the scores obtained for each putative contact residue. As an example, Figure 1 shows a protein 3D structure with highlighted the predicted interacting sites localized on surfaces.

**Table 3.** ISPRED 2.0 at work

| Method | Q2 | Q[+] | P[+] | Q[-] | P[-] | C |
|---|---|---|---|---|---|---|
| SPPIDER* | 0.74 | 0.60 | 0.64 | - | - | 0.42 |
| ISPRED2.0 | 0.66 | 0.54 | 0.7 | 0.77 | 0.62 | 0.32 |
| ISPRED2.0^ +2nd Net | 0.68 | 0.58 | 0.72 | 0.77 | 0.65 | 0.36 |
| ISPRED2.0# +2nd Net | 0.67 | 0.58 | 0.76 | 0.71 | 0.65 | 0.35 |

* * from [14]. ^with a second cascaded neural network. #trained on the 435 and tested on  S149. ISPRED 2.0 is trained on S626 and tested on S149.

**Table 4.** SPPIDER tested on randomly selected protein sets

| Dataset* | Q2 | Q[+] | P[+] | Q[-] | P[-] | C |
|---|---|---|---|---|---|---|
| #1 | 0.69 | 0.50 | 0.43 | 0.76 | 0.81 | 0.25 |
| #2 | 0.67 | 0.50 | 0.44 | 0.74 | 0.79 | 0.24 |
| #3 | 0.69 | 0.51 | 0.43 | 0.76 | 0.81 | 0.26 |
| #4 | 0.70 | 0.53 | 0.46 | 0.77 | 0.82 | 0.29 |
| #5 | 0.66 | 0.53 | 0.48 | 0.72 | 0.76 | 0.24 |
| AVERAGE | 0.68 | 0.51 | 0.45 | 0.75 | 0.80 | 0.26 |

*see Data set description

**Table 5.** Comparison of SPPIDER and ISPRED2.0 performance

| Method | Q2 | Q[+] | P[+] | Q[-] | P[-] | C |
|---|---|---|---|---|---|---|
| SPPIDER* | 0.68 | 0.54 | 0.47 | 0.74 | 0.79 | 0.28 |
| ISPRED2.0° | 0.66 | 0.45 | 0.43 | 0.74 | 0.78 | 0.21 |
| ISPRED2.0[+] | 0.66 | 0.49 | 0.44 | 0.73 | 0.77 | 0.22 |
| ISPRED2.0[#] +2nd Net | 0.67 | 0.53 | 0.45 | 0.73 | 0.79 | 0.25 |

*tested on S417. °trained on S626, tested on S417. [+]trained on S435, tested on S467. [#]trained on S435, tested on 467.

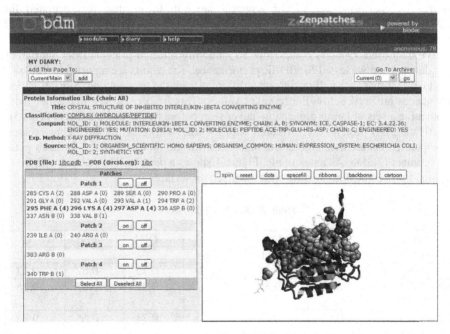

**Fig. 1.** Example of prediction of the protein 1ibc (inhibited interleukin—1beta conerting enzyme), obtained with ISPRED 2.0 and stored in the ZenPatches database (http://gpcr.biocomp.unibo.it/biodec/). In red are the true positives, in green false negatives and in yellow false negative predictions.

# 7  Conclusions

As a final consideration we can observe the recent methods focusing on predictions of interacting sites in proteins have exploited a large collections of chemico-physical properties for generating the input code. By this different rules have been described and obtained from different data bases. This obviously hampers a direct comparison of the different methods including ours. Even though standard rules are generally accepted when implementing machine learning approaches, different inputs apparently end up in different scoring indexes.

In this paper we focused on the specific problem of the data bases. We addressed the question of how two highly performing predictors can be compared when they have been trained with different criteria and data bases. We observe that when data bases are randomly selected the performance is rather similar suggesting that possible improvements described in literatures [14] may be due to over-fitting on a specific data base. It is therefore necessary to standardize both training and testing sets for implementing predictors that later on may be benchmarked on more rigorous criteria.

In any case the actual state of the art predictors seem to have reached an upper limit scoring performance that at least in our case on a more abundant data set of examples. A previous implementation of ISPRED [10], trained and tested on a less abundant data set scored with a 26% correlation index. This value, although obtained in 202, is similar to the most recent ones shown in Table 2. Our experiments indicate that an upper bound to the scoring indexes is possibly the one declared by SPPIDER, that we have shown is wuite sensible to different data sets, as opposite to ISPRED 2.0. Also the method seems not be an issue as recently demonstrated with a SVM based method that scored as high as SPPIDER (with a coefficient correlation of 33%) on a similar data set [5].

The problem of how to treat multiple interfaces then emerges. According to a molecular view of protein-protein interaction it should be considered that one protein may be even involved in different protein complexes, although in a transient way. Then the question poses as to whether we really are in the position of correctly describing an interacting surface, when it is possible that presently the PDB collects structural information only of those complexes whose life time is compatible with the crystallization procedures. When the same or a similar protein are interacting with different partners, are these patches to be included or excluded? According to a molecular view of protein-protein interaction it should be considered that one protein may be even involved in different protein complexes, although in a transient way. Then the question poses as to whether we really are in the position of correctly discriminating false positives. It may be possible that considering protein complexes solved with atomic resolution  we are missing other important features of protein surfaces that may help us in improving the gap among protein interacting networks and molecular recognition at the protein level. For the time being we can rely on putative interaction sites as those collected in the ZenPatches database (http:// gpcr.biocomp.unibo.it/biodec/) that can help in supporting experimental large scale data and at the molecular level in designing further experiments for validation studies.

## Acknowledgements

RC acknowledges the receipt of the following grants: FIRB 2003 LIBI--International Laboratory of Bioinformatics and the supportto the Bologna node of the Biosapiens Network of Excellence project within the European Union's VI Framework Programme (contract number LSGH-CT-2003-503265).

## References

1. Chothia, C., Janin, J.: Principles of protein-protein recognition. Nature 256, 705–708 (1975)
2. Jones, S., Thornton, J.M.: Prediction of protein-protein interaction sites using patch analysis. J. Mol. Biol. 272, 133–143 (1997)
3. Zhou, H., Qin, S.: Interaction site prediction for protein complexes: a critical assessment. Bioinformatics 21, 2469–2501 (2007)
4. De Vries, S.J., Bonvin, A.M.: How proteins get in touch: interface prediction in the study of biomolecular complexes. Curr. Protein Pept. Sci. 9, 394–406 (2008)
5. Ezkurdia, I., Bartoli, L., Fariselli, P., Casadio, R., Valencia, A., Tress, M.L.: Progress and challenges in predicting protein-protein interaction sites. Briefings in Bioinformatics 10(3), 233–246 (2009)
6. Fu, H.: Protein-Protein Interactions: Methods and Applications (Methods in Molecular Biology). Humana Press Inc., New York
7. Katchalski-Katzir, E., Shariv, I., Eisenstein, M., Friesem, A.A., Aflalo, C., Vakser, I.A.: Molecular surface recognition: determination of geometric fit between proteins and their ligands by correlation techniques. Proc. Natl. Acad. Sci. USA 89, 2195–2199 (1992)
8. Berezin, C., Glaser, F., Rosenberg, J., Paz, I., Pupko, T., Fariselli, P., Casadio, R., Ben-Tal, N.: ConSeq: the identification of functionally and structurally important residues in protein sequence. Bioinformatics 20, 1322–1324 (2004)
9. Nooren, I., Thornton, J.M.: Structural characterization and functional significance of transient protein-protein interactions. J. Mol. Biol. 325, 991–1018 (2003)
10. Fariselli, P., Pazos, F., Valencia, A., Casadio, R.: Prediction of protein-protein interaction sites in heterocomplexes with neural networks. Eur. J. Biochem. 269, 1356–1361 (2002)
11. Fariselli, P., Zauli, A., Rossi, I., Finelli, M., Martelli, P.L., Casadio, R.: A neural network method to improve prediction of protein-protein interaction sites in heterocomplexes. In: IEEE International Workshop on Neural Network on Signal Processing 2003, Toulouse, France, pp. 33–41. IEEE Press, Los Alamitos (2003)
12. Ofran, Y., Rost, B.: ISIS: interaction sites identified from sequence. Bioinformatics 23, 13–16 (2007)
13. Rost, B.: Prediction in 1D: Secondary structure, membrane helices, and accessibility. In: Bourne, P., Weissig, H. (eds.) Structural bioinformatics. Wiley-Liss, Hoboken (2002)
14. Porollo, A., Meller, J.: Prediction-based fingerprints of protein-protein interactions. Proteins 66, 630–645 (2007)
15. Dong, Q., Wang, X., Lin, L., Guan, Y.: Exploiting residue-level and profile-level interface propensities for usage in binding sites prediction of proteins. BMC Bioinform. 8, 147 (2007)
16. Chung, J., Wang, W., Bourne, P.E.: Exploiting sequence and structure homologs to identify protein-protein binding sites. Proteins 62, 630–640 (2006)

17. Li, J.J., Huang, D., Wang, B., Chen, P.: Identifying protein-protein interfacial residues in heterocomplexes using residue conservation scores. Int. J. Biol. Macromol. 38, 241–247 (2006)
18. Wang, B., Chen, P., Huang, D., Li, J.J., Lok, T.M., Lyu, M.R.: Predicting protein interaction sites from residue spatial sequence profile and evolution rate. FEBS Lett. 580, 380–384 (2006)
19. Bordner, A.J., Abagyan, R.: Statistical analysis and prediction of protein-protein interfaces. Proteins 60, 353–366 (2005)
20. Li, M., Lin, L., Wang, X.L., Liu, T.: Protein-protein interaction site prediction based on conditional random fields. Bioinformatics 23, 597–604 (2007)
21. Zhou, H.X., Shan, Y.: Prediction of protein interaction sites from sequence profile and residue neighbor list. Proteins 44, 336–343 (2001)
22. Chen, H., Zhou, H.: Prediction of interface residues in protein-protein complexes by a consensus neural network method: test against NMR data. Proteins 61, 21–35 (2005)

# The Equivalence between Biology and Computation

John K. Heath

School of Biosciences, University of Birmingham, UK
j.k.heath@bham.ac.uk

**Abstract.** A major challenge in computational systems biology is the articulation of a biological process in a form which can be understood by the biologist yet is amenable to computational execution. Process calculi have proved to especially powerful computational tools for modelling and reasoning about biological processes and we have previously described, and implemented, a Narrative approach to describing biological models which is a biologically intuitive high level language that can be translated into executable process calculus programs. Here we discuss an extension to the narrative approach which attempts to directly link biological data with Narrative primitives by suggesting an equivalence relationship between a string (the amino acid sequence) and a process. We outline future challenges in applying this approach more generally.

## 1 Introduction and Motivations

There are multiple incentives for a convergence between computing and biology [1]. Classically this has been the adoption by biologists of computational tools to store, search, visualise and interrogate large sets of biological data. However the relationship can be taken further by the aspiration to realise a biological process in the form of a computer program. The predicted benefits of this challenge are many. A pragmatic motivation for the biologist is that it could accelerate the pace of biological discovery, help design better laboratory experiments and reason more clearly about biological processes [2].

But there are greater wins to be had. A program that faithfully emulated biology could be used to study processes which cannot practically be investigated in the biologists laboratory. Examples would include biological processes that take place over very long periods of time such as pathway evolution, aging and degeneration, or the accumulation of mutations in the lifetime of a tumour. Programs could be used to study the outcomes of experiments that can be conceived but cannot be executed in the laboratory. Programs could be used to even identify experiments that biologist cannot conceive.

Computer programs that behaved like biological processes could revolutionise the development of new therapies by allowing exhaustive exploration of treatment scenarios without recourse to laboratory investigation and will be essential to realise the goal of personalised medicine: the program can be employed to predict the outcome of therapy.

P. Degano and R. Gorrieri (Eds.): CMSB 2009, LNBI 5688, pp. 18–25, 2009.

Finally programs that behaved like biology could be employed to design new biological processes with specified properties including, for example, biology that behaved like a computer.

Realisation of this challenge has outcomes for computer science in that it demands programs that do not simply emulate the system but in fact are the system. This calls for the development of computational techniques which exactly and formally behave like a biological process. This represents true convergence of the two disciplines.

This paper reviews current progress towards this aim with a specific emphasis on the equivalence between biological data types and events and computational formalisms.

## 2   Algorthimic Systems Biology

The classical, and most widely employed, approach to modelling biological processes has been via the use of mathematical tools that express the process in the form of numerical values which are related to each other in the form of equations. The behaviour of the process can then be simulated by calculating the solution to the equations which plots the changing relationship between the numerical variables over time. In this fomalism the computer is a calculating device [3,4].

In the past few years a different use of computers has been proposed. This springs from the fundamental properties of computation in which the state of the machine changes dynamically according to a set of rules (algorithims). Thus in computation the system passes through state changes according to the instructions supplied. In its most general sense the algorthimic approach to biology defines the states of a biological system and the instructions that relate the states to each other. Thus the central questions in this light are how is a biological state defined and what are the relationships between these states in the biological system. Thus in order to program the computing device to behave like a biological process we require a method which maps biological properties onto computational primitives. Such a mapping has to be rigorously specified and therefore requires some form of operational semantics which exists in the realm of biological data types.

### 2.1   Process Calculi

One of the most popular manifestations of the algorithmic approach to biology has been the use of process calculi which are set of approaches developed to study concurrent communicating systems [5]. The application of process calculi to biological systems was first advocated by Regev et al [6] who drew attention to the formal similarity between core concepts in process calculus languages and biological processes. These are: the representation of biological components as processes; the interaction between biological entities as a communication between processes and that the consequence of a communication event results in a change of state (ie the spawning of a new process). Process calculi provide a rich and expressive operational semantics for describing concurrent communicating systems with a small number of operational primitives and rigorous logical rules for manipulating the primitives allowing the computer scientist to reason about the processes embodied in the program. A number of different process calculus approaches have now been applied to the study of biological systems [3]. Of key importance it has been possible to demonstrate for a number of biological processes that the process calculus approach yields programs that exhibit realistic biological behaviours which can be successfully employed to reason about biological system and successfully predict the outcome of real (or hypothetical) biological interventions [2]. These outcomes are sufficiently encouraging to consider the further convergence between process calculus concepts and their biological counterparts.

## 2.2  Biological Narratives

In considering how to pursue the convergence of computer science and biology we need to consider the current process by which a biological scheme is executed in the form of a program. We shall here confine our analysis to the study of signal transduction pathways which have so far been the major class of biological problem investigated by process calculus tools.

Biologists traditionally express their hypotheses in the form of informal diagrams. These are informal in the following sense: they employ a variety of notational styles which can be ambiguous or confusing to the non biologist (although various formal notations have been proposed there is no current universal standard [7]). The diagrams are abstractions in that they are generally designed to emphasise some particular feature of the pathway and components that are not relevant to the immediate problem are often, for the sake of clarity, eliminated. It is therefore frequently confusing for the non- biologist to find seemingly the same process articulated by quite distinct diagrams. Finally diagrams are inherently static representations and dynamic features of pathway behaviour are difficult to represent. For these reasons the computer scientist can find the translation of a biological process into a computational formalism confusing and ambiguous.

The more productive approach to date has been true collaboration between the computer scientist and biologist in which a dialogue results in the formulation of the computer program in an appropriate language for execution. The dialogue is required because few (if any) biologists have the training or motivation to formulate their ideas in the form of specific computational-orientated semantics. The risk here is that the process of translation results in loss of information for successful articulation of the biological pathway.

We [8,9] have developed an approach which aims to address these issues by developing a semantic system which is aimed to be intuitive to the biologist, sufficiently expressive to faithfully articulate the biological process and yet readily translatable into a process calculus language. The basic components of the language are molecules (components) which are located in specified locations (compartments). The components can interact (for example bind or unbind) undergo modifications (eg phosphorylation/ dephosphorylation) and move between compartments. As a result of interactions components can change in number (ie synthesis or degradation). These basic elements (which map directly onto the underlying process calculus primitives) are then formulated as a Narrative of events. The Narrative thus represents the instructions that specify the biological process under investigation. In accord with the process calculus approach steps in the Narrative can be concurrent and exhibit dependencies: for example a binding event may be dependant on a prior phosphorylation event or relocation between compartments may require specific complexes to be formed.

A number of pathways have been formulated in a Narrative form [2,9] and been shown not only to produce outputs which resemble real world biological data, but, by removal of components in the narrative or alteration in rate parameters, permit the biologist to examine the consequences of perturbations on the behaviour of the model. The advantage of a high level, biologically intuitive and formally rigorous language is that models can be specified by the biologist without them needing to understand the details of the underlying computational execution. The availability of translators from the Narrative language to different process calculus languages should also allow the same model to be executed by different process calculus languages.

# 3   Equivalence between Biology and Computation

The Narrative approach illustrates a method of articulating a biological process in a way which is formally rigorous for computation but intuitive to the biologist. However there is still a gap between the realm of biological data and the computational model as the approach still requires the biologist to interpret their laboratory data in the form of a Narrative description. We now suggest an extension to the narrative approach in which the basic elements of the model are direct representations of laboratory data  leading to a formal equivalence between biological data and computational execution. For this purpose we need to understand the type of data generated in a typical biology experiment. Again by way of illustration we shall confine our discussion to the analysis of signal transduction pathways.

## 3.1   Components

In the computational model the component is simply a name with specified properties. However the name points to a real biological molecule with experimentally defined attributes. The corresponding component in the biological domain is a string: the amino acid sequence of the protein. Replacing the component with the string therefore directly connects the biological entity with the computational primitive. This opens up two avenues: the ability to specify the component in the model with verified biological data and a direct connection between model primitives and experimental biological data.

Thus for example in the model of Guerrerio et al [9] gp130 refers to a real biological molecule with known and verifiable biological attributes. These are stored

in various biological databases such as Uniprot [10] (http://www.uniprot.org).
Thus replacing the name with a database pointer UniProtKB/Swiss-Prot P40189
automatically connects biological knowledge with the computational formalism.
The information collected in the biological database contains all the details required for specification of the component in the model such as subcellular location (or compartment), sites of post translational modification (ie potential state
changes), domain architecture and primary sequence. In fact direct exploitation
of biological database information to define a component in a model in most
cases will provide much more information than the modeller would require.

The amino acid (or nucleotide) sequence present in the database also provides
the means for biological verification. Thus the existence of a component in the
biological domain can be verified by unique barcodes either in the form of the
component itself or by sequence tags added by genetic engineering techniques
[11]. The unique biological identifier can also be extended to components that
exist in certain states for example specific phosphorylated forms of the protein
can be identified by phospho specific antibodies [12]. In addition using for example fusions with intrinsically fluorescent proteins or antibody localisation the
location of the component can also be experimentally verifiable. There is thus
a direct map between the component in the model and the experimental data
generated by the biological investigator. By this means it becomes possible to
directly connect an observable eg the amount or location of a component and
its system state to the model.

## 3.2   Complexing Rules

The formation and destruction of molecular complexes is central to most signal
transduction processes. From the computational perspective these are processes
spawned from the merger of two previous processes. In the biological data the
existence of complexes can be detected by a variety of experimental techniques.
These include co-immunoprecipitation, co-localisation by microscopy or via proteomics techniques such as two-hybrid interactions or mass spectroscopy [13].
These data types have also been curated in biological protein/ protein interaction databases databases such as String [14] (http://string.embl.de/). Thus
a binding event in the model can be linked to the existence of an experimentally verified binding event in the database as a concatenation of two (or more)
strings. In addition as the model is run computationally it will generate outputs which refer to a potential biological experiment of the form does a complex
between component A and component B exist at time T in compartment X?
Thus connecting the biological data and the computational primitive allows the
model to yield outputs which describe the outcomes of a hypothetical biological
experiment.

## 3.3   Compartments and Space

A notion of confined space was introduced in the narrative approach as an extension to the original specifications of Regev et al [15] in order to more correctly

reflect biological reality in which the sequestration of biological processes, for example inside membrane confined spaces such as the nucleus or the golgi, plays a key role in the process under investigation [review]. Some biologically orientated process calculus languages such as Bio-ambients [15] have been specifically devised with compartments in mind and used to model signalling processes based upon directed trafficking of components between membrane-bound compartments. Indeed we have been able to show in modelling the gp130/Jak/STAT pathway using the Narrative-based process calculus approach that the rate of movement between cellular compartments (in this case nucleus and cytoplasm) is a key determinant of pathway behaviour [9]. We have also been able, using ambients, to model the trafficking of cell surface receptors into different cytosolic compartments which compares well with biological data [16]. Extensions to this approach in which trafficking of receptors is reactive to changes in the states of components in the model (eg dependent upon a phosphorylation event) are under development (M.G. Vigliotti, S.van Bakel, J. K. Heath unpublished).

These results are encouraging but need to be further developed if a full equivalence between biological datasets and computational models is to be achieved. In particular to fully articulate the richness of biological data some explicit notation for spatial localisation is required.

Experimental biological investigations of signalling pathways increasingly exploit the power of live cell imaging [17]. In this technique proteins are visualised by fusion to an endogenously fluorescent reporter protein such as Green Fluorescent Protein (GFP) and tracked dynamically in individual cells by confocal light microscopy. The biological data in this case is in the form of a video and is analysed a single cell level which is especially appropriate for modelling as each cell is in essence an individual execution of the model rather than averaged over many cells as would be the case for methods such as western blotting. The development of computational techniques for extracting and classifying data from live cell imaging is a rapidly developing area of computer science.

The application of this general approach is dramatically transforming our biological understanding of the spatial architecture of cell signalling pathways. Thus the commonly modelled ras/raf/mapk pathway is usually modelled as a connected set of binding and phosphorylation events occurring within a single compartment [18]. The imaging data reveals however that this module is highly dynamic with components of the pathway shuttling between different compartments as the signal is propagated [19]. There are also important problems in biology such as the molecular machines controlling cell motility or mitosis - in which molecular movement in time and space is the essence of the problem.

## 4    Conclusions

Process calculi concepts and approaches are now established as very well suited to modelling biological pathways by the algorithimic approach. An obstacle to more widespread adoption has been the language gap between computational and biological formalisations. High level, biologically intuitive languages as illustrated by the Narrative approach undoubtedly help to bridge this gap. However biological investigation is ultimately driven by data and formalisms for directly mapping biological data types into executable programs is the next step in the convergence of biology and computing.

## Acknowledgements

Research in the authors laboratory is supported by Cancer Research UK and the FP6 Endotrack programme.

## References

1. Priami, C.: Algorithmic Systems Biology An Opportunity for Computer Science. Communications of the ACM 52, 80–88 (2009)
2. Kwiatkowska, M.Z., Heath, J.K.: Biological pathways as communicating computer systems. J. Cell Sci. (in press, 2009)
3. Fisher, J., Henzinger, T.A.: Executable cell biology. Nat. Biotechnol. 25(11), 1239–1249 (2007)
4. Hunt, C.A., Ropella, G.E., Park, S., Engeleberg, J.: Dichotomies between computational and mathematical models. Nature Biotechnology 26, 737–738 (2008)
5. Milner, R., Parrow, J., Walker, D.: A calculus of mobile processes, Pts 1 and 2. Information and computation 100(11), 1–40 (1992) (print)
6. Regev, A., Silverman, W., Shapiro, E.: Representation and simulation of biochemical processes using the pi-calculus process algebra. In: Pacific Symposium Biocomput., pp. 459–470 (2001)
7. Matsuoka, Y., Ghosh, S., Kitano, H.: Consistent design schematics for biological systems: standardization of representation in biological engineering. J. R Soc. Interface (June 3, 2009)
8. Guerriero, M.L., Heath, J.K., Priami, C.: An automated translation from a narrative language for biological modelling into process algebra. In: Calder, M., Gilmore, S. (eds.) CMSB 2007. LNCS (LNBI), vol. 4695, pp. 136–151. Springer, Heidelberg (2007)
9. Guerriero, M.L., Dudka, A., Underhill-Day, N., Heath, J.K., Priami, C.: Narrative-based computational modelling of the Gp130/JAK/STAT signalling pathway. BMC Syst. Biol. 3, 40 (2009)
10. Jain, E., Bairoch, A., Duvaud, S., Phan, I., Redaschi, N., Suzek, B.E., Martin, M.J., McGarvey, P., Gasteiger, E.: Infrastructure for the life sciences: design and implementation of the UniProt website. BMC Bioinformatics 10, 136 (2009)
11. Puig, O., Caspary, F., Rigaut, G., Rutz, B., Bouveret, E., Bragado-Nilsson, E., Wilm, M., Sraphin, B.: The tandem affinity purification (TAP) method: a general procedure of protein complex purification. Methods 24(3), 218–229 (2001)
12. Chung, A.S., Chin, Y.E.: Antibody array platform to monitor protein tyrosine phosphorylation in mammalian cells. Methods Mol. Biol. 527, 247–255 (2009)
13. Albeck, J.G., MacBeath, G., White, F.M., Sorger, P.K., Lauffenburger, D.A., Gaudet, S.: Collecting and organizing systematic sets of protein data. Nat. Rev. Mol. Cell Biol. 7(11), 803–812 (2006)
14. Jensen, L.J., Kuhn, M., Stark, M., Chaffron, S., Creevey, C., Muller, J., Doerks, T., Julien, P., Roth, A., Simonovic, M., Bork, P., von Mering, C.: STRING 8–a global view on proteins and their functional interactions in 630 organisms. Nucleic Acids Res. 37(Database issue), D412–D416 (2009)

15. Regev, A., Panina, E.M., Silverman, W., Cardelli, L., Shapiro, E.: Bioambients: An abstraction for biological compartments. Theoretical Computer Science, 141–167 (2004)
16. Heath, J.K., Khan, I., van Bakel, S., Vigliotti, M.G.: Modelling intracellular fate of FGF receptors with BioAmbients. In: Proceedings of Int. Workshop Quantitative Aspects of Programming Languages (QAPL 2008). ENTCS, vol. 220, pp. 181–197 (2008)
17. Sabouri-Ghomi, M., Wu, Y., Hahn, K., Danuser, G.: Visualizing and quantifying adhesive signals. Curr. Opin. Cell Biol. 20(5), 541–550 (2008)
18. Calder, M., Gilmore, S., Hillston, J.: Modelling the influence of RKIP on the ERK signalling pathway using the stochastic process algebra PEPA. In: Priami, C., Ingólfsdóttir, A., Mishra, B., Riis Nielson, H. (eds.) Transactions on Computational Systems Biology VII. LNCS (LNBI), vol. 4230, pp. 1–23. Springer, Heidelberg (2006)
19. Lu, A., Tebar, F., Alvarez-Moya, B., Lpez-Alcal, C., Calvo, M., Enrich, C., Agell, N., Nakamura, T., Matsuda, M., Bachs, O.: A clathrin-dependent pathway leads to KRas signaling on late endosomes en route to lysosomes. J. Cell Biol. 184(6), 863–879 (2009)

# BlenX4Bio – BlenX for Biologists

Corrado Priami[1,2], Paolo Ballarini[1], and Paola Quaglia[1,2]

[1] CoSBi, Italy
[2] University of Trento, Italy
{priami,ballarini,quaglia}@cosbi.eu

**Abstract.** We introduce BlenX4Bio, a high-level interface for the programming language BlenX. BlenX4Bio allows biologists to write BlenX programs without having any programming skills. The main elements of a biological model are specified by filling in a number of tables. Such tables include descriptions of both static and dynamic aspects of the biological system at hand and can then be automatically mapped to BlenX programs for simulation and analysis by means of the CoSBi Lab software platform. In this paper we illustrate the main characteristics of BlenX4Bio through examples taken from biology textbooks.

## 1 Introduction

It is well accepted that modern biology, and mainly systems biology, has to have modeling techniques in its toolset. The ever increasing complexity of the modelled systems pose a challenge to biological modelling and the classical equation-based (both mathematical and chemical) modeling approaches, in particular, proved to suffer from the combinatorial explosion of the size of the model.

The recent field of algorithmic systems biology [12] proposes computer science as a foundation for systems biology and as a means which, relying on basic principles of programming language theory, allows for the specification of complex systems in a concise and precise manner. Algorithms force modelers and biologists to think about the mechanisms that govern the behavior of the system they are studying. Algorithms are high-level expressions of a system's behaviour which describe the states in which a system can be and the conditions under which a system evolves from one state to another. Computer science relies on programming languages to unambiguously describe and execute algorithms. Furthermore, concurrency must be a core design principle of any foundational formalism for modern biology, given the importance of simultaneous interactions in biological phenomena. Concurrent programming languages can easily and efficiently express the mechanistic rules that propel algorithmic systems biology as they already proved to be effective in modeling, analyzing and programming large computer networks.

Algorithmic systems biology must equip concurrent programming with an high-level interface simplified to the extent that biologists can use it without knowing the basics of programming. As a consequence we designed BlenX4Bio so to hide all programming based details from its users. BlenX4Bio is strongly

P. Degano and R. Gorrieri (Eds.): CMSB 2009, LNBI 5688, pp. 26–51, 2009.
© Springer-Verlag Berlin Heidelberg 2009

inspired by molecular cell biology and by the narrative language which was introduced in [6,4] as a biologists-friendly modeling formalism. BlenX4Bio is targeted and optimized for the BlenX language [3] from which it inherits the modeling philosophy.

The use of programming languages to model biological systems is an emerging field that enhances current modeling capabilities (richness of aspects that can be described as well as easiness, composability and reusability of models) [13]. The metaphor which inspires this idea is one where biological entities are represented as programs being executed simultaneously and the interaction of two entities is represented by the exchange of a message between the programs representing those entities in the model [15]. The biological entities involved in the biological process and the corresponding programs in the abstract model are in a one-to-one correspondence, thus coping by construction with the combinatorial explosion of variables needed in the equational approach to describe the whole set of states through which a single component can pass.

Programming language-based modeling is sometimes erroneously confused with rule-based modeling or with markup languages like SBML. The confusion with rule-based modeling arises due to the word *programming* because the advocates of rule-based modeling usually state that they program the behavior of biological systems through rules resembling both chemical equations and term-rewriting systems. The association of programming language-based modeling and SBML-like formalisms arises instead due to the common word *language* in the two families of formalisms. SMBL is an XML-based markup language whose overall goal is to develop an open standard that will enable simulation software to communicate and exchange models, ultimately leading to the ability for researchers to run simulations and analyses across multiple software packages (http://xml.coverpages.org/sbml.html).

BlenX4Bio is not related to model exchange; it is a tabular interface suitable to biologists to program in BlenX. SBML has been originally defined by merging the basic features of the formalisms supported by BioSpice [1], DBSolve [10], E-Cell [17], Gepasi [9], Jarnac [7], StochSim [11], and Virtual Cell [18]. Although more and more features have been added to SBML in successive versions and releases, it retains the main philosophy of chemical and mathematical modeling of the first design thus making it philosophically different from BlenX4Bio that instead is strongly affected by the key points of Beta-binders [14] and its evolution into BlenX [3] and of process algebras in general [5].

The paper is organized as follows. The next section recalls the basics of BlenX and introduces the main features of BlenX4Bio. Section 3 defines the details of the tables to be filled in for building a BlenX4Bio model by relying on examples taken from textbooks.

## 2   BlenX and BlenX4Bio

This section first briefly recalls the main characteristics and the inspiring principles of the programming language BlenX. Then we abstract the main parts of a BlenX program corresponding to the relevant information to be organized to define high-level models of biological phenomena. The proposed interface to BlenX is based on a tabular format that resembles database tables or spreadsheet pages, both being natural tools for biologists. The organization and relationships among tables are then discussed in the BlenX4Bio paragragh.

## 2.1   BlenX

BlenX is a programming language based on Beta-binders [14]. A program is a collection of boxes (representing biological components) with typed and dynamically varying interfaces running in parallel. The status of an interface (binder) can be available/active, hidden (for instance when a three-dimensional structure changes and it hides a domain of a protein from further interactions) or complexed (when the binder is bound to the binder of another entity). Furthermore any interface is equipped with a type that describes its properties. Primitives are available to dynamically hide or unhide binders, to create new binders and to change the type of a binder to vary its properties. This feature is not available in other process calculi.

The main step of computation for BlenX is the interaction between boxes or their binding/unbinding. Interaction is sensitivity-based: quantities associated to binders stochastically determine the possibility of interaction between two entities. We release the key-lock interaction mechanism based on the notion, implemented in all process calculi, of exact complementarity of channel names (in fact this is an inheritance from computer science modelling where two programs can interact only if they know the exact address of the interacting partners). This allows us to avoid any global policy on the usage of names for interaction between components, i.e. we do not need centralized authorities that decide how to name entities or interfaces.

BlenX supports a one-to-one correspondence between biological components and boxes specified in the model: a biological component that can be in $n$ different states is just a box in our approach, and it is different from mathematical/chemical modelling where $n$ variables are needed to represent $n$ different states.

BlenX allows users to describe complexes and their dynamic generation: BlenX users can define complexes (set of components bound together through specific interfaces) or can generate complexes during the simulation relying on the specification of the components and on the complexation affinities of their interfaces. This feature highly reduces the number of components to be specified in the initial state because the products of interactions are generated by the execution and they need not to be described initially. The high parallelism of the execution then shows all the possible scenarios. BlenX modeling improves over other process-calculi based approaches and also releases the assumption of mathematical/chemical modelling that imposes the specifications of all the species, complexes and their states reached throughout all the simulation (variables) already in the initial state (set of equations).

BlenX can also specify the dynamics of systems through events: global conditions can be expressed by the amount of components or complexes in the system at a given time, by the simulation time or simulation steps. Conditions trigger the enabling of actions called events that are then stochastically selected including them in the set of standard interaction-enabled actions. Events are used to remove or inject entities from/into the system, to join two entities into a single one, or to split an entity in two entities. BlenX events are essential for programming perturbations of the systems triggered by conditions emerging during the simulation phase, and for observing how the overall behaviour is affected by them. An example could be the knock-out of a gene at a given time. Another possible use of events is to program sensitivity analysis of the system into the model driven by dynamic conditions emerging during simulation. Events are essential to develop an in-silico lab. No other process calculi-based tool supports events in the manner described above.

## 2.2  BlenX4Bio

BlenX4Bio is a high-level, tabular interface for the BlenX programming language. A BlenX4Bio model consists of a number of tables that describe all the relevant aspects that characterise biological phenomena.

The BlenX4Bio tables are *Compartments, Components, Complexes, Binding Affinities, Dynamics, Translocation Affinities* and *Parameters* (see the tabs in Fig. 1). The first three tables store the *static information* of a biological model. The tables *Binding Affinities* and *Translocation Affinities* represent the capabilities of the entities populating the system to interact with each other and, respectively, to change location. This information is given in terms of binding sites so that the user does not need to specify through rules or equations all the possible interactions between components. The *Dynamics* table provides information for driving the evolution of the model during simulation. Dynamics of systems is either expressed through simple constrained sentences like in the narrative language [6], or directly through BlenX code. Finally, the *Parameters* table is used to store all the quantitative information, functions and predicates introduced in the other tables.

The seven BlenX4Bio tables, although structurally different, share a number of commonalities which we describe here. Every element of each table can be addressed uniquely through its *name* that is made up of an alphanumeric string defined by the user and a numerical identifier (reported in parentheses after the name) assigned by the system to disambiguate elements with the same name. To exemplify the naming, see the definition of the two mitochondria in Fig. 3: they can be distinguished only by their numerical identifiers. We stress that replication of entries is, in fact, very common in biological modelling at least to model the existence of several compartments of the same kind (e.g. several lysosomes, mitochondria, and vesicles may be part of a cellular model) or to model the ubiquity of species (e.g., some species may exist simultaneously in different compartments).

Every BlenX4Bio table contains also three textual fields to describe the element under definition, to add references to the entry and to keep modeler's note. Furthermore, all the numerical parameters are collected into the *Parameters*

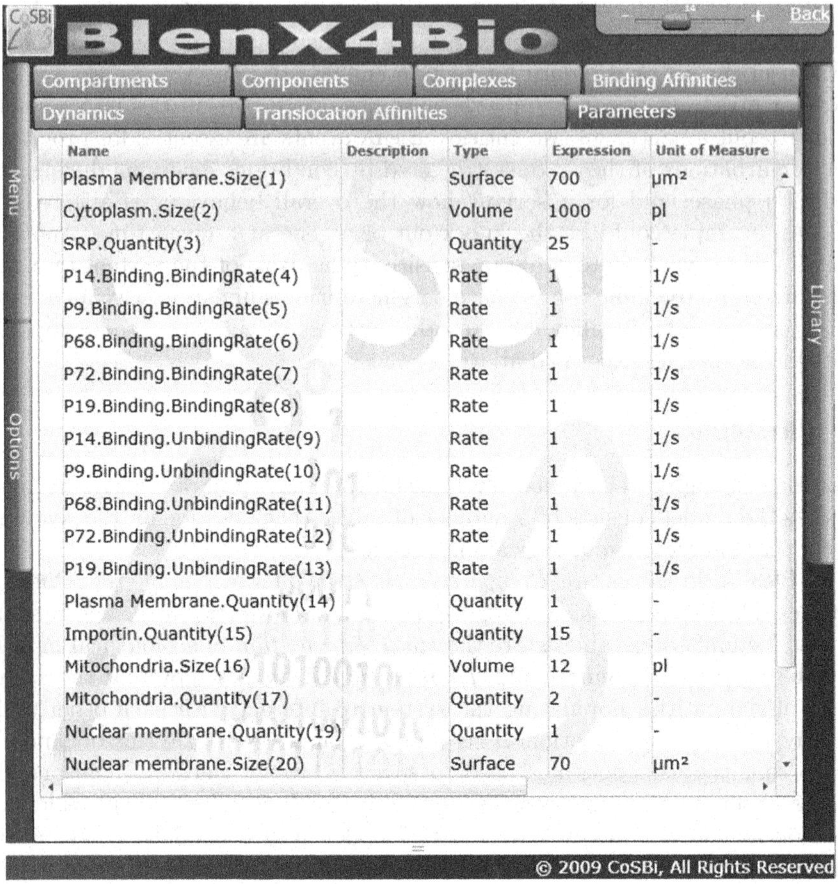


| Name | Description | Type | Expression | Unit of Measure |
|---|---|---|---|---|
| Plasma Membrane.Size(1) | | Surface | 700 | μm² |
| Cytoplasm.Size(2) | | Volume | 1000 | pl |
| SRP.Quantity(3) | | Quantity | 25 | - |
| P14.Binding.BindingRate(4) | | Rate | 1 | 1/s |
| P9.Binding.BindingRate(5) | | Rate | 1 | 1/s |
| P68.Binding.BindingRate(6) | | Rate | 1 | 1/s |
| P72.Binding.BindingRate(7) | | Rate | 1 | 1/s |
| P19.Binding.BindingRate(8) | | Rate | 1 | 1/s |
| P14.Binding.UnbindingRate(9) | | Rate | 1 | 1/s |
| P9.Binding.UnbindingRate(10) | | Rate | 1 | 1/s |
| P68.Binding.UnbindingRate(11) | | Rate | 1 | 1/s |
| P72.Binding.UnbindingRate(12) | | Rate | 1 | 1/s |
| P19.Binding.UnbindingRate(13) | | Rate | 1 | 1/s |
| Plasma Membrane.Quantity(14) | | Quantity | 1 | - |
| Importin.Quantity(15) | | Quantity | 15 | - |
| Mitochondria.Size(16) | | Volume | 12 | pl |
| Mitochondria.Quantity(17) | | Quantity | 2 | - |
| Nuclear membrane.Quantity(19) | | Quantity | 1 | - |
| Nuclear membrane.Size(20) | | Surface | 70 | μm² |

**Fig. 1.** BlenX4Bio registers all the quantitative information into the *Parameters* table for future updating and manipulation

table and each of them has an associated reliability value, i.e. a measure of how much the numerical datum is trusted. All the parameters can be changed and can be referred through naming conventions. This is an important feature that allows the user to dynamically change the values of the parameters, thus implementing in-silico experiments. Finally all the quantities must be equipped with a suitable unit of measure. BlenX4Bio implements integrity checks of the model under development to avoid ambiguities or inconsistencies in the set of relationships defined through the entries of the tables. Furthermore, BlenX4Bio supports the generation of BlenX code that can then be simulated through the BWB tool [2].

Although the main purpose of BlenX4Bio is completely different from the one of SBML-like formalisms [16], we conclude this section by commenting on the main differences between the current BlenX4Bio version and the most recent official version of SBML, namely: SBML level 2 version 4 release 1 (which is documented at http://sbml.org/Documents/Specifications). Since BlenX4Bio is an interface of the programming language BlenX, it naturally allows for modelling of sites-sensitive reactions (i.e. binding sites, phosphorylation sites, etc. are explicitly modelled in BlenX4Bio as attributes of BlenX4Bio components and sites sensitive reactions are straightforwardly expressed referring to components' sites). Furthermore, because of the binders-based semantics of BlenX, with BlenX4Bio reactions can be expressed either in terms of pairs of reactants (and their sites) or in terms of pair of sites (with no reference to components). The latter modality is a unique feature of BlenX4Bio (i.e. BlenX) which, prevents the specification of a biological model from suffering of the infamous combinatorial explosion issue.

SBML Level 2, on the other hand, does not support modelling of components' sites: SBML biochemical species are elementary objects (i.e. sites cannot be represented) which makes SBML 2 not suitable to model sites-sensitive reactions. Even though an improved third level of SBML is under development, which comprises the so called "Multi-state multi-component species" package designed to allow, modeling of sites-sensitive reactions, at present a stable version is not yet available hence it is difficult to understand how the two approaches compare. Another difference between BlenX4Bio and SBML 2 regards the ability to represent translocation of species between compartments. BlenX4Bio is designed to support fine-grained modelling of actual translocation mechanisms (as they are described in biology textbooks), which is: mechanisms based on transmembrane channels and translocation enablers. With SBML 2, on the other hand, translocation can be modelled only in a very abstract manner (i.e. by reactions occurring between species located in adjacent compartments): channels based mechanisms cannot be described with SBML. We also mention the ability of BlenX4Bio to describe arbitrarily large complexes like polymers without the need of introducing an arbitrary number of species as it would be the case in SBML. Finally, the possibility of using boolean-valued predicates to define state-based dynamics of models which is available with BlenX4Bio, is also not easy to render in SBML.

## 3  BlenX4Bio Modeling

This section illustrates the elements of the BlenX4Bio language in detail by adopting the following schema: first we describe the biological phenomenon of interest (e.g., *the spatial architecture of biological compartments* of a cell, or *the transportation mechanisms* driving inter-membrane translocations of biochemical species) by means of specific examples, then we find a corresponding unifying abstraction, and eventually we show how such abstractions/examples are represented in BlenX4Bio. In doing so, we provide details of the structure of every BlenX4Bio table.

### 3.1  Spatial Architecture of Cell Compartments

From a topological perspective a cell can be seen as an architecture of physical membrane-bounded locations (i.e. compartments whose boundary are determined by a membrane). Each compartment is a container of substances and the role it plays in regulating cell life depends on the substances that it contains. The content of compartments changes over time as a result of cell regulation, hence substances continuously migrate from one location of the cell to another in response to internal and/or external stimuli. The spatial architecture of compartments, therefore, is an essential aspect of cellular biology which calls for adequate modelling support. BlenX4Bio supports modelling of the spatial architecture of compartments via the *Compartments* table. We illustrate modelling of compartments architecture with BlenX4Bio by means of a concrete example: the compartments architecture of Eukaryote cells, which is described next.

The architecture of eukaryotes is made of a *plasma membrane* of about 700 $\mu m^2$ that surrounds the cell interior which is partitioned into smaller subcompartments called *organelles*. The interior, or *lumen*, of each organelle is enclosed by one or more *biomembranes* (globally amounting at 7000 $\mu m^2$) and contains a unique set of proteins that characterizes the functions of the organelle together with the membrane-bound proteins. The largest organelle is the *nucleus* that contains the DNA, the RNA synthetic apparatus, and a fibrous matrix. The part outside the nucleus is called *cytoplasm* and it contains all the other organelles. The aqueous part of the cytoplasm is called *cytosol* and it contains its distinctive proteins as well. The major organelles, besides the nucleus, that populate the cytoplasm are detailed below.

- *Endoplasmatic reticulum* (ER) has the largest membrane in the cell and is important for the synthesis of lipids, membrane proteins, and secretory proteins.
- *The Golgi complex* sorts secreted and membrane proteins.
- *Secretory vesicles* target proteins leaving the Golgi complex.
- *Lysosomes* are acidic organelles that contain degradative enzymes.
- *Mitochondria* are the main sites of ATP production.
- *Phagosomes* engulf in the cytoplasm large particles from the exterior of the cell like bacteria that are then delivered to lysosomes for degradation.
- *Endosomes* take up soluble macromolecules from the exterior of the cell and are then delivered to lysosomes.
- *Peroxisomes* degrade fatty acids and toxic compounds.
- *Chloroplasts* are the apparatus where photosynthesis takes place in plant cells.

Different types of eukaryotes are distinguished by the shape of the cell and the location of their organelles. The shape of a cell is determined by its *cytoskeleton*, a network of three kinds of filaments that mechanically support membranes. Filaments are classified according to their size into *actin filaments* (or microfilaments), *intermediate filaments*, and *microtubules*. Actin filaments provide the cell with structural support and motility; intermediate filaments provide the cell with support for the nuclear membrane and help cell adhesion into tissues formation; microtubules provide the cell with structural support, motility and cell polarity. The overall surface of the cytoskeleton (about 94000 $\mu m^2$) offers anchors to many membranes and proteins that bind to it, and works as a scaffold where many reactions take place.

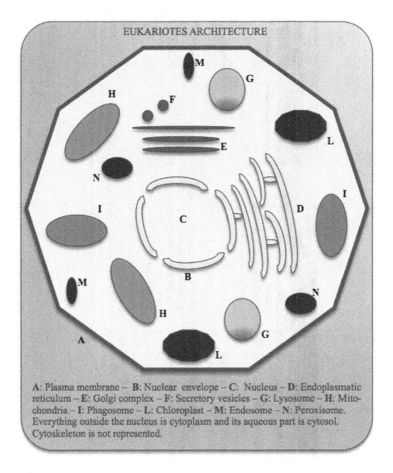

**Fig. 2.** The architecture of an eukaryotic cell that highlights its main compartments and sub-compartments

An abstract representation of the architecture of eukaryotes is depicted in Fig. 2. For more information on the architecture of cells see [8], chapter 5. The representation in BlenX4Bio of a part of the architecture in Fig. 2 is reported in Fig. 3.

The *Compartments* table always contains (even when not displayed) a global compartment termed *System*, which encloses all the other compartments. As a consequence, a BlenX4Bio model with no user-defined compartment is perfectly legal. Besides the columns for the textual description of references to the entry, and for the user's notes, the *Compartments* table contains the fields described below.

The column *Membrane* is a tag used to distinguish between two-dimensional (like, e.g., both the plasma membrane and the nuclear membrane in Fig. 2) and three-dimensional compartments (all the other compartments in the same figure).

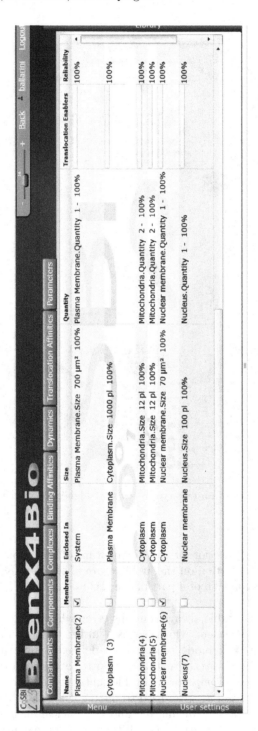

**Fig. 3.** BlenX4Bio partial architecture of an eukaryotic cell

To represent the hierarchical structure of compartments, for each newly introduced compartment, the user has to specify in which other compartment it is contained. This is the purpose of the *Enclosed In* column that offers a selection among the already defined compartments.

The *Size* of compartments is useful to compute the kinetic rates of reactions and hence to correctly manage simulations. As for all the other quantities, for the values inserted in the *Size* column it is necessary to specify an accompaining reliability measure.

The *Quantity* column denotes the number of available replicas of the same compartment. For instance, the system described in Fig. 3, contains two mitochondria.

The components populating compartments can move from one compartment to another. This process is called translocation and it is driven by facilitator complexes and proteins that control transmembrane movement of entities as well as protein sorting. Such macromolecules are classified in BlenX4Bio as *Translocation Enablers*. An example is *Importin* that translocates cargo proteins from the cytoplasm to the nucleus through the nuclear pore complex. The modeling of translocation scenarios is fully described below in Section 3.5.

## 3.2 Components

The components that are of interest for modeling a biological system constitute the list of elementary parts from which we start simulating and analyzing the system behavior. Cells are mainly a soup of molecules varying in size and functions floating in a water-based solution. Basic ingredients of the cell range from small molecules like ATP that stores chemical energy, hormones or neurotransmitters that transport signals, and monomers that are the basic bricks used to assemble larger molecules called polymers. The major class of polymers produced by a cell ([8], chapter 1) are: polysaccharides (whose forming monomers are sugars), proteins (whose forming monomers are aminoacids), and nucleic acids (whose forming monomers are called nucleotides).

In this presentation we take elementary molecules as the smallest entities of our model, namely we work at a level of abstraction where all the possible constituents of molecules are irrelevant and hence intentionally neglected. In the same way, if we had to consider protein interactions, we would not model the aminoacids forming those proteins. Also, if protein binding and unbinding were relevant events for the considered model, we would describe either the single proteins forming the macro-molecule or the complex as elementary components.

We exemplify here how to completely describe the relevant information of components in BlenX4Bio by describing the constituents of a *signal recognition particle* (SRP) represented in Fig. 4. Each component has a unique name as all the elements of the model, and every component is characterized by a list of sites that are used to interact, bind/unbind with the other components and complexes in the model. For instance, the *Importin* component has a site called *CargoReceptor* that is used to bind other proteins. The state of such site is

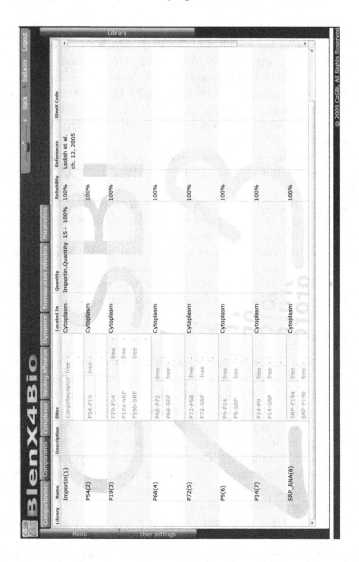

**Fig. 4.** Representation of elementary components in BlenX4Bio

*free*, meaning that it is available for binding. More generally, the current state of a site ranges over a set containing, e.g., *bound*, *phosphorylated*, *sumoylated*, *acetylated*. Another relevant information is the location where the component can be found, which can be used for managing the spatial structure of the system as well as the movement of objects. For example, *Importin* is located in the *Cytoplasm*.

Sites deserve some more discussion. Proteins can freely move within compartments or can be bound to membranes spanning over both surfaces (integral proteins) or just residing on one side of the membrane (peripheral proteins) ([8],

```
                                  BlenX
// Program File BlenX4Bio model CMSB09
[time = 10]

let Importin: bproc =
#(id_1, CargoReceptor)[ nil  ];

let P54: bproc =
#(id_2, P54-P19)[ nil  ];

let P19: bproc =
#(id_3, P19-P54),#(id_20, P19a-SRP),#(id_21, P19b-SRP)[ nil  ];

let P68: bproc =
#(id_7, P68-P72),#(id_22, P68-SRP)[ nil  ];

let P72: bproc =
#(id_9, P72-P68),#(id_23, P72-SRP)[ nil  ];

let P9: bproc =
#(id_11, P9-P14),#(id_24, P9-SRP)[ nil  ];

let P14: bproc =
#(id_13, P14-P9),#(id_25, P14-SRP)[ nil  ];

let SRP_RNA: bproc =
#(id_26, SRP-P19a),#(id_27, SRP-P19b),#(id_28, SRP-P68),
#(id_29, SRP-P72),#(id_30, SRP-P9),#(id_31, SRP-P14)[ nil  ];


run Importin || P54 ||  P19 ||  P68 ||  P72 ||  P9 ||  P14 ||  SRP_RNA
```

**Fig. 5.** BlenXcode corresponding to the BlenX4Bio description in Fig. 4

chapter 5). The single critical case to determine the location of a site (whether it is exposed either towards the outside of the membrane or towards the inside of the membrane) is relative to integral proteins. Indeed the exposition of the sites of integral proteins determines the set of components that can come in touch with the site due to the physical boundaries originated by the spatial hierarchical structure of cells. An example is the integral trans-membrane protein P in Fig. 6 that has three sites: one directed towards the outside of the membrane (S3) and two directed towards the inside of the membrane (S1 and S2). To correctly model these situations, we distinguished compartments into membranes (2D compartments) and normal ones (3D compartments). When a component is located in a compartment that turns out to be a membrane, the site description must be extended by a further field that says whether the site exposition is *Out* (towards the exterior of the membrane), *In* (towards the inner compartments delimited by the membrane) or *Within* (inside the membrane). For example Fig. 7 reports the description of the integral protein P drawn in Fig. 6.

We end this subsection by noting that the information stored in the *Components* table is almost in a one-to-one correspondence with the definition of the boxes in the BlenX program. Each component is mapped into a box and each site of a component is mapped into a corresponding interaction interface on the corresponding box. For example, the following portion of text

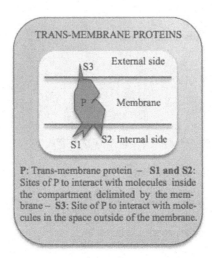

**Fig. 6.** Binding sites of trans-membrane proteins

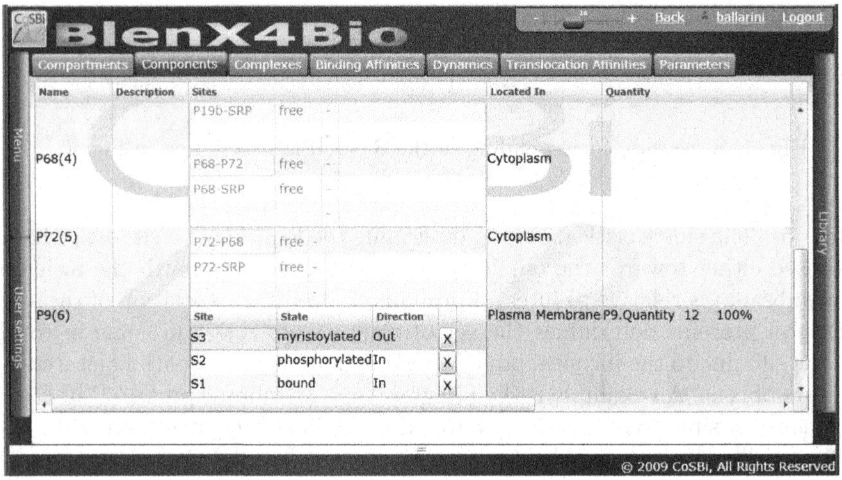

**Fig. 7.** Managing locations of sites in BlenX4Bio

```
let P19: bproc =
#(id_3, P19-P54),#(id_20, P19a-SRP),#(id_21, P19b-SRP)[ nil ]
```

in Fig. 5 corresponds to the description of P19 in Fig. 4. Indeed, the text is the BlenX declaration of a box named P19 with three interaction sites delimited within each #(...) and typed by P19-P54, P19a-SRP, and by P19b-SRP, respectively.

## 3.3   Complexes

Complexes are macromolecules formed by sets of entities bound together. Proteins can form complexes to accomplish specific functions. For instance the SRP that identifies the signal sequence exiting ribosomes to translocate the synthesizing protein to the ER is made up of a piece of RNA and the six proteins P9, P14, P19, P54, P68 and P72. Protein P54 binds to the signal sequence of the synthesizing protein; P9 and P14 interact with the ribosome; P68 and P72 are required for the translocation of the protein (see [8], chapter 16 and Fig. 12). All the proteins bind directly to the piece of RNA except for P54 that binds to P19. Furthermore, P68 and P72 are a dimer as well as P9 and P14.

Complexes are usually formed by bonds between domains of proteins with a high level of affinity/sensitivity. A suitable abstract representation for complexes is a graph whose nodes are the boxes we used as abstract representation of components and whose arcs represent the bonds connecting the affine domains of the elementary components. An abstract representation of SRP is reported in Fig. 8.

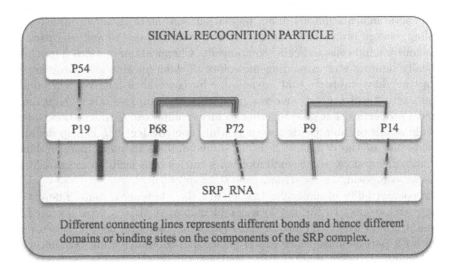

**Fig. 8.** Abstract representation of complexes through graphs of elementary components. The nodes represents te components and the arcs the bonds between components. Arcs have a type describing the affinity between the domains or binding sites that are connected.

To uniquely refer to a bond, it is enough to assign unique names to the involved components and to the binding sites (domains) of those components. In this way we can uniquely identify a bond as, e.g., the string "component1, site1; component2, site2". For instance if we name the RNA binding sites in Fig. 8, from left to right, SRP-P19a, SRP-P19b, SRP-P68, SRP-P72, SRP-P9 and SRP-P14; their complementary binding sites on the protein domains, from left to right, P19a-SRP, P19b-SRP, P68-SRP, P72-SRP, P9-SRP and P14-SRP;

the binding sites that form the bond between P19 and P54, from bottom to top, P19-P54 and P54-P19; the binding sites that form the bond between P68 and P72, from left to right, P68-P72 and P72-P68; the binding sites that form the bond between P9 and P14, from left to right, P9-P14 and P14-P9; then the column *Constituents* in Fig. 9 completely describes the SRP complex. For instance, the first line of the table describes the binding of P14 and SRP_RNA through the sites P14-SRP and SRP-P14.

The further information stored in the *Complexes* table defines the quantity of the complex available at the beginning of the simulation (in this case 25 SRP) associated with a reliability measure as for all the quantitative data.

As a final observation on the *Complexes* table, we note that the information stored in its fields is needed to define the initial environment defining the complexes that populates BlenX programs in the starting state.

### 3.4   Binding Affinities

The chemical interactions occurring in a cell are the essence of life: they implement genetic information handling, regulation, sensing of the environment and signaling, energy production and conversion, synthesis of the material used to continuously build cells and cells' components. Chemical interactions happen by physically binding and unbinding molecules of different size and structure as long as they have subparts that are suitable for covalent or non-covalent bonds. A main activity of the cell involves proteins with complementary shapes and chemical properties that glue together to form complexes. Covalent bonds usually hold atoms or aminoacids together, while non-covalent bonds help defining the structure of the complex. It is not only proteins to participate in binding/unbinding actions; some small molecules can serve as facilitators or building blocks for composing larger structures ([8], chapter 2).

The compatibility of molecules that enable their binding is quite different in biology with respect to the key-lock complementarity that is typically used in process algebras where, for instance, to enable a communication between processes both parties must share exactly the same communication channel with no possibility of approximation. Biological shapes and chemical properties allow instead interactions between molecules that are not perfectly complementary to each other, changing the strength of the bond or the probability of the interaction. We call this notion *affinity*, and model it in the BlenX4Bio *Binding Affinities* table (see Fig. 10 for an overview of its fields).

The sites (interfaces) through which an entity is potentially able to interact have been already identified in the definition of the corresponding entry in the *Components* table. Here, we only define affinities between sites. Each site is uniquely identified by the name of the component it belongs to, and by its proper the name. We list here pairs of sites and we associate to each pair a binding rate quantifying the affinity and an unbinding rate quantifying the reversibility of the binding. These quantities are then used to drive simulations.

We can further detail the binding/unbinding actions by specifying conditions that must be satisfied in order for the action to occur. For instance, P14 and

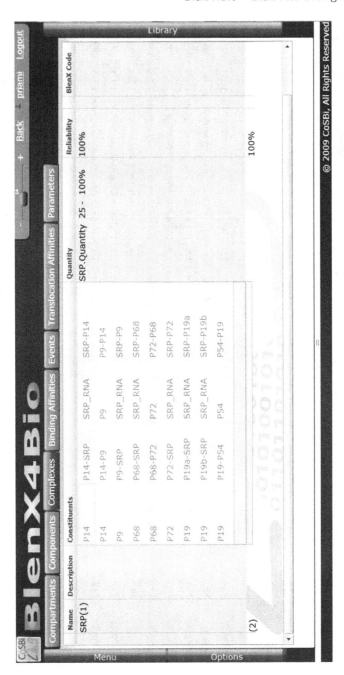

**Fig. 9.** Representation of the SRP complex in BlenX4Bio through the complete listing of the bonds identified by the names of the components, and of the sites that bind together. The four columns under *Constituents* refer, respectively, to the first component, the binding site of the first component, the second component, and the binding site of the second component.

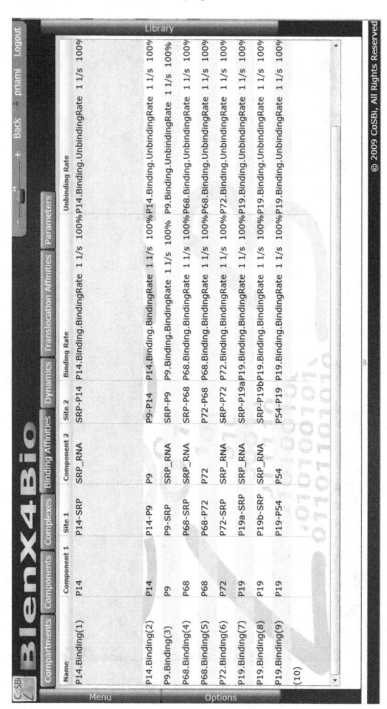

**Fig. 10.** Representation of the binding affinities in BlenX4Bio needed to build the SRP complex

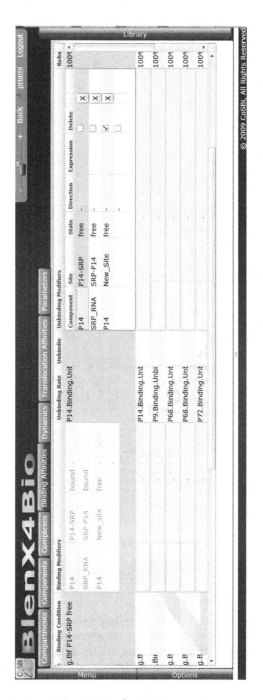

**Fig. 11.** Representation of the binding/unbinding conditions and of the modifiers

44      C. Priami, P. Ballarini, and P. Quaglia

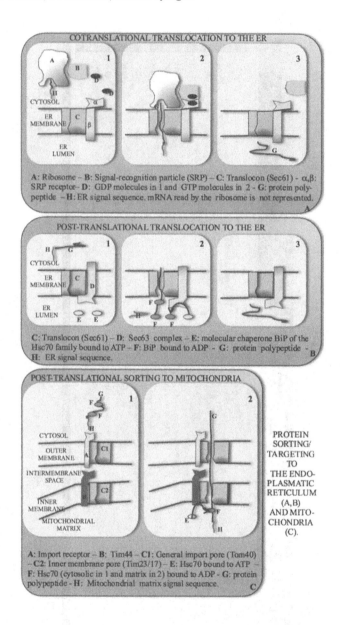

**Fig. 12.** The three boxes show the role of traslocation enablers in protein sorting. (A) The enablers are SRP and its receptor; (B) the enablers are the translocon and the signal region of the protein; (C) the enablers are the Import receptor and the signal region of the protein. In all the three cases, the activation of the translocation enablers by their mutual interaction open a channel in a membrane that allows one protein at time to move.

SRP_RNA can bind only if the P14-SRP site of P14 is *free*. Similarly, we can associate modifiers to binding/unbinding actions to manipulate the state of the sites as a consequence of binding/unbinding. Modifiers can also create or delete new sites (see Fig. 11). As an example, the binding of P14 and SRP_RNA creates a new free site New_Site that is deleted when P14 and SRP_RNA unbind.

We end this subsection by noting that the information stored in the *Binding Affinities* table is almost in a one-to-one correspondence with the affinities of the interaction interfaces of the boxes populating BlenX programs. Furthermore, binding modifiers and unbinding modifiers describe the actions that the binding/unbinding components have to take as a consequence of the formation/deletion of bonds.

## 3.5   Translocation Affinities

Half of the kinds of proteins of a mammalian cell have to be delivered to specific locations in order for the cell to function properly. The process of moving proteins around is usually called *protein targeting* or *protein sorting*. Protein sorting can mainly happen through *secretory pathways* or through *nonsecretory pathways* ([8], chapter 16).

Protein sorting is graphically illustrated in Fig. 12, where the uppermost two cartoons (A and B) show the use of secretory pathways. The secretory pathway initially targets secretory proteins to the endoplasmatic reticulum (ER)

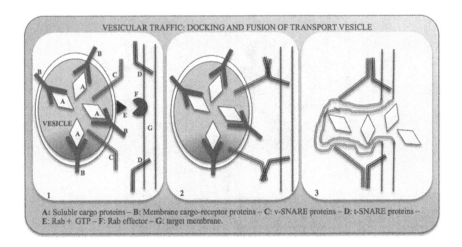

**Fig. 13.** Membrane cargo-receptors recruits the proteins to be transported by a vesicle. Once the recruitment is complete, the membrane forms a vesicle with a Rab+GTP receptor for docking the vesicle towards the right target membrane. Once the rab effector on the target membrane is identified and bound, the two SNARE counterparts tightly bind together allowing the fuse of the vesicle with the target membrane. As a consequence, the whole content of the vesicle is delivered in the compartment bound by the target membrane. The translocation enablers here are the v-SNARE and t-SNARE proteins.

membrane and translocated in the ER lumen (see Fig. 12 A and B). When translation is completed these proteins are further delivered to the lumen of Golgi and lysosomes or to plasma membranes via vesicles. A uniform principle drives this type of protein sorting: proteins are moved from one membrane-bounded compartment to another membrane-bounded compartment mediated by *transport vesicles* ([8], chapter 17). Vesicles collects *cargo proteins* in buds formed from the membrane of the source compartment and then fuse with the membrane of the target compartment by releasing in one shot all their content (see Fig. 13).

The nonsecretory pathway sorts proteins to the ER, mitochondria (see Fig. 12 C), chloroplasts, peorxisomas, and the nucleus. The main difference with respect to the situation described above is that these proteins are not secretory proteins and most of them are completely translated into the cytoplasm before their translocation. The mechanism of translocation is similar for all the cases mentioned so far apart from nuclear import/export ([8], chapter 12.3). Nuclear transport happens through the *nuclear pore complexes* (NPC) that form holes on the nuclear membrane. Macromolecules equipped with *nuclear localization signals* (NLS) bind to *importin/exportin* proteins that drive the macromolecules through the nuclear pores (see Fig. 14).

We classify all the above kinds of translocation phenomena into a single abstract translocation process suitable for modeling and mapping into the BlenX language. The abstract translocation phenomenon can be seen as a door that

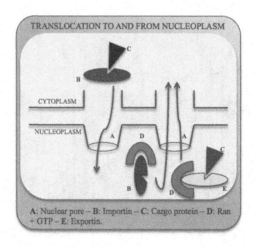

**Fig. 14.** A free importin binds to a cargo protein in the cytoplasm and the complex translocates into the nucleoplasm through the nuclear pore. The importin part of the complex interacts with Ran-GTP that causes a decrement of the affinity between the importin and the cargo protein. As a consequence the cargo protein is released in the nucleoplasm and the free importin migrates back to the cytoplasm. Similarly, cargo proteins can migrate from the nucleoplasm to the cytoplasm by forming a complex with an exportin and a Ran-GTP. The translocation enablers are here the importin and exportin proteins.

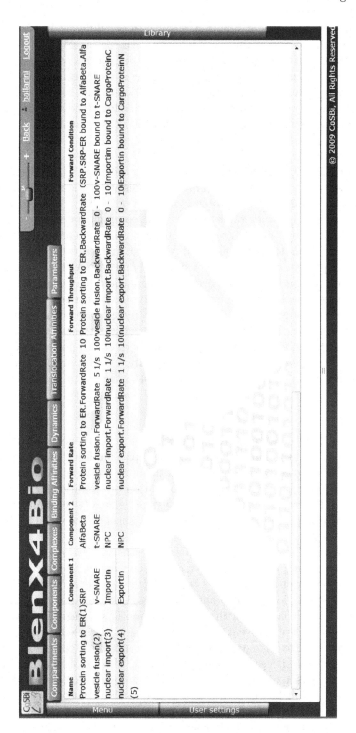

**Fig. 15.** BlenX4Bio model of translocations

can be opened by unlocking it simultaneously with two different keys. In the case of the nucleus the second enabler can coincide with the cargo protein that must complex with importin to enter the nucleus or with exportin to exit the nucleus. Once the door is open (e.g., nuclear pore, translocon Sec61, general import pores Tom40, Tim23/17, fuse of membranes in vesicular cases), its size determines how many entities can simultaneously step through the door (the throughput of the translocation channel). We distinguish the vesicular case in which all the content is translocated simultaneously from all the other cases discussed where a single entity at time can pass.

The BlenX4Bio table *Translocation Affinities* is the place where the above abstract representation of translocation phenomena is stored. For instance, Fig. 15 shows how the enablers of the protein sorting to ER (Fig. 12A) are modeled. The *Translocation Affinities* table specifies the components acting as enablers and some quantitative information on translocation. The enablers reported in Fig. 15 are the SRP in the cytoplasm and AlfaBeta, an integral protein of the ER membrane for protein sorting to ER; v-SNARE and t-SNARE for vesicular fusion with other compartments; Importin and the nuclear pore complex (NPC) to import cargo proteins into the nucleus; and finally Exportin and NPC to export cargo proteins from the nucleus. We further specify the rate of the translocation and its throughput, i.e. the speed and the bandwidth of the channel that allows translocation of components expressed in numbers of components per unit of time that traverse the channel. Finally we specify a condition that enables the opening of the channel and hence favours translocation.

## 3.6   Dynamics

So far we described how to represent the structural components of the system under investigation by defining the hierarchical space architecture and by listing the basic entities and complexes available in the initial state which is the starting point of the simulation. Furthermore, information on affinities for binding has been provided. Below, we discuss how the system behaves when time passes and how the simulation process deals with the static information defined in the tables already described.

The dynamics of systems is specified in the BlenX4Bio *Dynamics* table. All the entries have a unique name for referral purposes and are characterized by an algorithmic description of the action to be taken in order to make the system evolve. These descriptions are characterized by a conditional part and an action part that is enabled only when the side condition is satisfied. The expressions stating the dynamics of an event are specified in a constrained natural language similar to the narrative language [6]. Examples are reported in Fig. 16. Note that the dot notation is used to identify sub-parts of components. For instance, AlfaBeta.Alfa refers to the site Alfa of the component AlfaBeta (we do not describe here the full language because it is outside the scope of the present paper and we refer to the operative documentation of BlenX4Bio). Since the dynamics is quantitative and stochastic in nature, we must associate every entry of the *Dynamics* table with a rate.

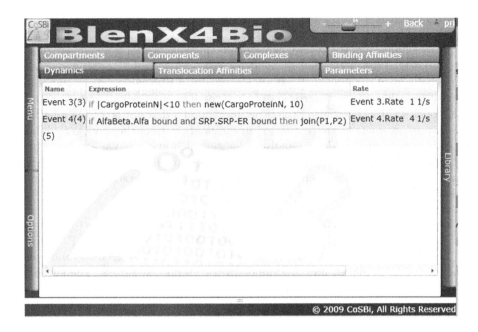

**Fig. 16.** BlenX4Bio model of system dynamics

We observe that the information stored in the *Dynamics* table defines the algorithm to be executed to simulate the temporal evolution of the described biological systems. This is the kernel of the BlenX program corresponding to the model specified through BlenX4Bio, and indeed the user can choose to specify the dynamics by directly writing BlenX code.

## 4    Conclusions and Future Work

The paper presented a tabular interface to the BlenX language that hides all the programming details from the users of BlenX4Bio. The CoSBi Lab platform supports the translation of BlenX4Bio models into BlenX programs ready to be simulated. In order to demonstrate the soundness of BlenX4Bio representation of biology we have considered basic aspects of cellular biology taken from biology textbooks and we provided evidence that BlenX4Bio is indeed a very useful means through which biologists can straightforwardly represent biological phenomena in a simple and intuitive manner. The key characteristics that distinguish BlenX4Bio from SBML or equational/chemical modeling have also been discussed.

We are currently working on extensions of the BlenX4Bio modelling approach. In fact, if on one hand BlenX4Bio allows the modeler to specify the biological system to be studied, on the other hand the actual analysis of the modeled systems requires some additional information to be provided. We envisage such additional information consisting of two parts: the aspects of the system's behaviour which is relevant to *look at* and the ways in which the specified model could be *perturbed*.

In our extension of BlenX4Bio, experiments describe in a formal manner which are the parts of the model that the modeler wishes to perturb in order to analyse the behaviour of the modelled system. Essentially, a BlenX4Bio experiment consists of two elements: a description of the (possibly conditional) perturbations the model is subject to during the experiment, and a description of the relevant *behavioural characteristics* (expressed in terms of *queries*) the modeller wishes to concentrate on in order to analyse the modelled system. We are currently extending the CoSBi Lab software to support the *experiments specification* extension of BlenX4Bio and hence move towards a fully integrated modeling and in-silico experimental platform for biologists.

**Acknowledgements.** The authors thank A. Csikasz-Nagy, L. Dematté, M. Forlin, M.L. Guerriero, J. Heath, A. Ihekwaba, A. Inga, R. Larcher, P. Lecca, R. Mardare, T. Mazza, I. Mura, A. Palmisano, N. Phuong, D. Prandi, A. Quattrone, A. Romanel, and J. Zamborsky for useful discussions. The authors thank F. Benedetti and the CoSBi development team for the implementation of BlenX4Bio.

# References

1. BioSpice home page, http://biospice.sourceforge.net/
2. Dematté, L., Priami, C., Romanel, A.: The Beta Workbench: a computational tool to study the dynamics of biological systems. Brief. Bioinform. 9(5), 437–449 (2008)
3. Dematté, L., Priami, C., Romanel, A.: The BlenX Language: A Tutorial. In: Bernardo, M., Degano, P., Zavattaro, G. (eds.) SFM 2008. LNCS, vol. 5016, pp. 313–365. Springer, Heidelberg (2008)
4. Guerriero, M.L., Dudka, A., Underhill-Day, N., Heath, J.K., Priami, C.: Narrative-based computational modelling of the gp130/jak/stat signalling pathway. BMC Systems Biology 3(40) (2009)
5. Guerriero, M.L., Prandi, D., Priami, C., Quaglia, P.: Process calculi abstractions for biology. In: Condon, A., Harel, D., Kok, J.N., Salomaa, A., Winfree, E. (eds.) Algorithmic Bioprocesses. Springer, Heidelberg (2009)
6. Heath, J.K., Guerriero, M.L., Priami, C.: An automated translation from a narrative language for biological modelling into process algebra. In: Calder, M., Gilmore, S. (eds.) CMSB 2007. LNCS (LNBI), vol. 4695, pp. 136–151. Springer, Heidelberg (2007)
7. Jarnac home page, http://www.sys-bio.org/software/jarnac.htm
8. Lodish, H., Berk, A., Matsudaira, P., Kaiser, C.A., Krieger, M., Scott, M.P., Zipursky, S.L., Darnell, J.: Molecular Cell Biology, 5th edn. Freeman, New York (2004)
9. Mendes, P.: Biochemistry by numbers: simulation of biochemical pathways with Gepasi 3. Trends Biochem. Sci. 22, 361–363 (1997)
10. Moehren, G., Markevich, N., Demin, O., Kiyatkin, A., Goryanin, I., Hoek, J., Kholodenko, B.N.: Temperature dependence of epidermal growth factor receptor signaling network can be accounted for using a kinetic model. Biochemistry 41(1), 306–320 (2002)

11. Le Novere, N., Shimizu, T.S.: StochSim: modelling of stochastic biomolecular processes. Bioinformatics 17, 575–576 (2001)
12. Priami, C.: Algorithmic systems biology. CACM 52(5), 80–88 (2009)
13. Priami, C., Quaglia, P.: Modeling the dynamics of bio-systems. Briefings in Bionformatics 5(3) (2004)
14. Priami, C., Quaglia, P.: Beta Binders for Biological Interactions. In: Danos, V., Schächter, V. (eds.) CMSB 2004. LNCS (LNBI), vol. 3082, pp. 20–33. Springer, Heidelberg (2005)
15. Regev, A., Shapiro, E.: Cells as computations. Nature 419, 343 (2002)
16. SBML home page, `http://sbml.org/`
17. Tomita, M., Hashimoto, K., Takahashi, K., Shimizu, T.S., Matsuzaki, Y., Miyoshi, F., Saito, K., Tanida, S., Yugi, K., Venter, J.C., Hutchison, C.A.3.: E-CELL: software environment for whole-cell simulation. Bioinformatics 15(1), 72–84 (1999)
18. Virtual Cell home page, `http://www.ibiblio.org/virtualcell/`

# Modelling Biological Clocks with Bio-PEPA: Stochasticity and Robustness for the *Neurospora crassa* Circadian Network

Ozgur E. Akman[1], Federica Ciocchetta[2], Andrea Degasperi[3],
and Maria Luisa Guerriero[4]

[1] Centre for Systems Biology at Edinburgh, The University of Edinburgh,
Edinburgh EH9 3JU, Scotland, UK
[2] The Microsoft Research - University of Trento Centre for Computational
and Systems Biology, Trento, Italy
[3] Laboratory for Foundations of Computer Science, The University of Edinburgh,
Edinburgh EH8 9AB, Scotland, UK
[4] Department of Computing Science, The University of Glasgow,
Glasgow G12 8QQ, Scotland, UK

**Abstract.** Circadian clocks are biochemical networks, present in nearly all living organisms, whose function is to regulate the expression of specific mRNAs and proteins to synchronise rhythms of metabolism, physiology and behaviour to the 24 hour day/night cycle. Because of their experimental tractability and biological significance, circadian clocks have been the subject of a number of computational modelling studies.

In this study we focus on the simple circadian clock of the fungus *Neurospora crassa*. We use the Bio-PEPA process algebra to develop both a stochastic and a deterministic model of the system. The light on/off mechanism responsible for entrainment to the day/night cycle is expressed using discrete time-dependent events in Bio-PEPA.

In order to validate our model, we compare it against the results of previous work which demonstrated that the deterministic model is in agreement with experimental data. Here we investigate the effect of stochasticity on the robustness of the clock's function in biological timing. In particular, we focus on the variations in the phase and amplitude of oscillations in circadian proteins with respect to different factors such as the presence/absence of a positive feedback loop, and the presence/absence of light. The time-dependent sensitivity of the model with respect to some key kinetic parameters is also investigated.

## 1 Introduction

Circadian clocks are oscillatory gene networks developed by living organisms in order to adapt to the 24-hour day/night cycle. In general, the biochemical mechanisms regulating circadian rhythms are robust enough for approximately 24 hour oscillations to persist over a range of constant lighting and temperature conditions. Exposure to periodic external stimuli (e.g. light/dark or temperature cycles) has the effect of resetting these free-running oscillations so as to establish stable phase relationships with the forcing stimulus (circadian entrainment). This enables cyclic changes in the environment to

P. Degano and R. Gorrieri (Eds.): CMSB 2009, LNBI 5688, pp. 52–67, 2009.
© Springer-Verlag Berlin Heidelberg 2009

be anticipated, such as seasonal variations in the length of day (photoperiod) [1]. Circadian rhythms are present in nearly all eukaryotes, from mammals and plants, to insects and fungi. There is now detailed experimental data showing that these rhythms can be produced by networks of multiple, interlocked positive and negative feedback loops in which the protein product of a gene modulates expression of either its own transcript or that of another target gene in the network [2].

Several mathematical models have been proposed in recent years to describe the specific oscillation-generating mechanisms in a range of different organisms. These include the fruit fly *Drosophila melanogaster* [3,4], the plant *Arabidopsis thaliana* [5,6] and the mouse *Mus musculus* [7,8]. Here we focus on the fungus *Neurospora crassa*, which possesses one of the most comprehensively studied circadian networks [9]. In recent years, a number of mathematical models of the *Neurospora* clock have been developed, including continuous-deterministic models that are described in terms of ordinary differential equations (ODEs) [10,11,12,13,14], as well as discrete-stochastic models [15,16]. Such models have been used successfully to explore the relationship between the architecture of the *Neurospora* circadian network and the robustness of its function in biological timing.

Within this theme, our aim in this work is to investigate the effect of stochastic fluctuations on the performance of the *Neurospora* clock. While deterministic models are good approximations of real biochemical systems when the number of molecules is sufficiently high, at low copy numbers the effect of random fluctuations becomes significant and so stochasticity needs to be taken into account to obtain a faithful representation of the real biochemical system [17]. To explore the effect of these fluctuations on circadian timing in *Neurospora*, we implement a discrete-stochastic version of a continuous ODE model previously developed to investigate the entrainment of the clock by light and temperature [13,18]. We use the ODE representation of this clock to validate our stochastic model and to highlight the differences between deterministic and stochastic representations of the network. In particular, where previous stochastic studies have concentrated mainly on the unforced (free-running) *Neurospora* clock, modelling entrainment as a weak modulation of transcription [15,16], here we investigate how stochasticity affects the robustness of circadian oscillations for a more realistic model which explicitly incorporates elements of the light-signalling pathway [13,18]. We exploit discrete time-dependent events to represent light/dark cycles and analyse the behaviour of the system under different light conditions and in the absence of a core feedback loop. As part of this analysis, we use a novel sensitivity analysis method to determine the time within the circadian cycle at which a given phase marker is most responsive to parameter variations.

We use Bio-PEPA [19,20] as our modelling language. Bio-PEPA is a process algebra recently developed for modelling biochemical systems. Among its key strengths as a language for systems biology is the fact that it is equipped with different semantics, enabling both continuous-deterministic and discrete-stochastic representations of the same model description to be automatically generated. Another important feature of Bio-PEPA is that it permits the definition of generic rate laws. This allows the specification of complex kinetic formulae, such as those used in the ODE representation of the original *Neurospora* model (see Sect. 4 below). In addition, time-dependent events can be easily incorporated, enabling periodic external stimuli such as light/dark cycles to be represented in a straightforward manner.

The rest of the paper is structured as follows. The circadian clock of *Neurospora crassa* and the Bio-PEPA model of the clock are described in Sect. 2 and Sect. 4, respectively. Bio-PEPA is introduced in Sect. 3. In Sect. 5 the simulation and analysis results are presented. Finally, in Sect. 6 we report some concluding remarks.

## 2    The Circadian Clock of *Neurospora Crassa*

*Neurospora* exhibits a 22 hour rhythm in asexual spore formation (conidiation) when grown in constant darkness (DD). The conidiation rhythm is a key clock output which can be entrained by both light and temperature [21]. In natural 24 hour cycles of alternating light and dark (LD), the phase of entrainment (judged by the time of conidiation onset) coincides with the middle of the night in both long and short days, providing a simple, biologically relevant measure of circadian function [22,13,18].

The core, multi-loop genetic oscillator believed to underlie the conidiation rhythm is formed by the rhythmic gene *frequency (frq)* and the constitutively expressed gene *white collar-1 (wc-1)* [9]. The protein product of the *white collar-1* gene, WC-1, comprises the positive element of a central negative feedback loop, activating transcription of *frq*. The protein product of the *frq* gene, FRQ, is the negative element of the loop, interacting with *frq*-bound WC-1 to inhibit *frq* expression [23,24]. In addition to its role as a transcriptional inhibitor, FRQ positively regulates expression of WC-1, giving a positive feedback loop that interlocks with the central loop [25]. Light entrains the clock by promoting the binding of a flavin chromophore to WC-1, resulting in a light-activated form which enhances *frq* transcription [24].

A network diagram for the model of the core oscillator that we consider here is shown in Fig. 1. For the ODE representation of the model presented in [13,18], the repressive

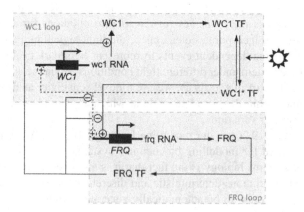

**Fig. 1.** A schematic representation of the gene network underlying the model of the *Neurospora* clock. WC-1* represents light-activated WC-1. The dashed lines indicate light-dependent gene-protein interactions.

action of FRQ transcription factor (hereafter called *PF*) on *frq*-bound WC-1 transcription factor (*PW*) and *frq*-bound light-activated WC-1 transcription factor (*PWL*) was assumed to occur through a noncompetitive inhibition process modelled using Hill kinetics. Hill kinetics were also used to describe the upregulation of WC-1 translation by *PF* as well as the light-dependent increase in the transcription of *wc-1* mRNA (*MW*), necessary to simulate loss-of-function *wc-1* mutants [13]. Michaelis-Menten kinetics were used to describe enzyme-mediated degradation of mRNA and protein, while the conversion of *PW* to *PWL* was modelled as a reversible first order mass-action reaction $PW \rightleftharpoons PWL$ with a light-dependent forward rate. The light input to the ODEs took the form of a smoothly differentiable function that switches rapidly between 0 and 1 at dawn ($t = t_{dawn}$), and from 1 back to 0 at dusk ($t = t_{dusk}$), modelling the lighting protocol commonly used in circadian experiments.

In order to obtain oscillatory behaviour, a delay was introduced into the central negative feedback loop by assuming that just-translated FRQ protein (*E1F*) is modified into a second intermediate protein (*E2F*) before being converted into transcription factor [13]. The conversion processes $MF \rightarrow E1F \rightarrow E2F \rightarrow PF$, which include translation of FRQ from *frq* mRNA (*MF*), were each modelled as first order mass-action reactions. Similarly, a delay was introduced into the positive feedback loop by introducing two intermediate WC-1 protein species (*E1W* and *E2W*), and describing the conversions $MW \rightarrow E1W \rightarrow E2W \rightarrow PW$ with first order kinetics.

The ODE representation of the model comprises 9 equations with 34 kinetic parameters. The parameters were fitted to gene and protein expression time series in DD and LD using a bipartite optimisation method developed for high-dimensional computational biology models [5]. This technique combines a random parameter search with simulated annealing to minimise a qualitative cost function that assesses the goodness-of-fit of the model to key experimental data sets [5,13]. For the ODE model, the best parameter set was taken to be that yielding the smallest cost function score following the application of the optimisation scheme to 50 million randomly distributed points in the 34-dimensional parameter space [18]. This optimal parameter set yielded a good fit to each of the target time series, and also reproduced the variation in entrainment phase with photoperiod observed experimentally [18]. We use the same parameter set here for the Bio-PEPA representation of the model detailed in Sect. 4.

## 3 Bio-PEPA

In this section we give a short description of Bio-PEPA [19,20], a language that has recently been developed for the modelling and analysis of biological systems. The main components of a Bio-PEPA system are the *species components*, describing the behaviour of each species, and the *model component*, describing the interactions between the species and initial amounts.

The syntax of the Bio-PEPA components is defined as:

$$S ::= (\alpha, \kappa) \text{ op } S \mid S + S \mid C \quad \text{with op} = \downarrow \mid \uparrow \mid \oplus \mid \ominus \mid \odot \qquad P ::= P \underset{\mathcal{L}}{\bowtie} P \mid S(x)$$

where $S$ is the *species component* and $P$ is the *model component*. In the prefix term $(\alpha, \kappa)$ op $S$, $\kappa$ is the *stoichiometry coefficient* of species $S$ in reaction $\alpha$, and the *prefix combinator* "op" represents the role of $S$ in the reaction. Specifically, $\downarrow$ indicates a *reactant*, $\uparrow$ a *product*, $\oplus$ an *activator*, $\ominus$ an *inhibitor* and $\odot$ a generic *modifier*. We can use the shorthand notations $(\alpha, \kappa)$ op and $\alpha$ op for $(\alpha, \kappa)$ op $S$ and $(\alpha, 1)$ op $S$, respectively.

The operator "+" expresses the choice between possible actions, and the constant $C$ is defined by an equation $C \overset{def}{=} S$. The process $P \bowtie_{\mathcal{L}} Q$ denotes synchronisation between components $P$ and $Q$; the set $\mathcal{L}$ determines those activities on which the operands are forced to synchronise, with $\bowtie$ denoting a synchronisation on all common action types. In the model component $S(\overset{*}{x})$, the parameter $x \in \mathbb{R}$ represents the initial concentration (or the number of molecules in a discrete-stochastic setting). The reader is referred to [19] for further details on the language and its semantics.

Recently Bio-PEPA has been extended to incorporate events [26], constructs that represent changes in the system due to some triggering conditions. This allows biochemical perturbations to the system to be represented, such as the timed introduction of reagents or the modulation of system components by external stimuli. A Bio-PEPA event has the form *(id, trigger, event_assignment, delay)*, where *id* is the event name, *trigger* is a mathematical expression involving the components of the Bio-PEPA model and/or time, *event_assignment* is a list of assignments causing some changes to elements in the system, and *delay* is either 0 *(immediate events)* or a positive real value *(delayed events)*.

A Bio-PEPA system representing a biochemical network consists of a set of sequential components, a model component, and context (defining information such as kinetics rates, parameters, locations, and events). Its formal definition is the following:

**Definition 1.** *A Bio-PEPA system $\mathcal{P}$ is a 8-tuple $\langle t, \mathcal{L}, \mathcal{N}, \mathcal{K}, \mathcal{F}_R, Comp, P, Events \rangle$, where: t is time, $\mathcal{L}$ is the set of locations, $\mathcal{N}$ is the set of (auxiliary) information for the species, $\mathcal{K}$ is the set of parameters, $\mathcal{F}_R$ is the set of functional rates, Comp is the set of species components, P is the model component and Events is the set of events.*

Bio-PEPA offers a formal representation of biochemical systems, on which different kinds of analysis can be carried out, through the defined mappings into continuous-deterministic and discrete-stochastic modelling languages. The Bio-PEPA language is supported by software tools which automatically process Bio-PEPA models and generate other representations in forms suitable for different kinds of analysis [19,27]. In particular, the generated simulation model can be executed using the Dizzy simulation tool [28], in which both stochastic simulation algorithms and differential equation solvers are implemented.

## 4   The Bio-PEPA Model of the Circadian Clock

In the following we provide an overview of the Bio-PEPA model for the circadian clock described in Sect. 2. The full model is reported in the Supplementary Material.

The clock is characterised by robust entrainment to light/dark cycles. Light entrains the clock by modulating general kinetic laws different from mass-action that abstract complex sequences of more elementary steps [13,18]. These features can be easily represented in Bio-PEPA using events and functional rates. From the Bio-PEPA description of the clock we can derive both the model for stochastic simulation and the related system of differential equations.

In order to derive a stochastic model, the continuous concentration values of ODE models must be translated into discrete numbers of molecules. In general, assuming concentrations are expressed in *molars* (M), the initial amounts must be multiplied by the factor $N_A \cdot V$ (where $N_A$ is the Avogadro number and $V$ is the volume of the compartment in which the reactions take place), and the kinetic parameters must be rescaled accordingly (see [29] for details). For the *Neurospora* model, as the compartment size and absolute concentration values are not known to any great accuracy, we follow the approach used in [15] and introduce a generic scaling factor $\Omega$ that modulates the number of molecules. Specifically, concentrations are turned into discrete numbers of molecules by multiplying them by $\Omega$, and the kinetic parameters are also rescaled by $\Omega$ (see the Supplementary Material for details).

**Reactions and kinetic laws.** Each reaction is associated with an action type and a functional rate, expressing the kinetic law. For instance, the transcription of $MF$ under upregulation by $PWL$ and inhibition by $PF$ is represented by the action type *transcription_MF_by_PWL* and the kinetic law:

$$transcription\_MF\_by\_PWL = \frac{a_1 \cdot PWL^n}{(1 + (PF/b_1)^g) \cdot (PWL^n + b_2{}^n)}$$

**Species.** Each biological species is abstracted by a Bio-PEPA species component. Below we report the definition of $PF$; the other species are described similarly.

$$PF \stackrel{def}{=} (transcription\_MF\_by\_PW, 1) \ominus + (transcription\_MF\_by\_PWL, 1) \ominus +$$
$$(transformation\_E2F\_to\_PF, 1) \uparrow + (degradation\_PF, 1) \downarrow +$$
$$(translation\_E1W\_by\_PF, 1) \oplus$$

$PF$ is involved in five reactions: it is an inhibitor of the transcription of $MF$ with and without the influence of light (first line), a product of the transformation from $E2F$ to $PF$, a reactant in the degradation of $PF$ (second line) and an activator of the translation of $E1W$ (last line). Note the use of shorthand notation in the definition of $PF$.

The full system is described in terms of the model component

$$MF(mf0) \bowtie_{*} E1F(e1f0) \bowtie_{*} E2F(e2f0) \bowtie_{*} PF(pf0) \bowtie_{*} MW(mw0) \bowtie_{*}$$
$$PW(pw0) \bowtie_{*} PWL(pwl0) \bowtie_{*} E1W(e1w0) \bowtie_{*} E2W(e2w0)$$

where the values in parenthesis are the initial values for the species.

**Events.** Entrainment by light/dark cycles is represented by events in Bio-PEPA. In the initial state the system is in dark conditions and, therefore, the transformation from the protein $PW$ to the form activated by light ($PWL$) is not possible. This is represented by setting the kinetic parameter $r_1$ for the transformation reaction $PW \rightarrow PWL$ equal to 0. At dawn, the reaction is suddenly activated and therefore $r_1$ is reset to its maximum value 5.1759. At dusk the reaction is deactivated again by resetting $r_1$ to 0. This periodic sequence of parameter changes is represented by the following set of immediate events

$$Events = [(dawn_i; t = t_{dawn} \cdot i; r_1 = 5.1759; 0),$$
$$(dusk_i; t = t_{dusk} \cdot i; r_1 = 0; 0), \ i = 1, 2, \ldots, D \ ]$$

where $D$ is the number of simulated days, and $t_{dawn}$ and $t_{dusk}$ are the times of the day at which dawn and dusk occur, respectively. By changing the values of $t_{dawn}$ and $t_{dusk}$ we can simulate the effect of changing the photoperiod ($t_{dusk} - t_{dawn}$), a key entrainment parameter for the *Neurospora* clock [22,18]. Here we focus on two conditions: constant darkness (DD) and alternating 12-hour cycles of light and dark (12:12 LD).

## 5   Model Analysis

In this section we present the validation of our model against the original ODE representation and we illustrate some analysis results. We use a version of the Dizzy simulator [28] developed at the University of Edinburgh [30], which extends the tool with sensitivity analysis techniques and additional simulation methods. The time-dependent events in the Bio-PEPA model are translated into time-dependent reaction rates in the Dizzy model (defined in terms of the step function *theta*, which is predefined in Dizzy), and we use the Gibson-Bruck stochastic simulation algorithm [31]. The choice of this algorithm is due to its efficiency (in simulation time) with respect to other stochastic simulators and to the fact it supports time-dependent rates.

### 5.1   Validation of the Model

As a preliminary step, we validate the Bio-PEPA model by comparing it against the original deterministic representation [18,13]. In Fig. 2(a) and Fig. 2(b) we show the comparison for the DD system and for 12:12 LD cycles respectively. In each graph we plot three time-series: the behaviour of the original model (dashed lines), the solution of the system of ODEs generated by the Bio-PEPA model (solid lines), and the average behaviour over 10 stochastic simulation runs with scaling factor $\Omega = 10000$ (points). The variables plotted are the clock outputs *frq* mRNA ($MF$), *wc-1* mRNA ($MW$), total FRQ protein ($FP = E1F + E2F + PF$), and total WC-1 protein ($WP = E1W + E2W + PW + PWL$).

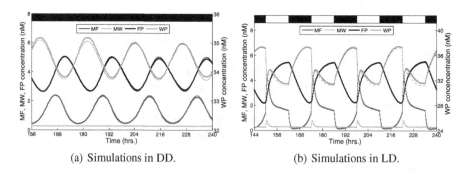

(a) Simulations in DD.                          (b) Simulations in LD.

**Fig. 2.** Comparisons of the original ODE model of the *Neurospora* clock with the deterministic and stochastic Bio-PEPA representations. Black bars represent lights-off and white bars lights-on.

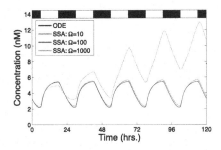

**Fig. 3.** The effect of increasing $\Omega$ on the LD system. Plotted are time-series showing oscillations in total FRQ protein $FP$ averaged over 1000 stochastic simulations for $\Omega=10$, 100 and 1000 (coloured lines). Each time series has been normalised by $\Omega$ to enable comparison with the $FP$ oscillations obtained from the deterministic model (black line).

The scaling factor $\Omega = 10000$ was chosen with the purpose of having a reasonably high number of molecules to minimise the effect of stochastic fluctuations [29]. Consequently, in both DD and LD systems, the time-series resulting from the deterministic and the stochastic analysis of the Bio-PEPA models are in very close agreement, with the stochasticity almost unnoticeable, despite the small number of simulation runs. Comparing the deterministic time-series generated by our model against that of the original model, we observe that in the DD system they are in perfect agreement, while a small difference can be observed in the LD system (especially for $WP$): the reason for this difference resides in the different ways in which the light switch is modelled: a smooth function in the original ODE model versus discrete events in the Bio-PEPA model. A similar agreement was obtained for different photoperiods (results not shown).

### 5.2 Effect of the Scaling Factor $\Omega$ on Stochasticity

Higher values of the scaling factor $\Omega$ correspond to larger molecular populations in the stochastic model, yielding smaller stochastic fluctuations [29]. We have seen in the previous section that for $\Omega = 10000$ the stochasticity is reduced to such a point that even with a small number of simulations runs, the average simulation behaviour is near-identical to the deterministic behaviour. As a consequence, the higher the scaling factor, the more regular the circadian oscillations will be, whereas we expect the effects of noise to be more evident with a smaller scaling factor. Figure 3 shows the average oscillations in total FRQ protein $FP$ for different values of $\Omega$ (10, 100 and 1000). We observe that for $\Omega = 10$ the average behaviour of the stochastic system differs significantly from the $FP$ oscillation in the deterministic system, yielding unstable oscillations that are inconsistent with the stable cycling of FRQ observed experimentally [9]. By contrast, for $\Omega = 100$ and $\Omega = 1000$, regular oscillating dynamics are obtained. We also note that the average oscillations for $\Omega = 100$ and $\Omega = 1000$ are very close to the deterministic solution, indicating that increasing $\Omega$ in this range only affects the variability about the average. We consider a scaling factor $\Omega = 1000$ in the remainder of the work. Similar results (with slightly higher variability) were obtained in all cases with $\Omega = 100$.

## 5.3  Investigating the Role of Positive Feedback

In this section, we study how the positive feedback loop of the *Neurospora* network affects its stochastic behaviour. A previous analysis of the deterministic model identified the upregulation of WC-1 production by FRQ transcription factor, controlled by the parameter $a_7$, as a key regulatory process in the light entrainment of the clock [18]. With $a_7$ at its nominal value $a_7^{WT} = 2.4695$, the ODE model yields the correct experimental responses to changes in photoperiod, with the phase of total FRQ protein $FP$ tracking the middle of the night in both long and short days [18]. Knocking out positive feedback by reducing $a_7$ to 0 destroys self-sustained oscillations in DD by pushing the deterministic model through a supercritical Hopf bifurcation [18]. This is in agreement with the loss of free-running conidiation rhythms reported in experiments [32]. The destruction of the DD limit cycle has a significant effect on the light responses of the model, yielding a system that is unable to respond to changes in photoperiod during long days. This suggests a role for the positive loop in promoting robustness against seasonal photoperiod changes [18]. Here, we compare the behaviour of the stochastic and deterministic models for DD and 12:12 LD cycles in the presence and absence of the positive loop, focusing on the resulting changes to the FRQ oscillation in each case.

**The Effect of Removing Positive Feedback on the DD System.** Figure 4(a) shows the difference between the deterministic and stochastic behaviour for the unperturbed network in DD. While the ODE model exhibits self-sustained FRQ oscillations, the average oscillation generated by the stochastic system damps to a constant value. This is a consequence of the individual realisations of the stochastic model going out of phase with each other, as can be seen in Fig. 4(c) and Fig. 4(d). This phase diffusion in the free-running system, characterised by a phase distribution spanning the full circadian cycle, agrees with previous stochastic analysis of circadian models [15].

Setting the positive feedback strength $a_7$ to 0 yields damped FRQ oscillations in the ODE model, as $a_7$ is below the Hopf bifurcation value (Fig. 4(b)). Individual realisations of the stochastic model, however, are still oscillatory, albeit with smaller amplitudes compared to the unperturbed network (Fig. 4(c)). Again, the average FRQ oscillation damps to a constant value as a consequence of phase diffusion.

The persistence of self-sustained oscillations when positive feedback is removed demonstrates that stochasticity can introduce greater robustness against modifications to the network architecture. This finding is consistent with models of the mammalian clock for which simulated knockouts that are arrhythmic in ODE implementations can become rhythmic when stochasticity is incorporated [8].

**The Effect of Removing Positive Feedback on the LD System.** Finally, we consider the 12:12 LD system and examine the effect of setting $a_7$ to 0 on the oscillatory behaviour of the model. Comparing Fig. 4 and Fig. 5, it is clear that for both the unperturbed system and the positive loop knockout, entrainment regularises the dynamics, markedly reducing the variability of oscillations compared to the free-running system (similar findings were reported for a model of the *Drosophila* clock in [15]). In both cases, there is relatively little phase diffusion, as evidenced by phase distributions that are concentrated about their corresponding deterministic values (Fig. 5(c)). Interestingly, although removal of positive feedback shifts the mean value of FRQ phase,

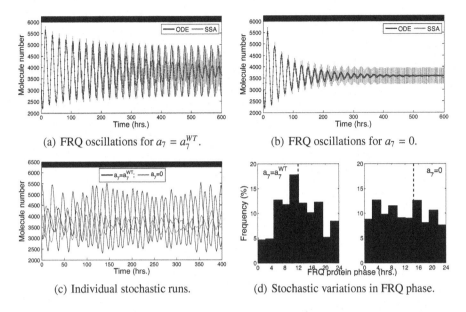

(a) FRQ oscillations for $a_7 = a_7^{WT}$.

(b) FRQ oscillations for $a_7 = 0$.

(c) Individual stochastic runs.

(d) Stochastic variations in FRQ phase.

**Fig. 4.** Changes to the FRQ protein oscillation $FP$ resulting from the removal of positive feedback in DD. Black lines in (a) and (b) denote the solution of the deterministic system, red points the average of 1000 stochastic simulations and red lines the corresponding standard deviations. In (d), the phase of FRQ protein was taken as the time at which $FP$ has decreased to its half-maximum value over the interval $576 \leq t \leq 600$. The histograms show the distribution of this phase marker over all 1000 runs of the stochastic model. Dotted lines denote the phase of FRQ in the corresponding deterministic systems.

consistent with the analysis of the deterministic system ([18]), the variation about the mean is unaffected (the standard deviation is 0.3720 for the unperturbed system and 0.3238 for the loop knockout). This demonstrates that the positive loop is able to buffer the clock against environmental variations (seasonal changes in photoperiod) without degrading its robustness to stochastic fluctuations in the chemical reactions comprising the oscillatory mechanism.

## 5.4 Sensitivity Analysis for the 12:12 LD System

Sensitivity Analysis (SA) aims to identify the relationships between the inputs and outputs of mathematical models of biochemical networks [33]. A key goal is the production of Sensitivity Indices (SI) that quantify these relationships, revealing which factors are the most influential with respect to model outcome. The most widespread SA method is "one-at-a-time" (OAT). Given a mathematical model with parameters set to those considered the most likely (also called *nominal parameters*), each parameter is perturbed individually by a fixed value or by a percentage of its nominal value, and the change in the output(s) of interest measured. OAT has seen widespread use in ODE models of biochemical interactions; this has included circadian networks for which a standard

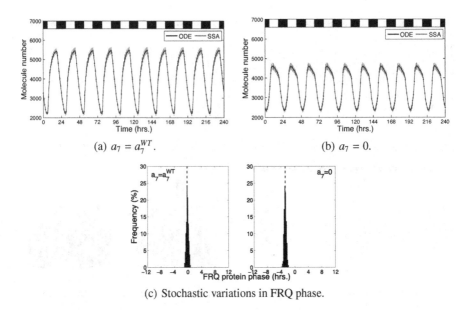

(c) Stochastic variations in FRQ phase.

**Fig. 5.** Changes to the FRQ protein oscillation $FP$ resulting from the removal of positive feedback in 12:12 LD cycles. Black lines in (a) and (b) denote the solution of the deterministic system, red points the average of 1000 stochastic simulations and red lines the corresponding standard deviations. The phase of the FRQ oscillation in (c) was taken as the time at which $FP$ decreases to its half-maximum value, identified recently as a molecular correlate of conidiation onset [22,13,18]. As in Fig. 4, the phase marker was computed over the interval $576 \le t \le 600$ for all stochastic realisations of the model. The dotted lines denote FRQ phase for the ODE model.

approach has been to compute the sensitivities of period and amplitude over one cycle of the oscillation [16,11,12].

In [34] this method has been extended to stochastic models. In this case, the output at a given time is not just a value representing the amount of a species as in ODEs; it is, instead, a set of possible values, obtained from independent stochastic simulations. The SA extension has been obtained by substituting the difference between perturbed and nominal output values employed in the traditional approach with a difference measure based on the density distribution surface of the output, estimated with a suitable number of simulations. An estimate of this density distance based on stochastic simulations can be obtained using histogram distance, as originally presented in [35]. This stochastic version of OAT therefore applies when one is interested in observing the change in the distribution of the amount of a particular species at a given time.

Here we apply the traditional approach to the means of the stochastic simulations and also consider the extended approach, based on histogram distance. These are indeed complementary, as the former does not incorporate any notion of stochastic variability, while the latter quantifies the likelihood of having the same distribution in both the perturbed and unperturbed systems. Moreover, a feature of the histogram distance is that its value will always be 2 when there is no overlapping of the distributions, making the traditional approach still necessary to determine sensitivities for large displacements.

Figure 6 summarises the result of the local sensitivity analysis obtained by changing a subset of parameters predicted to have a significant effect on entrained phase. Each parameter was incremented by 10%, and the results shown are averages over 1000 simulation runs for $\Omega = 100$ and $\Omega = 1000$. It can clearly be seen that both the density- and average-based sensitivity measures vary significantly over the circadian cycle for the parameters considered. For both measures, the most sensitive parameters are $a_4$, $d_1$ and $d_3$ representing the maximum rates of light-independent wc-1 transcription, frq degradation and wc-1 degradation respectively. All 3 parameters yield maximum sensitivities with respect to the average-based measure around dusk, when $FP$ is close to its peak value. By contrast, maximum sensitivities with respect to the density-based measure occur at the time when $FP$ decreases to its half-maximum value, a molecular correlate of conidiation onset [22,13,18]. This demonstrates that in terms of the average FRQ oscillation, the marker of entrained phase most responsive to evolutionary parameter variations is peak FRQ phase. However, when stochastic variations in the FRQ waveform are considered, the most responsive marker is the phase of the FRQ half-maximum. As it is the latter which correlates with physiological entrained phase for the *Neurospora* clock, this analysis suggests that stochastic fluctuations in FRQ expression

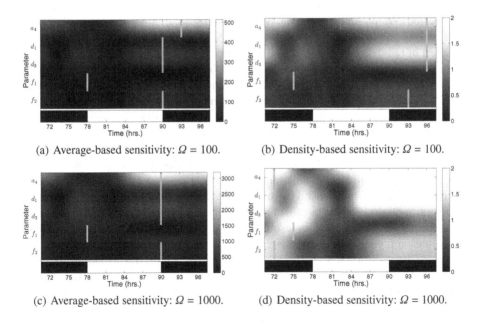

(a) Average-based sensitivity: $\Omega = 100$.

(b) Density-based sensitivity: $\Omega = 100$.

(c) Average-based sensitivity: $\Omega = 1000$.

(d) Density-based sensitivity: $\Omega = 1000$.

**Fig. 6.** Local sensitivity to parameter variation. Sensitivities were computed every 3 hours over one circadian cycle ($72 \leq t \leq 96$). The color gradient represents the difference in the amount of $FP$ between the nominal and modified parameter sets in each case (with increasing sensitivity going from black to white). In all panels, the light blue lines denote the cycle phase at which maximum sensitivity is attained for a given parameter. Because of the periodic behaviour of the system, qualitatively similar sensitivities are obtained over other 24h intervals (see Fig. S1 in the Supplementary Material).

may have been an important contributing factor to the selection of the half-maximum as a phase marker in nature.

As previously mentioned, average-based SA computes differences in terms of the average $FP$ value, while density-based SA also considers the stochastic variability. Therefore, the differences in the sensitivities when using different scaling factors can only be captured consistently by using density-based SA. For instance, similar average-based sensitivities are obtained for $\Omega = 100$ and $\Omega = 1000$ (cf. Figs. 6(a) and 6(c)), while the density-based sensitivities are in general smaller for $\Omega = 100$ as a consequence of the higher variance causing greater overlapping in the probability distributions of $FP$ (cf. Figs. 6(b) and 6(d)).

## 6  Conclusions

In this work we presented and studied a stochastic model of the circadian clock in *Neurospora crassa* under two different light conditions: constant darkness and 12:12 light/dark cycles. We used Bio-PEPA as our modelling language. This language allowed us to represent in a straightforward manner two features of the system: complex kinetic laws and time-dependent events representing cyclic light/dark conditions. The model was validated against an existing ODE representation describing key behaviours observed in laboratory experiments, including the variation of entrained phase with photoperiod. We presented some analysis results illustrating the differences between deterministic and stochastic representations of the clock. In particular, we investigated the effect of removing the positive feedback loop, previously identified as a significant factor in the determination of entrained phase. We found that while removal of positive feedback destroys self-sustained free-running (DD) rhythms in the deterministic system, oscillations with significant amplitude persist when stochastic fluctuations are considered, demonstrating the greater robustness of the oscillatory mechanism in the stochastic model. In addition, we showed that knocking out the loop has little effect on the stochastic variability of entrained phase for a given photoperiod, suggesting that positive feedback can be used to tune the phase-photoperiod relationship without introducing greater variation due to noise amplification.

Finally, we considered sensitivity analysis techniques in order to identify the most influential parameters on the circadian function of the 12:12 LD system. We focused on the variations in FRQ expression resulting from perturbations to 5 putatively sensitive parameters, applying a local sensitivity method at different time points within the 24-hour cycle. By using a novel stochastic sensitivity measure based on histogram distance, we found that the FRQ waveform is maximally sensitive at the time it reaches is half-maximum level, a molecular correlate for conidiation onset. We commented that this implicates stochasticity as a potential factor in the selection of this seemingly complex phase marker by evolution, rather than the phase of peak FRQ expression that is predicted to be maximally sensitive when variations in average FRQ level are considered. We conclude that while the local method we used only focuses around a specific point in the parameter space, it can still be informative, giving an idea about the impact of parameter changes on the behaviour of the system. In the future, we plan to apply some global methods in order to explore the full parameter space (or a meaningful subset of it) and to quantify the relationships between different parameters.

The use of stochastic simulation with our model merits some discussion, as it is characterised by some non-elementary reactions with complex kinetic laws, abstracting sets of interactions whose details are unknown. The use of Gillespie's stochastic simulation algorithm (or its variants, such as Gibson-Bruck [31]) in the case of general kinetic laws has been discussed by several authors [36,37,38]. Rao and Arkin [36] showed that this approach is valid for some specific kinetic laws, such as Michaelis-Menten and competitive inhibition. On the other hand, in [37] the authors demonstrated that this extension of Gillespie's algorithm is not always appropriate. Here, we applied stochastic simulation paying particular attention to the interpretation of the simulation results and to their validation: in Sect. 5 we showed that the behaviour we obtain using our stochastic model is in agreement with the known behaviour of the system, and therefore we conclude that in this case the use of stochastic simulation is appropriate.

## Acknowledgements

The authors thank Jane Hillston and Andrew Millar for their helpful comments. The Centre for Systems Biology at Edinburgh is a Centre for Integrative Systems Biology (CISB) funded by BBSRC and EPSRC, reference BB/D019621/1. At the time of writing this paper, Federica Ciocchetta was a research fellow at the University of Edinburgh, supported by the U.K. Engineering and Physical Sciences Research Council (EPSRC) research grant EP/C543696/1 "Process Algebra Approaches to Collective Dynamics". Andrea Degasperi is supported by a University of Glasgow Lord Kelvin/Adam Smith scholarship. Maria Luisa Guerriero is supported by EPSRC grant EP/E031439/1 "Stochastic Process Algebra for Biochemical Signalling Pathway Analysis".

## Supplementary Material

The Bio-PEPA source files, the generated Dizzy files, some further information on the model and supplementary figure S1 are available online on the Bio-PEPA web page
`http://homepages.inf.ed.ac.uk/jeh/Bio-PEPA/ACDG_CMSB09_Supplement.zip` .

## References

1. Dunlap, J.P., Loros, J.L., DeCoursey, P.J.: Chronobiology: Biological Timekeeping. Sinauer, Sunderland (2003)
2. Young, M., Kay, S.: Time zones: a comparative genetics of circadian clocks. Nat. Rev. Genet. 2(9), 702–715 (2001)
3. Tyson, J., Hong, C., Thron, C., Novak, B.: A simple model of circadian rhythms based on dimerization and proteolysis of PER and TIM. J. Biophys. 77, 2411–2417 (1999)
4. Ueda, H., Hagiwara, M., Kitano, H.: Robust oscillations within the interlocked feedback model of *Drosophila* circadian rhythm. J. Theor. Biol. 210, 401–406 (2001)
5. Locke, J., Kozma-Bognar, L., Gould, P., Fehér, B., Kevei, E., Nagy, F., Turner, M., Hall, A., Millar, A.: Experimental validation of a predicted feedback loop in the multi-oscillator clock of *Arabidopsis thaliana*. Mol. Sys. Biol. 2, 59 (2006)

6. Zeilinger, M., Farré, E., Taylor, S., Kay, S., Doyle, F.: A novel computational model of the circadian clock in Arabidopsis that incorporates PRR7 and PRR9. Mol. Sys. Biol. 2(60) (2006)

7. Leloup, J., Goldbeter, A.: Toward a detailed computational model for the mammalian circadian clock. Proc. Natl. Acad. Sci. USA 100, 7051–7056 (2003)

8. Forger, D., Peskin, C.: Model based conjectures on mammalian clock controversies. Theor. Biol. 230(4), 533–539 (2004)

9. Loros, J., Dunlap, J.: Genetic and molecular analysis of circadian rhythms in *Neurospora*. Annu. Rev. Physiol. 63, 757–794 (2001)

10. Leloup, J., Gonze, D., Goldbeter, A.: Limit cycle models for circadian rhythms based on transcriptional regulation in *Drosophila* and *Neurospora*. J. Biol. Rhythms 14(6), 433–448 (1999)

11. Francois, P.: A model for the *Neurospora* circadian clock. Biophys. J. 88(4), 2369–2383 (2005)

12. Ruoff, P., Loros, J., Dunlap, J.: The relationship between FRQ-protein stability and temperature compensation in the *Neurospora* circadian clock. Proc. Natl. Acad. Sci. USA 102(49), 17681–17686 (2005)

13. Akman, O., Locke, J., Tang, S., Carré, I., Millar, A., Rand, D.: Isoform switching facilitates period control in the *Neurospora crassa* circadian clock. Mol. Sys. Biol. 4, 64 (2008)

14. Hong, C., Jolma, I., Loros, J., Dunlap, J., Ruoff, P.: Simulating dark expressions and interactions of *frq* and *wc-1* in the *Neurospora* circadian clock. Biophys. J. 94(4), 1221–1232 (2008)

15. Gonze, D., Halloy, J., Goldbeter, A.: Robustness of circadian rhythms with respect to molecular noise. Proc. Natl. Acad. Sci. USA 99(2), 673–678 (2002)

16. Smolen, P., Baxter, D., Byrne, J.: Reduced models of the circadian oscillators in *Neurospora crassa* and *Drosophila melanogaster* illustrate mechanistic similarities. OMICS 7(4), 337–354 (2003)

17. McAdams, H., Arkin, A.: Stochastic mechanisms in gene expression. Proc. Natl. Acad. Sci. USA 94(3), 814–819 (1997)

18. Akman, O., Rand, D., Brown, P., Millar, A.: Robustness from flexibility in the fungal circadian clock (submitted, 2009)

19. Ciocchetta, F., Hillston, J.: Bio-PEPA: a Framework for the Modelling and Analysis of Biological Systems. Theoretical Computer Science (to appear, 2009)

20. Ciocchetta, F., Hillston, J.: Bio-PEPA: an extension of the process algebra PEPA for biochemical networks. In: Proc. of FBTC 2007. ENTCS, vol. 194, pp. 103–117 (2008)

21. Vitalini, M., de Paula, R., Park, W., Bell-Pedersen, D.: The rhythms of life: circadian output pathways in *Neurospora*. J. Biol. Rhythms 21(6), 432–444 (2006)

22. Merrow, M., Boesl, C., Ricken, J., Messerschmitt, M., Goedel, M., Roenneberg, T.: Entrainment of the *Neurospora* circadian clock. Chronobiol. Int. 23(1-2), 71–78 (2006)

23. Merrow, M., Franchi, L., Dragovic, Z., Gorl, M., Johnson, J., Brunner, M., Macino, G., Roenneberg, T.: Circadian regulation of the light input pathway in *Neurospora crassa*. EMBO J. 20(3), 307–315 (2001)

24. Froehlich, A., Loros, J., Dunlap, J.: Rhythmic binding of a WHITE COLLAR-containing complex to the *frequency* promoter is inhibited by FREQUENCY. Proc. Natl. Acad. Sci. USA 100(10), 5914–5919 (2003)

25. Cheng, P., Yang, Y., Liu, Y.: Interlocked feedback loops contribute to the robustness of the *Neurospora* circadian clock. Proc. Natl. Acad. Sci. USA 98(13), 7408–7413 (2001)

26. Ciocchetta, F.: Bio-PEPA with events. Transactions on Computational Systems Biology (to appear, 2009)

27. Bio-PEPA Home Page, http://www.biopepa.org/

28. Dizzy Home Page, http://magnet.systemsbiology.net/software/Dizzy
29. Gillespie, D.: Exact stochastic simulation of coupled chemical reactions. J. Phys. Chem. 81(25), 2340–2361 (1977)
30. Dizzy Edinburgh version, http://homepages.inf.ed.ac.uk/stg/software/Dizzy/
31. Gibson, M., Bruck, J.: Efficient Exact Stochastic Simulation of Chemical Systems with Many Species and Many Channels. The Journal of Chemical Physics 104, 1876–1889 (2000)
32. Schafmeier, T., Káldi, K., Diernfellner, A., Mohr, C., Brunner, M.: Phosphorylation-dependent maturation of *Neurospora* circadian clock protein from a nuclear repressor toward a cytoplasmic activator. Cell 20(3), 297–306 (2006)
33. Heinrich, R., Schuster, S.: The Regulation of Cellular Systems. Chapman and Hall, Boca Raton (1996)
34. Degasperi, A., Gilmore, S.: Sensitivity Analysis of Stochastic Models of Bistable Biochemical Reactions. In: Bernardo, M., Degano, P., Zavattaro, G. (eds.) SFM 2008. LNCS, vol. 5016, pp. 1–20. Springer, Heidelberg (2008)
35. Cao, Y., Petzold, L.: Accuracy limitations and the measurements of errors in the stochastic simulation of chemically reacting systems. J. Comput. Phys. 212(1), 6–24 (2006)
36. Rao, C.V., Arkin, A.P.: Stochastic chemical kinetics and the quasi-steady-state assumption: Application to the Gillespie algorithm. J. Chem. Phys. 118(11), 4999–5010 (2003)
37. Bundschuh, R., Hayot, F., Jayaprakash, C.: Fluctuations and Slow Variables in Genetic Networks. Biophys. J. 84, 1606–1615 (2003)
38. Cao, Y., Gillespie, D., Petzold, L.: Accelerated Stochastic Simulation of the Stiff Enzyme-Substrate Reaction. Journal of Chemical Physics 123, 144917–144929 (2005)

# Quantitative Pathway Logic
# for Computational Biology*

Michele Baggi[1], Demis Ballis[2], and Moreno Falaschi[1]

[1] Dip. di Scienze Matematiche e Informatiche
Pian dei Mantellini 44, 53100 Siena, Italy
{baggi,moreno.falaschi}@unisi.it
[2] Dip. Matematica e Informatica
Via delle Scienze 206, 33100 Udine, Italy
demis@dimi.uniud.it

**Abstract.** This paper presents an extension of Pathway Logic, called
Quantitative Pathway Logic (QPL), which allows one to reason about
quantitative aspects of biological processes, such as element concentrations and reactions kinetics. Besides, it supports the modeling of inhibitors, that is, chemicals which may block a given reaction whenever
their concentration exceeds a certain threshold. QPL models can be specified and directly simulated using rewriting logic or can be translated
into Discrete Functional Petri Nets (DFPN) which are a subclass of Hybrid Functional Petri Nets in which only *discrete* transitions are allowed.
Under some constraints over the anonymous variables appearing in the
QPL models, the transformation between the two computational models
is shown to preserve computations. By using the DFPN representation
our models can be graphically visualized and simulated by means of well
known tools (e.g. Cell Illustrator); moreover standard Petri net analyses (e.g. topological analysis, forward/backward reachability, *etc.*) may
be performed on the net model. An executable framework for QPL and
for the translation of QPL models into DFPNs has been implemented
using the rewriting-based language Maude. We have tested this system
on several examples.

## 1 Introduction

Pathway Logic (PL) [20] is a symbolic approach to the modeling and analysis
of biological systems which is based on rewriting logic [14]. Rewriting logic is
a logical framework which allows one to easily formalize systems. Within this
framework, the states of a system are represented as elements of an algebraic data
type, specified by an equational theory, while its behavior is modeled via rewrite
rules describing local transitions between states. The process of application of
rewrite rules generates computations, which —in the case of biological systems—
correspond to pathways.

* This work has been partially supported by the Italian MUR under grant
RBIN04M8S8, FIRB project, Internationalization 2004.

P. Degano and R. Gorrieri (Eds.): CMSB 2009, LNBI 5688, pp. 68–82, 2009.

A PL model is an *executable* rewriting logic specification representing a biological process which consists of an equational part where biological molecules, their states and locations are declared. Such entities may be decorated by metadata such as entity synonyms and categories, protein families, *etc.* Such information can be retrieved by standard databases (e.g. HUGO [7], UniProt/Swiss-Prot [22]). A collection of rewrite rules specifying individual reaction steps establishing how the system may evolve. In the case of signal transduction, rewrite rules represent processes such as activation, phosphorylation, complex formation, or translocation. As an example, consider the following PL rewrite rule labeled C.reloc of a given PL model.

```
rl[C.reloc] {CLm | clm A} {CLi | cli B} {CLc | clc C} =>
            {CLm | clm A} {CLi | cli B [C - reloc]} {CLc | clc}
```

The left hand-side of the rule represents a generic cell with three locations: the membrane (location tag CLm) containing element A, the inside of the membrane (location tag CLi) containing element B and the cytoplasm (location tag CLc) containing element C. The rule specifies that when such a configuration is detected, then C is relocated in CLi. Note that Pathway Logic provides a very simple way to add spatial information over the biological element describing a cell as a set of locations containing elements. Unfortunately, such a description is only qualitative, since the specification is not equipped with quantitative information regarding the elements into play.

PL models are specified using Maude [5], a rewriting-logic-based language. Maude specifications representing the PL models are executable, therefore they give a direct support to model simulation. Moreover, using Maude built-in search capabilities, PL models can be naturally queried in order to find pathways of interest that may (may not) be activated. Finally, Maude specifications can be translated into equivalent Petri net (PN) representations by means of the Pathway Logic Assistant (PLA) [21]. The benefit of such PN representations generated by PLA is twofold. On the one hand, it allows one to verify PL models reusing efficient model-checking techniques which have been already developed for PNs; on the other hand it provides a visual and interactive denotation of the model.

**Our contribution.** Although Pathway Logic may be very useful to model biological processes and provides a simple way to express the systems dynamics, it has some important limits: (a) PL only supports qualitative modeling of the biological events of interests. As a matter of fact, it provides no explicit way to add quantitative information to the models such as element concentrations in cell locations, levels of production as well as consumption of elements occurring in a reaction, reaction thresholds, *etc.* (b) PL does not provide adequate capabilities to express inhibitory actions occurring in biological reactions, which are very common e.g. in regulatory networks. Basically, an element acts as an inhibitor in a reaction if its concentration over a given threshold prevents the reaction to take place. Instead, the inhibitor let the reaction take place whenever its concentration is under the considered threshold.

In this paper we provide an extension of Pathway Logic called *Quantitative Pathway Logic* (QPL for short) with the aim of overcoming the mentioned PL limits and obtaining a more precise approach to modeling while keeping the possibility to compute with and analyze these complex systems. More specifically, QPL efficiently integrates quantitative data (such as element concentrations, reaction thresholds, production and consumption rates) into PL models. Besides, it allows one to model reaction inhibitors.

To manage the different aspects of biological systems, we equip QPL specifications with two equivalent computational models following and adapting the PL approach of [21]. This allows one to adopt different representations with different expressive capabilities for handling the complexity of the systems under examination. On the one hand, QPL specifications can be directly formalized and executed by using the Maude rewriting logic formalism. In this way, both model simulation and model search can take advantage of quantitative information to yield more accurate results. On the other hand, QPL models can be translated into an extension of the classical Petri nets called Discrete Functional Petri Nets (DFPN), which are basically Hybrid Functional Petri Nets (HFPN) [12] in which only discrete transitions are authorized. By using such a representation our models can be graphically visualized, simulated, and analyzed by means of well known tools (e.g. Cell Illustrator [13]).

**Plan of the paper.** The rest of the paper is organized as follows. In Section 2 we briefly recall some necessary notions about rewriting logic and the Maude language. Section 3 presents the QPL rewriting logic formalism for modeling biological processes. Besides, we show how to specify, simulate, search and model-check QPL models using the Maude language. In Section 4, we provide an alternative representation of QPL models which is based on DFPNs. We also provide a transformation between the two representations that, under certain constraints, is shown to preserve the computations. Section 5 describes the prototypical implementation of our methodology. In Section 6 we discuss related work and we draw some conclusions.

## 2    Preliminaries

Rewriting logic [14] is a simple computational logic very well suited as a semantic framework within which many different models of computation, systems and languages can be naturally modeled. A *rewrite theory*, that is, a theory in rewriting logic is a triple $\mathcal{R} = (\Sigma, E, R)$, where $(\Sigma, E)$ is the equational theory *modulo which* we rewrite, and $R$ is a set of (possibly conditional) labeled rules. $\Sigma$, called the *signature*, specifies the operators and the type structure of $\mathcal{R}$ as usual, while $E$ is a set of (possibly conditional) equations. Rules in $R$ are of the form $(label : t_0 \Rightarrow t_1 \text{ if } c)$ where $t_0$, $t_1$ are terms, the rule left-hand side (lhs) and the rule right-hand side (rhs) respectively, and $c$ is an optional boolean term representing the rule condition. Variables may appear in rules and equations and are denoted by lower letters, while operators are denoted by identifiers whose first character is a capital letter. A *context* $C$, is a term with a single hole, denoted by [ ], used to indicate the location of a rewrite application. $C[t]$ is the result of placing $t$ in the hole of $C$. A *substitution* $\sigma$ is a finite mapping from variables to terms, $\sigma(t)$ is the result of applying $\sigma$ to term $t$.

Rewrite proceeds modulo the equational theory $E$ and it is accomplished by performing *pattern matching modulo* the equational theory. More precisely, given an equational theory $E$, a term $u$ and a term $v$, we say that $u$ *matches* $v$ *modulo* $E$ (or that $u$ *E-matches* $v$) via substitution $\sigma$ if there exists a context $C$ such that $C[\sigma(u)] =_E v$, that is, $C[\sigma(u)]$ and $v$ are equal modulo the equational theory $E$. Given a term $t$ and rule $r = (label : t_0 \Rightarrow t_1 \text{ if } c)$, we say that $t$ *rewrites* to $t'$ via $r$ (in symbols $t \xrightarrow{r} t'$), if there exists a substitution $\sigma$ s.t. $t_0$ E-matches $t$ via $\sigma$, $t' = C[\sigma(t_1)]$ and $\sigma(c)$ holds (i.e. it is equal to *true* modulo E). A computation over $\mathcal{R}$ is a sequence of rewrites of the form $\mathcal{R} \vdash s_0 \xrightarrow{r_1} s_1 \ldots \xrightarrow{r_k} s_k$, with $r_1, \ldots, r_k \in R$.

Rewrite theories can be encoded in Maude [5], a high-performance reflective language supporting both equational and rewriting logic programming, which is particularly suitable for developing domain-specific applications. Maude specifications can be executed and verified using some powerful language built-in operators such as `rew` (which generates a computation starting from an initial term), `search` (which generates all the possible computations starting from a given initial term), and `modelCheck` (which supports model checking w.r.t. the Linear Temporal Logic (LTL) [6]).

# 3   Quantitative Pathway Logic

A QPL model is a rewrite theory $\mathcal{Q} = (\Sigma, E, R)$ which models biological systems. A QPL model is naturally divided in two parts: the *equational part* and the *rules part*. The former allows one to represent the cellular states, while the latter specifies the system dynamics.

**The Equational Part.** This part corresponds to equational theory $(\Sigma, E)$ of the QPL model $\mathcal{Q}$. It provides sorts and operators useful to model molecular components and more in general all the entities involved in a biological system.

As in the standard PL framework, the main sorts for entities include `Chemical`, `Protein`, `Complex`, which are all subsorts of sort `Thing` specifying a generic entity. Cellular compartments are identified by sort `Location` which provides location names to each compartment, while `Modification` is a sort used to classify Post-transactional protein modifications, which are defined by the operator `[ - ]` (e.g. the term `[EgfR - act]` represents the epidermal growth factor Egf receptor in an activated state).

Besides that, we provide a special sort `QThing` which is represented by the pair $(Thing, \mathbb{R}^+)$, where $\mathbb{R}^+$ specifies the sort for the non-negative real numbers. The sort `QThing` is employed to manage entity concentrations (e.g. `(Erk, 3.3)` might model the fact that the concentration of the the Mitogen-Activated Protein Kinase Erk is 3.3 units). We call *occurrence* any term of sort `QThing`. A *soup* is a set of occurrences and cellular compartments which is identified by type `Soup`.

Now, a *cell state* is represented by a term of the form `[cellType | locs]`, where `cellType` specifies the type of cell and `locs` represents the contents of a cell organized by cellular compartments (or locations). Each location is modeled by a term of the form { locName | comp }, where locName is a name identifying the location (e.g. CLm may represent the cell membrane location), and comp is a soup in that location.

**The Rules part.** Given a QPL model $\mathcal{Q} = (\Sigma, E, R)$, rules part is specified via the set of rewrite rules $R$, which contains rewrite rules formalizing individual reaction steps. In the case of signal transduction, rewrite rules represent processes such as activation, phosphorylation, complex formation, or translocation. Basically, as in PL, QPL rewrite rules transform a cell state into another via pattern matching modulo an equational theory. Moreover, such a transformation in QPL can take advantage of the quantitative information associated with the entities into play. In this context, it is very easy to define promoters (entities enabling a reaction when their concentration is over a certain threshold), inhibitors (entities blocking a reaction when their concentration is over a certain threshold), tests (entities not consumed by a reaction), and reaction rates modeled via consumption and production functions. Let us see some examples.

*Example 1 (Promoters and tests).* Consider a reaction modeled by the following rewrite rule.

ex1 : $\{\text{CLi} \mid \text{cli} (A, a)\}\{\text{CLm} \mid \text{clm} ([B\text{-}GDP], b)(D, d)\} \Rightarrow \{\text{CLi} \mid \text{cli} (A, a/2.0)(C, a/2.0 + b)\}$
$$\{\text{CLm} \mid \text{clm} (D, d)\} \text{ if } a \geq 3.5.$$

The rule states that, if we detect a cell state in which (1) an entity A with concentration a is inside the cell membrane (location CLi); (2) entities [B-GDP] (i.e. entity B bounded to a molecule of Guanosine diphosphate) and D with concentrations b and d are on the border of the cell membrane (location CLm); then, A promotes the reaction whenever its concentration is greater or equal than 3.5 units and a new cell state is generated in which Reactants A and [B-GDP] are consumed: the former is consumed according to the consumption function (a/2.0) and the latter is completely consumed. Entity C is produced inside the membrane according to the production function a/2.0+b, while D is a test whose concentration is left unchanged (its presence is necessary for the reaction to take place but there is no need to consume it).

Note that production functions may depend on concentrations of several reactants. Instead, we assume that the consumption function of a reactant A must depend only on the concentration of A. Inhibitory behaviors are easily modeled by means of conditional rules. For this purpose, we first define the auxiliary function checkInhibitors$(s, (i_1, t_1) \ldots (i_n, t_n))$ which takes in input a soup s, and a list of occurrences $(i_j, t_j)$ and returns true if there does not exist any occurrence $(i_j, c_j)$ in s such that $c_j \geq t_j$. Now, an inhibitor $i_j$ is an entity located in some compartment s which prevents a reaction to take place whenever its concentration $c_j$ is over a certain threshold $t_j$. This amount to saying that a set of inhibitors $(i_1, \ldots, i_n)$ can block a reaction iff checkInhibitors$(s, (i_1, t_1) \ldots (i_n, t_n)) = $ false. Therefore, we can model a reaction containing inhibitors $i_1$ and $i_2$ by a rewrite rule of the form $t \Rightarrow t'$ if checkInhibitors(s,($i_1,t_1$)($i_2,t_2$)).

*Example 2 (Inhibitors).* Consider a reaction modeled by the following rewrite rule.

ex2 : {CLi | cli (A, a)} $\Rightarrow$ {CLi | cli (C, a)} if checkInhibitors(cli, (B, 4.0))

The rule states that, when there exists a cell state in which an entity A with concentration a appears inside the cell membrane (location CLi), then the cell state is transformed by consuming completely the reactant A and producing a units of entity C provided that there is no inhibitor B with concentration greater than 4.0 inside the cell membrane (i.e. in the location CLi). Note that $B$ is not consumed by the reaction.

## 3.1 Simulation and Analysis of QPL Models

Quantitative Pathway Logic models are rewrite theories, hence *executable* specifications, since they describe system states and provide rules specifying the way in which states may change. In other words, we can directly exploit the Maude rewrite engine to run our models. In particular Maude supports the *forward simulation* which is the first kind of analysis that can be carried out given such an executable specification. It consists in running the model from a given initial state for a fixed number of steps or until a steady state has been reached. It is very useful for initial exploration of the transition graph but not suitable for understanding the dynamics of systems having infinite behaviors. Maude is also equipped with *forward search* facilities, which allow one to perform a breath-first search of all rewrite paths generated for a given initial state. If the specification is finite, such a search will find all possible outcomes from a given initial state. Moreover, the search can be constrained to find only states satisfying a given property or until a fixed number of rewrite steps.

To execute both forward simulation and forward search we need to provide an initial state. Initial states (called *dishes*) are encoded in our rewriting framework by terms of the form PD(out cellstate), where cellstate represents a cell state and out specifies a soup of ligands and other molecular components in the cells surroundings which may interact with the cell.

*Example 3.* Consider the following dish

PD({{(Egf, 2.0)}[HMEC|{CLo|empty}{CLm|(EgfR, 1.0)(PIP2, 3.0)}
{CLi|([Hras − GDP], 4.0)(Src, 5.0)}
{CLc|(Gab1, 5.0)(Grb2, 6.0)(Pi3k, 2.0)(Plcg, 7.0)(Sos1, 1.0)}]).

The dish above contains the cell state of a cell of type HMEC which is made up of three locations CLo, CLi and CLc. Each location contains a soup of occurrences. Besides, the ligand Egf occurs in the cell surroundings with concentration 2.0.

Now, given an initial state and a QPL model, we can analyze the behavior of the system by means of Maude forward simulation and search capabilities. Let us see some examples.

*Example 4.* Assuming that a QPL model for the signal transduction network of the epidermal growth factor receptor (EgfR) is given, we may run the model starting from an initial cell state by means of the Maude `rewrite` command (abbreviated `rew`), that is, we may explore the behavior of the specified system for different initial cell states. For instance, the Maude query

```
rew [100] PD((Egf,2.0) [HMEC | {CLm | (EgfR,1.0) (PIP2,3.0)}
                                {CLi | ([Hras-GDP],4.0) (Src,5.0)}])
```

asks Maude to perform at most 100 rule applications (i.e. rewrite steps) to rewrite the given initial state and returns the final cell state we reached. It is also possible to rewrite without specifying an upper bound on the number of rule applications. Since the model may be non-deterministic (i.e. there might be several computations starting from the same initial state), Maude selects only one of such computations by means of a predetermined rewrite strategy. Therefore, the returned cell state may represent only one of the possible system behaviors.

Maude also provides the `frew` which allows one to implement user-defined rewrite strategies. As explained in Example 4, the forward simulation explores only one possible model behavior. To analyze all possible model dynamics we may employ the Maude `search` feature as shown in the following example.

*Example 5.* Assuming the same QPL model of the previous example, we want to know whether —starting from a dish containing a given concentration of the ligand Egf— it is possible to produce the Mitogen-Activated Protein Kinase Erk (as described in [20]) with a concentration greater than 3.6 units. Such analysis can be modeled by means of the following Maude forward search query:

```
search [1]
 PD((Egf,2.0) [HMEC | {CLo | empty} {CLm | (EgfR,1.0)(PIP2,3.0)}
   {CLi | ([Hras-GDP],4.0)(Src,5.0)}
   {CLc | (Gab1,5.0)(Grb2,6.0)(Pi3k,2.0)(Plcg,7.0)(Sos1,1.0)}])
    =>+ PD(out:Soup [HMEC|cyto:Soup {CLi|(Erk,k)}]) such that k > 3.6 .
```

The term before the right arrow denotes the initial state, while the one after the arrow specifies the pattern of the state we are looking for along with the condition that Erk has to appear with a concentration higher than 3.6.

Finally, also Maude *Model checking* [6] can be employed to analyze QPL models. Model checking enlarges the set of properties which can be investigated. While `search` only concerns with properties of individual states, model checking deals with properties of computations (i.e. pathways). In this context, the model-checker is typically asked to check the assertion that there is no computation starting from the given initial state satisfying the property of interest; thus a path can be extracted from a counterexample, if one is found.

*Example 6.* By using Maude `modelCheck` function we can easily verify whether, starting from a given initial state `istate` containing Egf, it is possible to activate the entity Src with a concentration greater than 4.5 units without having previously produced the entity Rala-GDP. To this purpose, we define a parametric property *entAct(e,n)*, which is satisfied when the CLi location contains an occurrence of entity e with a concentration higher than n. The property might be specified as follows eq PD(out:Soup [HMEC|cyto:Soup {CLi|cli:Soup(e,k)}]) |= entAct(e,n) = k > n . Now, we can model the desired analysis by means of the following Maude model-checking query:

```
red modelCheck(istate,
  []~(<> entAct([Src-act],4.5) /\
      (entAct([Src-act],4.5) |-> entAct(RalaGDP,0.0))))
```

The query is expressed in linear temporal logic and consists of a conjunction of two sub-queries, the former asks for the activation of Src with a concentration of 4.5 units and the latter asks for a sequence of events (expressed by the |-> operator) where the production of Rala strictly follows the activation of Src.

# 4   Representing QPL Models via DFPNs

QPL models can be represented by means of Discrete Functional Petri Nets (DFPN) which are restricted HFPNs[12] able to model quantitative aspects of a given system by means of functional discrete transitions. The advantage of such an alternative formalism is twofold: on the one hand, graphical representations are naturally derived from DFPNs, which allow one to visualize the model of interest and to graphically interact with it using common tools available on the market (e.g. Cell Illustrator); on the other hand, analysis of DFPNs can profit from well-known techniques which have been already developed for the Petri net settings. In what follows, we formalize DFPNs by borrowing terminology and notation from [12].

## 4.1   Discrete Functional Petri Nets

**Definition 1 (Marking).** *Given a finite set of places $P$, a* marking *of $P$ is a mapping $M : P \to \mathbb{R}^+$. Given $p \in P$, $M(p)$ is called the* mark *of $p$. We denote the set of all possible markings of $P$ by $\mathcal{M}$. Given two markings $M$ and $M'$ of $P$, we say that $M \geq M'$ iff for each $p \in P$, $M(p) \geq M'(p)$. Moreover, we say that $M$ and $M'$ are* incomparable *if $M \nleq M'$ and $M' \nleq M$.*

Let $\mathcal{M}$ be the set of all markings of the set of places $P$, we denote the set of all functions mapping a marking of $P$ into a non-negative real number as $\mathcal{F}_P = \{f \mid f : \mathcal{M} \to \mathbb{R}^+\}$. Functions in $\mathcal{F}_P$ are called *update* functions.

**Definition 2 (Discrete Functional Petri Nets).** *A* Discrete Functional Petri Net *(DFPN) is a triple $\mathcal{P} = (P, T, C)$ where $P = \{p_1, \ldots, p_n\}$ is a non-empty finite set of* places*, $T = \{t_1, \ldots, t_m\}$ is a non-empty finite set of* transitions *such that $P \cap T = \emptyset$, and $C$ is a tuple $(PT, TP, a, w, u)$ defined as follows:*

- *$PT \subseteq P \times T$ and $TP \subseteq T \times P$. Elements in $PT$ (resp. $TP$) are called* input connectors *(resp.* output connectors*). Each connector has a* connector type *which is given by a function $a : PT \cup TP \to \{\text{process}, \text{test}, \text{inhibitor}, \text{output}\}$. Input connectors whose type is* process *(resp.* inhibitor*,* test*) are also called* process *(resp.* inhibitor*,* test*) connectors.*

- $w : PT \to \mathbb{R}^+$ *is a mapping called* threshold labeling *assigning non-negative real numbers to input connectors.*
- $u : PT \cup TP \to \mathcal{F}_P$ *is a partial mapping called* update labeling *such that $u(c)$ is defined iff $a(c) \in \{\texttt{process}, \texttt{output}\}$ (i.e. mapping $u$ assigns update functions to process and output connectors only).*

Connectors in a DFPN are labeled by threshold and update labeling. More specifically, threshold labeling puts non-negative real numbers (i.e. *thresholds*) on input connectors $(p, t)$ and are used to fix the minimum threshold on the mark of place $p$ which is required to enable/disable transition $t$. Update labeling decorates both process and output connectors with update functions and is employed to change the marks of places involved with transitions firing.

The enablement relation of a transition $t$ in a DPFN depends on the type of the input connectors $(p, t)$. Basically, an inhibitor connector enables transition $t$ when the mark of $p$ is under a certain threshold; process and test connectors enable transition $t$ whenever the mark of $p$ is over the threshold. More formally,

**Definition 3.** *Let $\mathcal{P} = (P, T, C)$, where $C = (PT, TP, a, w, u)$, be a DFPN. Given a transition $t \in T$ and a marking $M \in \mathcal{M}$ of $P$, we say that $t$ is* enabled *in $M$ iff for each input connector $c = (p, t) \in PT$ the following conditions hold:*
*(1) $M(p) < w(c)$ if $a(c) = \texttt{inhibitor}$     (2) $M(p) \geq w(c)$ if $a(c) \neq \texttt{inhibitor}$. Otherwise transition $t$ is said to be* disabled *in $M$. We denote the set of all the transitions which are enabled in $M$ by $\mathcal{E}(M)$.*

Now, given a DFPN $\mathcal{P}$, we can define computations over $P$ as follows.

**Definition 4.** *Let $\mathcal{P} = (P, T, C)$, where $C = (PT, TP, a, w, u)$, be a DFPN. Let $M, M' \in \mathcal{M}$ be two markings of $P$ and $t$ be a transition in $T$ such that $t \in \mathcal{E}(M)$. Then, $M$ evolves into $M'$ using $t$ in $\mathcal{P}$ (in symbols, $\mathcal{P} \vdash M \overset{t}{\mapsto} M'$) iff $M' = DFPN\_One\_Step(\mathcal{P}, M, t)$, where $DFPN\_One\_Step(\mathcal{P}, M, t)$ is a function defined as follows:*

$DFPN\_One\_Step(\mathcal{P}, M, t)$
　　**for each** $(p, t) \in PT$ *with* $a(p, t) = \texttt{process}$:   $M'(p) \leftarrow M(p) - u(p, t)(M)$
　　**for each** $(t, p) \in TP$:   $M'(p) \leftarrow M(p) + u(t, p)(M)$
　　**return** $M'$

A computation *in $\mathcal{P}$ is a (possibly infinite) sequence $\mathcal{P} \vdash M_0 \overset{t_0}{\mapsto} M_1 \overset{t_1}{\mapsto} M_2 \overset{t_2}{\mapsto} \dots$ Marking $M_0$ is called* initial *marking.*

Roughly speaking, a marking $M$ can evolve into a marking $M'$ by firing an enabled transition $t$ producing $M'$ from $M$ in the following way: for each process connector $(p, t)$ the mark $M(p)$ is consumed by applying the update function labeling $(p, t)$; for each output connector $(t, p)$, the mark $M(p)$ is increased by applying the update function labeling $(t, p)$; for each inhibitor/test connector $(p, t)$, the mark $M(p)$ is left unchanged

Transitive ($\mapsto^+$) and transitive and reflexive ($\mapsto^*$) closures of relation $\mapsto$ are defined in the usual way. Note that several transitions can be enabled at the same time in a DFPN producing non-deterministic computations.

## 4.2   Translating QPL Models into DFPNs

Given a QPL model expressed by a rewrite theory $\mathcal{Q} = (\Sigma, E, R)$, we can easily derive a DFPN having the same model behavior which will be denoted by $\mathcal{P}_\mathcal{Q}$. Basically, the translation procedure produces a DFPN transition for each rule $r$ belonging to $R$. Let us see how the translation of a single rule into a transition proceeds. Let $r = (l : t \Rightarrow t'$ if $c)$, we define $\mathcal{OL}(r) = \{(e, q, loc) \mid$ occurrence $(e, q)$ appears in $t$ in location $loc\}$, $\mathcal{OC}(r) = \{(e, q, loc) \mid$ entity $e$ appears in predicate `checkInhibitors` in $c$ associated to location $loc\}$,     and $\mathcal{OR}(r) = \{(e, q, loc) \mid$ occurrence $(e, q)$ appears in $t'$ in the location $loc\}$[1]. First of all, for each $(e, q, loc) \in \mathcal{OL}(r) \cup \mathcal{OR}(r) \cup \mathcal{OC}(r)$, we define a place $\langle e, loc \rangle$ whose marking $M(\langle e, loc \rangle)$ is represented by $q$. Intuitively, each place in the resulting DFPN will model a given entity $e$ appearing in a compartment $loc$, while the concentration $q$ provides information regarding the mark of the place.

Then, we generate the transition $l$, where $l$ is the label of the rule $r$ under examination. Transition $l$ is connected to the generated places via input/output connectors in the following way. For each $(e, q, loc) \in \mathcal{OL}(r)$ we generate an input connector $con = (\langle e, loc \rangle, l)$ whose type $a(con)$ depends on the role of the entity $e$ in the original rule $r$ (i.e. is $e$ a process or a test?). For each $(e, loc) \in \mathcal{OC}(r)$ we generate an input connector $con = (\langle e, loc \rangle, l)$ whose type $a(con) = $ `inhibitor`. For each $(e, q, loc) \in \mathcal{OR}(r)$ if $a(\langle e, loc \rangle, l) \neq$ `test` we generate an output connector $con = (l, \langle e, loc \rangle)$ whose type $a(con) = $ `output`. Finally, thresholds and update functions are defined according to the expressions appearing in the rule condition.

Note that rule's variables not representing concentrations are not encoded into the net $\mathcal{P}_\mathcal{Q}$. Such variables have only an auxiliary purpose in the QPL model. Indeed, they are used to enable pattern matching, which is an evaluation mechanism not employed in DFPNs, and hence they do not have a counterpart in the resulting net model[2]. For further details, the complete description of the translation method can be found in [3]. Figure 1 shows DFPN obtained by the the translation of rules `ex1` and `ex2` of Examples 1 and 2.

A PL dish, that is an initial state for a QPL model, is naturally encoded into an initial marking for the resulting DFPN in which we assign a given mark to (some of) the places $\langle e, loc \rangle$ of the net.

## 4.3   Model Equivalence

Given a QPL model $\mathcal{Q}$, the resulting DFPN $\mathcal{P}_\mathcal{Q}$ is equivalent to $\mathcal{Q}$ in the sense that computations are preserved. In other words, any computation over $\mathcal{Q}$ is mapped to a computation over $\mathcal{P}_\mathcal{Q}$ and vice versa. The equivalence holds under certain conditions we need to enforce on the translation. In the remainder of

---

[1] As usual locations are identified by their location name.

[2] Due to lack of space and for the sake of readability, we have not treated the case of entity variables which can range over a finite set of values (e.g. a variable representing a class of ligands which may interact with a given receptor). A full explanation of such a case can be found in the technical report [3].

$H = (P, T, C)$
$P = \{\texttt{<A,CLo>}, \ \texttt{<C,CLo>}, \ \texttt{<[B-GDP],CLm>}, \ \texttt{<D,CLm>}, \ \texttt{<A,CLi>}, \ \texttt{<B,CLi>}, \ \texttt{<C,CLi>}\}$
$T = \{\texttt{ex1, ex2}\}$
$C = \{In, Out, a, w, u\}$
$In = \{(\texttt{<A,CLo>},\texttt{ex1}), \ (\texttt{<[B-GDP],CLm>},\texttt{ex1}), \ (\texttt{<D,CLm>},\texttt{ex1}), \ (\texttt{<A,CLi>},\texttt{ex2}),$
$\qquad\quad (\texttt{<B,CLi>},\texttt{ex2})\}$
$Out = \{(\texttt{ex1},\texttt{<C,CLo>}), \ (\texttt{ex2},\texttt{<C,CLi>})\}$

$$a(c) = \begin{cases} \text{process} & \text{if } c \in \{(\texttt{<A,CLo>},\texttt{ex1}), \ (\texttt{<[B-GDP],CLm>},\texttt{ex1}), \ (\texttt{<A,CLi>},\texttt{ex2})\} \\ \text{test} & \text{if } c = (\texttt{<D,CLm>},\texttt{ex1}) \\ \text{inhibitor} & \text{if } c = (\texttt{<B,CLi>},\texttt{ex2}) \\ \text{output} & \text{if } c \in \{(\texttt{ex1},\texttt{<C,CLo>}), \ (\texttt{<ex2,C,CLi>})\} \end{cases}$$

$$w(c) = \begin{cases} 3.5 \text{ if } c = (\texttt{<A,CLo>},\texttt{ex1}) \\ 4.0 \text{ if } c = (\texttt{<B,CLi>},\texttt{ex2}) \end{cases}$$

$$u(c) = \begin{cases} \texttt{A\_CLo\_q}/2.0 & \text{if } c = (\texttt{<A,CLo>},\texttt{ex1}) \\ \texttt{[B-GDP]\_CLm\_q} & \text{if } c = (\texttt{<[B-GDP],CLm>},\texttt{ex1}) \\ \texttt{A\_CLo\_q}/2.0 + \texttt{B\_CLm\_q} & \text{if } c = (\texttt{ex1},\texttt{<C,CLo>}) \\ \texttt{A\_CLi\_q} & \text{if } c = (\texttt{<A,CLi>},\texttt{ex2}) \\ \texttt{A\_CLi\_q} & \text{if } c = (\texttt{ex2},\texttt{<C,CLi>}) \end{cases}$$

**Fig. 1.** DFPN encoding of rules ex1 and ex2

this section, we provide a brief explanation of this result. We refer the reader to the technical report [3] for more details and the complete proofs. Our approach follows and adapts the approach presented in [21] by Talcott and Dill for establishing the equivalence between Pathway Logic and standard Petri nets.

Let $\mathcal{Q}$ be a QPL model and $\mathcal{P}_{\mathcal{Q}}$ the DFPN obtained from $\mathcal{Q}$. Let $\mathcal{S}_{\mathcal{Q}}$ be the set of all possible (ground) cell states of $\mathcal{Q}$ and $\mathcal{M}_{\mathcal{Q}}$ be the set of all possible markings of $\mathcal{P}_{\mathcal{Q}}$. We define a mapping $s2m \colon \mathcal{S}_{\mathcal{Q}} \to \mathcal{M}_{\mathcal{Q}}$ which maps a cell state $s \in \mathcal{S}_{\mathcal{Q}}$ into a marking $M_s \in \mathcal{M}_{\mathcal{Q}}$ such that, for each entity $e$ appearing in a location $loc$ with concentration $q$, $M_s(\langle e, loc \rangle) = q$. We define the inverse mapping of $s2m$ by $m2s \colon \mathcal{M}_{\mathcal{Q}} \to \mathcal{S}_{\mathcal{Q}}$, and thus we have $m2s(s2m(s)) = s$ for any cell state $s \in \mathcal{S}_{\mathcal{Q}}$. Since cell states are terms which can be built upon contexts and holes, we extend $s2m$ to contexts and holes in such a way that $s2m(C[t]) = s2m(t) \cup s2m(C)$. Therefore, given a place $\langle e, loc \rangle$ of $\mathcal{P}_{\mathcal{Q}}$, which is obtained from a cell state $s = C[t] \in \mathcal{S}_{\mathcal{Q}}$, where $C$ is a context with a hole at location $loc_C$, the mark of $\langle e, loc \rangle$ w.r.t. the mapping $s2m(C[t])$ is defined as

$$s2m(C[t])(\langle e, loc \rangle) = \begin{cases} s2m(t)(\langle e, loc \rangle) & \text{if } loc = loc_C \\ s2m(C)(\langle e, loc \rangle) & \text{otherwise} \end{cases}$$

In order to guarantee the computational equivalence between QPL models and their DFPN counterparts, we need to enforce a constraint over the rules of QPL models. Let us see an example illustrating some issues related to the model translation which may arise.

*Example 7.* Consider the two following QPL rules:

r1 : {CLm | (A, a) {CLc | cyto (B, b)}} ⇒ {CLm | (A, a) ([B − act], b) {CLc | cyto}}

r2 : {CLm | clm (A, a) {CLc | cyto (B, b)}} ⇒ {CLm | clm (A, a) ([B − act], b) {CLc | cyto}}

where `cyto` and `clm` are "anonymous" variables matching any other component located in the cytoplasm or cell membrane respectively.

Consider now the state $s = \{CLm|(A, 3.3)(C, 4.7)\{CLc|(B, 0.1)(D, 0.9)\}\}$. The rule r2 applies to $s$ but r1 does not because it lacks variable `clm` which enables the pattern matching between $s$ and the lhs of r1 on the location Clm. On the other hand, r1 and r2 are translated into transitions t1 and t2 which are equal modulo renaming of the transition labels. In particular, t1 and t2 connect the same places via the same input/output connectors. Therefore, t1 is enabled whenever t2 is, and vice versa. Clearly, this fact generates a discrepancy between the computational behaviors of the QPL model and the corresponding DFPN.

To avoid the situation presented in Example 7, we basically forbid rules like r1 from being specified in QPL models.

**Definition 5.** *Let $\mathcal{Q} = (\Sigma, E, R)$ be a QPL model. Then $\mathcal{Q}$ is well-specified iff for each rewrite rule $r \in R$, each location Loc appearing in the lhs or rhs of $r$ contains a variable loc.*

Notice that the QPL model of Example 7 is *not* well-specified.

The equivalence result between the two computational models is stated by the following theorem.

**Theorem 1.** *Let $\mathcal{Q} = (\Sigma, E, R)$ be a well-specified QPL model and $\mathcal{P}_\mathcal{Q}$ be the DFPN obtained from $\mathcal{Q}$. Let r2t be a function mapping each rule $r \in R$ into the corresponding transition r2t(r) of $\mathcal{P}_\mathcal{Q}$. Then, the following result hold:*

$$\mathcal{Q} \vdash s_0 \xrightarrow{r_1} s_1 \ldots \xrightarrow{r_k} s_k \Leftrightarrow \mathcal{P}_\mathcal{Q} \vdash s2m(s_0) \xrightarrow{r2t(r_1)} s2m(s_1) \ldots \xrightarrow{r2t(r_k)} s2m(s_k)$$

*where $s_i \in \mathcal{S}_\mathcal{Q}$, $r_j \in R$*

## 5    Implementation

We implemented the framework presented so far in a prototypical system written in Maude, which is publicly available along with some examples at http://users.dimi.uniud.it/~michele.baggi/qpl. Basically, our system extends the Pathway Logic data structures and mechanisms to cope with quantitative information. Our system allows one to specify and analyze Quantitative Pathway Logic models by means of Maude language features. In particular, it is possible to exploit Maude built-in operators to easily express queries to simulate, search and model-check the models under examination. We tested the prototype on several small/medium size biological systems (e.g. the circadian rhythm in Drosophila) achieving rather promising results. In the future, we plan to provide a thorough experimental evaluation on real-size case studies assisted

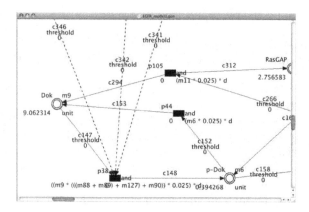

**Fig. 2.** Cell Illustrator screenshot of the EgfR pathway model

by the biologists of our group. The prototype is also equipped with a model translator which allows one to automatically derive the corresponding DFPN from the given QPL model. On the net representation, we can apply well-known Petri net analysis methodologies such as topological analysis for relevant subnet detection, backward/forward reachability analysis, *etc.* For more information about DFPN analyses, please refer to our technical report [3]. Moreover, we provide the possibility to export DFPN descriptions in the *Cell System Markup Language* (CSML) [17], an XML format for modeling biopathways which covers widely used data formats, e.g. CellML[11], SBML[8]. Then, CSML representations of DFPNs can be imported in Cell Illustrator [13], which is a software tool by means of which we can visualize and graphically interact with the DFPN models. In Figure 2 a screenshot of the Cell Illustrator application with a small fragment of the EgfR pathway model is shown. Here, discrete transitions are identified by solid rectangles, while continuous places are represented by double circles. Dashed arrows and solid arrows stand for process connectors and test connectors, respectively; finally, labels identify thresholds and update functions.

## 6    Related Work and Conclusions

There are many different computational models of biological processes, depending on the aspects we want to focus on. Our approach to biopathway modeling basically exploits rewrite theories [14] and Petri nets [16] with the aim of taking advantage of the benefits conveyed by both these formalisms. The Petri net formalism allows one to model networks of reactions describing processes as well as process execution with a graphical representation that naturally corresponds to conventional representation of biochemical networks. Standard Petri nets have a limited expressiveness (e.g. they cannot describe quantitative aspects of the model or capture reaction inhibition effects). To this respect, many powerful variants of the Petri net formalism have been developed along with a plethora of techniques and tools for system specification and analysis (e.g. see [4,9,10]).

Nonetheless the specification of large biological systems involving a huge number of states and transitions might result in a rather complex task, since Petri nets make system state and state changes explicit through marking of places and transitions which are required to be individually specified.

On the other hand, rule-based formalisms directly model states of molecular components and state changes by means of rules in which rule patterns may subsume several distinct states. This implies that rule-based specifications are in general more concise and easy to write than Petri nets. As already explained, Pathway Logic (PL) [20] represents biological processes by means of rewriting logic theories. PL allows one to only model the qualitative aspects of the processes. A recent extension of the Pathway Logic [1] has been proposed to represent and reason about semiquantitative and probabilistic aspects of biological processes. Basically, this approach annotates reaction rules with affinity information that can be used to implement distinct simulation strategies which can also include timing information. Although this approach improves expressiveness of standard PL, it only handles a semi-quantitative modeling of biological processes which does not allow us to define complex reaction kinetics. In contrast, QPL fully supports quantitative information such as element concentrations, reaction enabling thresholds, production and consumption rates in a concise and quite intuitive way. Besides, from the QPL model, we can automatically generate a graphical DFPN representation which can be displayed and intuitively manipulated via Cell Illustrator.

The extension we propose is the first step towards the development of a general PL formalism for the full specification of hybrid models (specifying both discrete and continuous components) and their stochastic counterparts. As future work, we intend to (i) add probabilistic behaviors to our models in the style of [1], and (ii) endow our specification language with timing information following the approach of [18], where hybrid systems are formalized by means of suitable rewrite theories.

# References

1. Abate, A., Bai, Y., Sznajder, N., Talcott, C., Tiwari, A.: Quantitative and Probabilistic Modeling in Pathway Logic. In: 7th IEEE International Conference on Bioinformatics and BioEngineering, pp. 922–929. IEEE Xplore, Los Alamitos (2007)
2. Abdulla, P.A., Čerāns, K., Jonsson, B., Tsay, Y.: Algorithmic analysis of programs with well quasi-ordered domains. Information and Computation 160(1-2), 109–127 (2000)
3. Baggi, M., Ballis, D., Falaschi, M.: Applications to Systems Biology of Quantitative Pathway Logic. Technical report (2009), http://users.dimi.uniud.it/~michele.baggi/qpl/BBF09tr.pdf
4. Chaouiya, C., Remy, E., Thieffry, D.: Petri net modelling of biological regulatory networks. Journal of Discrete Algorithms 6(2), 165–177 (2008)
5. Clavel, M., Durán, F., Eker, S., Lincoln, P., Martí-Oliet, N., Meseguer, J., Talcott, C.: The Maude 2.0 System. In: Nieuwenhuis, R. (ed.) RTA 2003. LNCS, vol. 2706, pp. 76–87. Springer, Heidelberg (2003)

6. Eker, S., Meseguer, J., Sridharanarayanan, A.: The maude LTL model checker and its implementation. In: Ball, T., Rajamani, S.K. (eds.) SPIN 2003. LNCS, vol. 2648, pp. 230–234. Springer, Heidelberg (2003)

7. Eyre, T.A., et al.: The HUGO Gene Nomenclature Database, 2006 updates. Nucleic Acids Research 34(Database issue), 319–321 (2006)

8. Gauges, R., Rost, U., Sahle, S., Wegner, K.: A model diagram layout extension for SBML. Bioinformatics 22(15), 1879–1885 (2006)

9. Genrich, H.J., Küffne, R., Voss, K.: Executable Petri net models for the analysis of metabolic pathways. International Journal on Software Tools for Technology Transfer 3(4), 394–404 (2001)

10. Gilbert, D., Heiner, M., Lehrack, S.: A Unifying Framework for Modelling and Analysing Biochemical Pathways Using Petri Nets. In: Calder, M., Gilmore, S. (eds.) CMSB 2007. LNCS (LNBI), vol. 4695, pp. 200–216. Springer, Heidelberg (2007)

11. Lloyd, C.M., Lawson, J.R., Hunter, P.J., Nielsen, P.F.: The CellML Model Repository. Bioinformatics 24(18), 1367–2123 (2008)

12. Nagasaki, M., Doi, A., Matsuno, H., Miyano, S.: A versatile petri net based architecture for modeling and simulation of complex biological processes. Genome Informatics 15(1), 180–197 (2004)

13. Nagasaki, M., Doi, A., Matsuno, H., Miyano, S.: Genomic Object Net: A platform for modelling and simulating biopathways. Applied Bioinformatics 2, 181–184 (2004)

14. Martí-Oliet, N., Meseguer, J.: Rewriting Logic: Roadmap and Bibliography. Theoretical Computer Science 285(2), 121–154 (2002)

15. Minsky, M.L.: Computation: finite and infinite machines. Prentice-Hall, Inc., Upper Saddle River (1967)

16. Murata, T.: Petri Nets: Properties, analysis and applications. Proc. IEEE 77(4), 541–580 (1989)

17. Nagasaki, M., Saito, A., Li, C., Jeong, E., Miyano, S.: Systematic reconstruction of TRANSPATH data into Cell System Markup Language. BMC Systems Biology 2(1) (2008)

18. Ölveczky, P.C., Meseguer, J.: Specification of real-time and hybrid systems in rewriting logic. Theoretical Computer Science 285(2), 359–405 (2002)

19. Reinhardt, K.: Reachability in Petri Nets with Inhibitor Arcs. Electronic Notes in Theoretical Computer Science 223, 239–264 (2008)

20. Talcott, C.: Pathway logic. In: Bernardo, M., Degano, P., Zavattaro, G. (eds.) SFM 2008. LNCS, vol. 5016, pp. 21–53. Springer, Heidelberg (2008)

21. Talcott, C., Dill, D.L.: Multiple representations of biological processes. In: Priami, C., Plotkin, G. (eds.) Transactions on Computational Systems Biology VI. LNCS (LNBI), vol. 4220, pp. 221–245. Springer, Heidelberg (2006)

22. The_uniprot_consortium. The Universal Protein Resource (UniProt). Nucleic Acids Research 35(Database issue) (2007)

# A Prize-Collecting Steiner Tree Approach for Transduction Network Inference

Marc Bailly-Bechet[1,2], Alfredo Braunstein[1], and Riccardo Zecchina[1]

[1] Microsoft TCI Research, Dipartimento di Fisica,
Politecnico di Torino, Corso Duca degli Abruzzi 24, 10129 Torino, Italy
[2] Université de Lyon, F-69000, Lyon, CNRS, UMR5558,
Laboratoire de Biométrie et Biologie Evolutive, F-69622, Villeurbanne, France
mbailly@biomserv.univ-lyon1.fr, alfredo.braunstein@polito.it,
riccardo.zecchina@polito.it

**Abstract.** Into the cell, information from the environment is mainly propagated via signaling pathways which form a transduction network. Here we propose a new algorithm to infer transduction networks from heterogeneous data, using both the protein interaction network and expression datasets. We formulate the inference problem as an optimization task, and develop a message-passing, probabilistic and distributed formalism to solve it. We apply our algorithm to the pheromone response in the baker's yeast *S. cerevisiae*. We are able to find the backbone of the known structure of the MAPK cascade of pheromone response, validating our algorithm. More importantly, we make biological predictions about some proteins whose role could be at the interface between pheromone response and other cellular functions.

## 1 Introduction

Living cells need to react to a wide spectrum of changes –physical, chemical or biological– in their environment [1]. Conversely the cell reactions span from the activation of small-scale processes, *e.g.* synthesis of precise molecular components or excretion of others, to complex changes in the global cellular state, such as the diauxic shift or pheromone response and mating [2] in yeast. In order for the cell to survive, these changes must be tightly regulated. One type of regulation occurs through signaling cascades, which represents how the information propagates inside a cell, from receptor proteins to transcription factors and other effector proteins. At the molecular level, this information transits by activation or inactivation of specific signaling proteins. Activation mechanisms include a variety of protein-protein interactions such as conformation changes, or dimerization [3]; one of the most studied is the well-known phosphorylation-dephosphorylation system provided by kinases [4]. Known signaling cascades show desirable properties, from a system point of view: they act as low-pass filters, ensuring an adequate cell response only when external stimuli are above the molecular noise level [5], but they also provide signal amplification [6]. Recently, it was also shown that information could travel in both directions on signaling

P. Degano and R. Gorrieri (Eds.): CMSB 2009, LNBI 5688, pp. 83–95, 2009.

cascades, due to chemical equilibrium shifts in the cascades[7]. These properties can be used by the cell to tune the signal propagation and therefore the response to the environment.

The intersection of the signaling pathways forms the transduction network, whose nodes are proteins and whose edges represent protein interactions transmitting information. Due to the many interconnections between different signaling cascades in the transduction network, a precise regulation of the cross-talk between different pathways is necessary. One way to ensure pathway specificity in answer to a given signal is the usage of scaffold proteins which will specifically bind to other members of a given signaling cascade, increasing specificity of the response [8,9]. On the other hand, one way to diffuse signal to many pathways is to root them all to the same activator protein. The complexity of these cross-interactions has allowed evolution to shape these pathways so as to be very efficient in sensing and adjusting to the environment, but makes them very difficult to study independently and even to identify precisely. In this work we tackle the issue of transduction network inference from proteomics and transcriptomics data. Phosphoproteomics works have been led to reconstruct these cascades, but are still very expensive and time consuming. At the algorithmic level, this problem has been widely studied, mainly by inference of linear cascades. In this context Scott *et al.* [10] developed an algorithm based on color-coding to infer linear subparts of the transduction network, and found with good accuracy the known MAPK kinases cascades. White *et al.* [11] made a step forward by looking for transduction networks as a superposition of shortest paths on the protein interaction network (PIN). The focus of their method is to unbias the solution tree from the high connectivity bias, as often hubs of the PIN tend to be over-represented in the inferred networks, as a consequence of their high inbetweenness. Other works about the inference of transduction network include [12], who introduced a Steiner tree formalism to recover this network based on expression data and an existing PIN. This formalism states that the transduction network is a subtree of the global protein interaction network which contains all proteins of a given subset, named terminals, defined by the user. This subset is composed of proteins known to be part of the signaling network, or selected via another criterion such as expression level. The problem is then to reconstruct such a tree, respecting also other combinatorial constraints such as *e.g* small tree size or fixed tree depth. This last approach was recently developed by [13], who used an integer linear programming relaxation to find subnetworks involved in signal transduction, improving the algorithmic performance.

The previously cited approaches have been effective at finding already known signaling cascades, but made few predictions, mainly because of the time constraints of the available techniques in the field of combinatorial inference problems. Indeed the Steiner tree problem [14] is NP-hard and classical algorithms allowing to solve it at the probabilistic level are slow. Here we provide a new method from the class of message-passing algorithms to infer a Steiner tree from a weighted graph, which directly applies to infer transduction networks from a PIN and expression data. From a numerical point of view, message-passing algorithms are probabilistic and distributed, allowing for a very fast resolution of inference problems [15], even for large networks. Moreover, our algorithm does not need *a priori* selected terminals (i.e proteins of interest), and compute the transduction network as a whole, instead of a sum of linear subparts, as was done in previous works. This results in a high exploratory power of combinatorial effects that could uncover biologically meaningful cross-talk. We apply it to the pheromone response in *S. cerevisiae*; results show that we are able to reconstruct accurately known pathways, to infer how the signal propagates in other signaling cascades of the cell, and to make functional predictions about a new group of genes implied in the pheromone response.

## 2   Material and Methods

The rationale of our model is that the transduction network is a subtree of the PIN, which should be composed with links of the PIN corresponding to real protein interactions, and proteins being of biological relevance for the biological process under study. Indeed, protein-protein interactions detected in proteomic assays contain a high fraction of false positives [16], creating the need to take into account in our model the statistical confidence we have for each link of the PIN. As proteomics data are still scarce, whether expression data are nowadays available in huge quantities, we hypothesized, as was previously done by [13,17,18,10,12], that genes being differentially expressed during the activation of the signaling pathway encode proteins being necessary for the signaling response itself, and employed expression data to measure the relative importance of each protein in a given environmental context. Therefore we could model the transduction network inference as an optimization problem, given weights for every edge of the PIN, to represent the propensity of the edge to be a false positive, and prizes for the nodes, proportional to the level of differential expression of the corresponding genes in the expression data relative to the phenomenon under study.

### 2.1   Inference of the Transduction Network

In general terms, we are interested in finding a "minimal" sub-network that is connected to a given protein node, known as the root. We will model this problem as a Prize-Collecting Steiner Tree on Graphs problem (see e.g. [19,20]). Given a network $G = (V, E)$ with positive (real) weights $\{w_l : l \in E\}$ on edges and $\{w_n : n \in V\}$ on vertices, we are interested in finding the connected sub-network that minimizes the following quantity:

$$C = \sum_{links} w_l - \lambda \sum_{nodes} w_n \qquad (1)$$

It is easy to see that such network must be a tree (links closing cycles can be removed, lowering $C$). $\lambda$ is a parameter regulating the balance between optimization of the two terms of the sum.

This problem is known to be NP-Hard, implying that is unlikely that an algorithm that can efficiently solve any instance of the problem exists. To solve it we will use a small variation of an extremely efficient heuristics based on belief propagation developed on [14] that is known to be exact on many classes of random networks [14,21]. The algorithm iterates the following set of equations for the quantities $\{\psi_{ij}\}_{(ij)\in E}$ (called "messages") to a fixed point:

$$\psi_{ji}^{t+1}(d_j, p_j) = -c_{jp_j} + \sum_{(kj)\in E\backslash(ij)} \max_{f(d_k,p_k,d_j,p_j)\neq 0} \psi_{kj}^t(d_k, p_k) \qquad (2)$$

where $d_i \in \mathcal{D} = \{0, ..., D-1\}$, $p_i \in V(i) \cup \{\emptyset\}$, $\psi_{ji} : \mathcal{D} \times V(i) \cup \{\emptyset\} \to \mathbb{R}$ and $f_{ij}$ is a characteristic function that ensures the condition $p_i = j \Rightarrow p_j \neq \emptyset, d_j = d_i - 1$ defined as follows:

$$f_{ij} = g_{ij}g_{ji}$$
$$g_{ij} = (1 - \delta_{p_j,i}(1 - \delta_{d_i,d_{j-1}}))(1 - \delta_{p_j,i}\delta_{p_i,\emptyset})$$

On a fixed point, the following quantities ("field") are computed:

$$\psi_j(d_j, p_j) = \sum_{(kj)\in E} \max_{f(d_k,p_k,d_j,p_j)\neq 0} \psi_{kj}^t(d_k, p_k)$$

Then a tree $T^*$ is built from the parenthood relations defined as follows: define $d_j^*, p_j^* = \arg\max_{d_j, p_j} \psi_j(d_j, p_j)$. Then if $p_j^* \neq \emptyset$, define the parent of $j$ as $p_j$. Otherwise, $j$ does not belong to $T^*$. With a minimal non-degeneracy assumption on the initial fields, it is relatively straightforward to verify that with variables $d_j^*, p_j^*$, $f_{ij} = 1 \forall (ij) \in E$ and this implies that $T^*$ is indeed a tree. It can be proved in some limit cases that the algorithm is optimal, and verified experimentally that it generally gives an excellent approximation to the optimal. For more details see [21].

## 2.2    Data Source and Definition of the Weights

The yeast protein interaction network (PIN) was built by combining data from two databases : DIP [22] and MIPS [23]. The combined network has 5217 nodes and 22637 edges. To define their weights, edges were divided in two categories: a high confidence one, containing links extracted from small-scale experiments or found many times; and a low confidence one, containing links found only once in a large-scale experiment. We defined the two corresponding weights so as to maximize the correlation of our weight set and the one of [24], giving a weight $w_l = 1$ for high confidence edges (24.9% of the PIN) and a weight $w_l = 1.74$ for low confidence edges. The choice of this weight set as a reference is based on the observation that it is one of the most reliable [25], and does not derive the weights from expression data.

We analyzed 56 expression datasets from [26]. We computed node prizes for each dataset in a classical way by taking $w_n = -\log(p_n)$, where $p_n$ is the $p$-value of differential expression of node $n$ in the corresponding microarray. Though, a high prize was attributed to genes having a significant $p$-value in the expression data.

The 56 datasets were analyzed independently with values of $\lambda$ ranging from 0.05 to 0.9. The chosen root was the receptor protein STE2 in datasets comparing cells submitted or not to pheromone $\alpha$ action. In datasets with an artificially overexpressed gene under GAL4 promoter control, this gene was chosen as the root. If the strain used contained deletions, the corresponding genes were removed from the PIN prior to inference. In datasets comparing deleted strains to wild type strains without exposition to pheromone, deleted genes were selected as roots.

### 2.3   Statistical Analyses

Functional homogeneity of the trees inferred for each expression microarray and each value of $\lambda$ was assessed by comparing the number of GO Slim annotations [27] shared by interacting proteins in the inferred Steiner trees, and random trees with same root and size, with edges probabilistically weighted as in the real data or not. Random trees were generated 50 times and results were averaged.

Steiner proteins were defined as proteins present in the Steiner tree with $w_n < \frac{1}{\lambda}$: such proteins have a local cost to be added in the tree, which has to be compensated. Enrichment of the inferred trees in proteins of interest was estimated by comparison with random trees generated with permuted expression data, for $\lambda = 0.2$ (30 iterations).

## 3   Results

Our algorithm infers an organism transduction network, using as a support the PIN and expression data to find a Steiner tree maximizing the level of differential expression of its nodes (genes) *and* built preferentially with edges of high confidence. The free parameter $\lambda$ (see Mat. Meth.) regulates the balance between optimization on the edges and on the nodes, and therefore regulates the tree size. For each microarray given as input, the Steiner tree found is a representation of the transduction network activated in the corresponding condition. The Steiner trees representing transduction networks were inferred in 56 expression datasets from a study about pheromone response [26], with 7 different values of $\lambda$. A statistical description of the trees found is provided in Table 1. As expected, both the frequency of high-cost links selected and the average tree size increase with $\lambda$.

As an integrity check we analyzed the correlation between the tree size and the average prize $w_n$ of the nodes in the datasets, which is a direct measure of the numbers of genes differentially expressed on the microarray (Fig 1). As expected, the average tree size increases both with $\lambda$ and average node prize; indeed, this second dependence even seems linear, a property that could be interesting to detect anomalies in the inferred trees.

An averaged representation of the trees found for $\lambda = 0.2$ is given in Fig 2. Proteins usually found as members of the pheromone response pathway are present, such as FUS3, GPA1 or SST2; some missing intermediates appear for higher

**Table 1.** Statistical properties of the trees inferred. One can see the evolution of the average tree properties with increasing values of the parameter $\lambda$, notably the increase in average tree size and decrease in the fraction of Steiner proteins found. The global fraction of high-cost links in the PIN is 75.1%, notably higher than the fraction present in the inferred trees.

| $\lambda$ | Tree size (# prot) | Fraction of high-cost edges | Fraction of Steiner proteins |
|---|---|---|---|
| 0.05 | 1.5±1.1 | 0.034 | 0.471 |
| 0.1 | 9.7±15.8 | 0.058 | 0.295 |
| 0.2 | 85.0±123.1 | 0.273 | 0.248 |
| 0.3 | 173.2±222.8 | 0.345 | 0.233 |
| 0.5 | 337.3±363.7 | 0.389 | 0.213 |
| 0.7 | 478.5±450.6 | 0.404 | 0.198 |
| 0.9 | 612.7±516.2 | 0.407 | 0.188 |

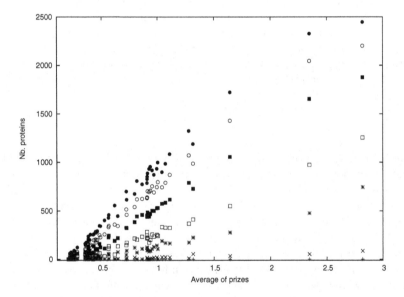

**Fig. 1.** This figure shows the strong correlation between the average node prize in each dataset (x-axis) and the number of proteins found in the inferred tree (y-axis). Different types of points correspond to different values of $\lambda$: vertical crosses $\lambda = 0.05$, diagonal crosses $\lambda = 0.1$, stars $\lambda = 0.2$, empty boxes $\lambda = 0.3$, filled boxes $\lambda = 0.5$, empty circles $\lambda = 0.7$, filled circles $\lambda = 0.9$. Note the linearity of the relation for each given value of $\lambda$.

values of $\lambda$. To assess the quality of the trees found, we computed the average number of shared GO Slim annotations between neighbors, and compared it to random trees, either weighted or not (Fig 3). the average number of common annotations is higher for low values of $\lambda$ (Fig 3), showing a clear functional enrichment of the Steiner trees. Topology and PIN weights only account for a part

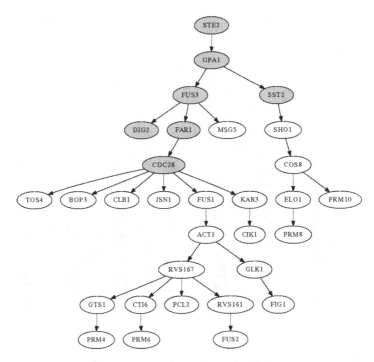

**Fig. 2.** This tree is formed by the superposition of all 56 Steiner trees found for $\lambda = 0.2$. Link intensity is proportional to the number of times the link was found, either in one sense or another; in case of links inferred in different directions, the orientation represented is the one mostly found. Links found in less than 30% of the trees are not shown for clarity. Grey nodes represent the proteins involved in the known pheromone pathway.

of this enrichment, shown by the simulations with random weighted trees, the rest being a combined consequence of both the proteins and the paths selected in the tree, which can thus be considered to represent biologically meaningful transduction networks. For high values of $\lambda$, this enrichment is not visible, and we will therefore focus on results at low $\lambda$.

Previous to analyses, a technical bias has to be accounted for. Due to differences in in-betweenness – or connectivity, see [11] –, certain proteins occur more or less often in the Steiner trees. Indeed, proteins with a high in-betweenness in the PIN tend to be frequently present in the Steiner trees, even if they are attributed a low prize. From a probabilistic point of view, including these proteins in the Steiner tree allows to gain access to proteins with a positive contribution to the global tree cost, enough to compensate for their own relative costs. One can see this trend in Fig 4: proteins selected more often have a high in-betweenness. Still, this correlation is only partial ($R^2 = 0.37$), and let ample space for other factors to explain presence of certain proteins in the final trees.

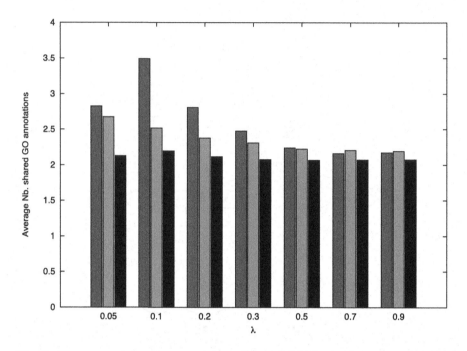

**Fig. 3.** Histogram of the average number of shared GO Slim annotations per link, on average on all 56 inferred trees at a given value of $\lambda$. First histogram represents the real values, second random weighted trees, and third random unweighted trees (See Mat. Meth.). Note the high differences for low values of $\lambda$.

**Table 2.** Properties of the 11 putative Steiner proteins found the most frequently in $\lambda = 0.2$ datasets. $k$ stands for connectivity and "In-bet." for in-betweenness.

| Gene name | Protein name | Frac. found (real data) | Frac. found (random data) | Ratio | $k$ | In-bet. ($\times 10^5$) |
|---|---|---|---|---|---|---|
| YBR160W | CDC28 | 0.66 | 0.57 | 1.2 | 227 | 15 |
| YDR388W | RVS167 | 0.52 | 0.31 | 1.7 | 121 | 4.9 |
| YHL048W | COS8 | 0.45 | 0.09 | 5.0 | 46 | 0.63 |
| YFL039C | ACT1 | 0.45 | 0.17 | 2.7 | 47 | 1.1 |
| YER118C | SHO1 | 0.43 | 0.06 | 7.0 | 42 | 1.5 |
| YJR091C | JSN1 | 0.43 | 0.44 | 1.0 | 293 | 25 |
| YCL040W | GLK1 | 0.43 | 0.003 | 144 | 6 | 0.25 |
| YBR159W | IFA38 | 0.41 | 0.09 | 4.8 | 101 | 1.9 |
| YGL181W | GTS1 | 0.41 | 0.09 | 4.5 | 43 | 1.5 |
| YPL181W | CTI6 | 0.41 | 0.02 | 20.3 | 26 | 0.26 |
| YMR059W | SEN15 | 0.34 | 0.007 | 47.5 | 57 | 1.4 |

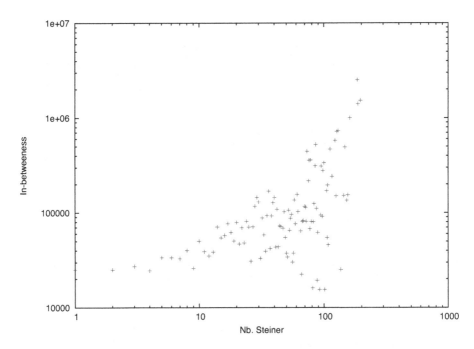

**Fig. 4.** Number of times a protein appears as Steiner (in all trees inferred) *vs* in-betweenness of the protein in the PIN. Note the correlation between them.

An interesting feature of our formalism is the definition of *Steiner proteins, i.e* proteins present in the Steiner trees without being highly differentially expressed. These proteins form bridges between groups of proteins with a positive contribution to the optimization criterion, and they could not be discovered by analyzing only the expression levels in the microarray, as they do not differ significantly from the background. Is is the combination of information from the PIN structure and expression data that unveil them. In the following analyses we focus on the Steiner proteins that appear at low values of $\lambda$, *i.e* those less distinguishable from the background expression.

In order to quantitatively measure the significance of Steiner proteins, we did a bootstrap experiment by generating Steiner trees for random expression data, obtained by permutations of the real datasets. Then, we compared the frequency of occurrence of proteins found very often as Steiner proteins in the real data to their frequency of occurrence in this randomized data; the ratio of these quantities was then used to assess the biological significance of the putative Steiner proteins (see Table 2), a high ratio meaning biologically meaningful inference and a low one typical of an artifact due to PIN topology and high in-betweenness bias.

Proteins with such a high ratio have an average in-betweenness (see Table 2). Using this table, one can easily see that the proteins CDC28, JSN1 and RVS167 should not be accounted as Steiner proteins, based on the ratio value and their very high in-betweenness. To get better insights about these proteins and their

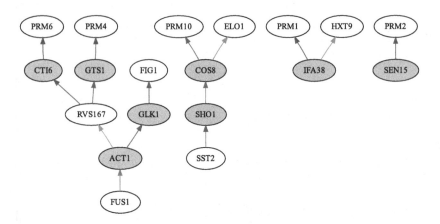

**Fig. 5.** Main first-order interactions of the proteins identified as *Steiner proteins*, $\lambda =$ 0.2. Steiner proteins are shown in grey. Link intensity is proportional to the number of times the link is found when protein is considered as Steiner. Only links found in more than 50% runs are shown.

implication in the pheromone response as Steiner proteins, we looked which partners they interact with. The partners found in more than half of the trees are represented in Fig 5. Many interactants are membrane proteins, in particular PRM proteins [28]. One interesting feature is that the Steiner proteins COS8 and SHO1 seem to be strongly interacting, as ACT1 and GLK1 do either. We detail these two cases in the following paragraphs.

GLK1 give access to the FIG1 protein, a membrane protein which has already been implicated in the pheromone response and in particular in cell fusion [29]. Another protein implicated in the glucose metabolism, GTS1, is inferred as a Steiner protein. As both these proteins are interacting with the actin protein ACT1, one could hypothesize a cross-talk between pheromone response, glucose metabolism and cytoskeleton structure. If role of the cytoskeleton in mating is quite well-known, implication of the glucose metabolism is not, but could be a sign of a global regulation of the cellular state previous to mating, as GTS1 is also a known regulator of transcription.

COS8 is found at the end of the SST2-SHO1 cascade. SST2 is the regulator of desensitization of the pheromone pathway, while SHO1 is the main cell osmosensor and initiates various signaling pathways. The subtree found behind COS8 is composed of membrane proteins and the fatty-acid elongase ELO1. Moreover, COS8 interacts often with proteins involved in sphingolipid synthesis, such as LAC1 and AUR1 (not shown in Fig 5 because they occur less than 50% of the time). Finally, the main cascade leading to IFA38, a beta-keto reductase implicated in fatty acid metabolism and also found as Steiner protein, is indeed passing by COS8. The multiple interactions of COS8 with these proteins, either membrane-spanning or located in the ER, allows to hypothesize that COS8 plays a role in the secretory pathway – probably in relation with sphingolipid synthesis– during pheromone response. Interestingly, COS8 is one but a member of a very conserved gene family [30], and finding the function of COS8 could help to understand the role of the entire family.

## 4  Discussion

In this work we presented an efficient strategy to infer structure relations from sparse gene expression information and protein-protein interaction probabilities. This approach is based on statistical physics principles, is scalable (completely parallelizable) and is expected to be well-suited for large networks. The scheme is highly efficient (the computation time scales as $D|E|$ where $|E|$ is the number of edges of the protein network, and it normally suffices to take $D = O(logN)$ to achieve optimal values). This property allowed us to explore values of the parameter $\lambda$ and a large number of pathways very quickly, much faster than it would have taken with complete algorithms and other available heuristics.

The main drawback of the approach resides in its input limitations, that is, it cannot infer new interactions between proteins and must follow the structure of the PIN given as input. This makes it difficult to apply our method to organisms where the PIN is unknown or poorly described, which is the case for many organisms, such as human. However, this issue is becoming obsolete with the rise of new experimental techniques in the proteomic fields. Moreover, there are bioinformatic solutions that could be used in order not to be limited by this problem. First, one could add in the PIN very high cost edges on a set of putative interactions. Analysis of the Steiner trees with increasing values of $\lambda$, may allow to see, among the added edges, which are selected more frequently by the algorithm, and thereby discriminate between them. Second, as the methods to infer protein-protein interactions based solely on sequence become more efficient (see *e.g.* [31]), it should be possible to develop an integrated framework where protein interactions are inferred numerically before applying our methodology.

Our methodology, while using state-of-the-art computational techniques, is able to infer quantitatively which Steiner proteins could play a role in a given context, as represented by expression data. The network representation allows a clear interpretation: specific interactions are predicted in defined conditions. This type of predictions is easy to confirm experimentally by double-hybrid and genetic experiments, making our methodology an invaluable input for wet labs. Collaborations have indeed been started to experimentally test our predictions. Moreover, our algorithm could be made still more efficient, by including genetic or regulatory interactions in the base network or searching for protein complexes instead of protein interactions. Developments in this sense, coupled with experimental validations of our predictions, will finally allow the development an integrated message-passing framework for systems biology, in direct contact with experimental data and labs.

# Acknowledgements

We thank S. Fortunato for the computation of the in-betweenness, and J.M. François for help in the biological analyses. M.B.B and A.B acknowledge a fellowship from Politecnico di Torino and a combined Microsoft research/Politecnico di Torino funding.

# References

1. Elston, T.C.: Probing pathways periodically. Sci. Signal 1(42) (2008); pe47
2. Dohlman, H.G., Slessareva, J.E.: Pheromone signaling pathways in yeast. Sci. STKE 2006(364) (December 2006); cm6
3. Luttrel, L.: Transmembrane signalling by g protein couple receptors. Methods Mol. Biol. 332, 3–49 (2006)
4. Chen, R.E., Thorner, J.: Function and regulation in MAPK signaling pathways: lessons learned from the yeast Saccharomyces cerevisiae. Biochem. Biophys. Acta 1773(8), 1311–1340 (2007)
5. Thattai, M., van Oudenaarden, A.: Attenuation of noise in ultrasensitive signaling cascades. Biophys. J. 82(6), 2943–2950 (2002)
6. Kholodenko, B.N.: Cell-signalling dynamics in time and space. Nat. Rev. Mol. Cell Biol. 7(3), 165–176 (2006)
7. Ventura, A.C., Sepulchre, J.A., Merajver, S.D.: A hidden feedback in signaling cascades is revealed. PLoS Comput. Biol. 4(3), e1000041 (2008)
8. Locasale, J.W., Chakraborty, A.K.: Regulation of signal duration and the statistical dynamics of kinase activation by scaffold proteins. PLoS Comput. Biol. 4(6), e1000099 (2008)
9. Bashor, C.J., Helman, N.C., Yan, S., Lim, W.A.: Using engineered scaffold interactions to reshape MAP kinase pathway signaling dynamics. Science 319(5869), 1539–1543 (2008)
10. Scott, J., Ideker, T., Karp, R.M., Sharan, R.: Efficient algorithms for detecting signaling pathways in protein interaction networks. J. Comput. Biol. 13(2), 133–144 (2006)
11. White, A., Ma'yan, A.: Connecting seed lists of mammalian proteins using steiner trees. Nature Precedings (2008)
12. Scott, M.S., Perkins, T., Bunnell, S., Pepin, F., Thomas, D.Y., Hallett, M.: Identifying regulatory subnetworks for a set of genes. Mol. Cell Proteomics 4(5), 683–692 (2005)
13. Zhao, X.M., Wang, R.S., Chen, L., Aihara, K.: Uncovering signal transduction networks from high-throughput data by integer linear programming. Nucleic Acids Res. 36(9), e48 (2008)
14. Bayati, M., Borgs, C., Braunstein, A., Chayes, J., Ramezanpour, A., Zecchina, R.: Statistical mechanics of steiner trees. Phys. Rev. Lett. 101(3), 037208 (2008)
15. Mézard, M., Parisi, G., Zecchina, R.: Analytic and algorithmic solution of random satisfiability problems. Science 297(5582), 812–815 (2002)
16. Gavin, A.C., Bosche, M., Krause, R., Grandi, P., Marzioch, M., Bauer, A., Schultz, J., Rick, J.M., Michon, A.M., Cruciat, C.M., Remor, M., Hofert, C., Schelder, M., Brajenovic, M., Ruffner, H., Merino, A., Klein, K., Hudak, M., Dickson, D., Rudi, T., Gnau, V., Bauch, A., Bastuck, S., Huhse, B., Leutwein, C., Heurtier, M.A., Copley, R.R., Edelmann, A., Querfurth, E., Rybin, V., Drewes, G., Raida, M.,

Bouwmeester, T., Bork, P., Seraphin, B., Kuster, B., Neubauer, G., Superti-Furga, G.: Functional organization of the yeast proteome by systematic analysis of protein complexes. Nature 415(6868), 141–147 (2002)

17. Chang, W.C., Li, C.W., Chen, B.S.: Quantitative inference of dynamic regulatory pathways via microarray data. BMC Bioinformatics 6, 44 (2005)

18. Steffen, M., Petti, A., Aach, J., D'haeseleer, P., Church, G.: Automated modelling of signal transduction networks. BMC Bioinformatics 3, 34 (2002)

19. Johnson, D., Minkoff, M., Phillips, S.: The prize collecting steiner tree problem: theory and practice. In: Proceedings of the eleventh annual ACM-SIAM symposium on Discrete algorithms, Society for Industrial and Applied Mathematics, Philadelphia, PA, USA, pp. 760–769 (2000)

20. Lucena, A., Resende, M.G.C.: Strong lower bounds for the prize collecting Steiner problem in graphs. Discrete Applied Mathematics 141(1-3), 277–294 (2004)

21. Bayati, M., Braunstein, A., Zecchina, R.: A rigorous analysis of the cavity equations for the minimum spanning tree. Journal of Mathematical Physics 49(12), 125206 (2008)

22. Xenarios, I., Rice, D.W., Salwinski, L., Baron, M.K., Marcotte, E.M., Eisenberg, D.: DIP: the database of interacting proteins. Nucleic Acids Res. 28(1), 289–291 (2000)

23. Güldener, U., Oesterheld, M., Pagel, P., Ruepp, A., Mewes, H.W., Stümpflen: Mpact: the MIPS protein interaction resource on yeast. Nucleic Acids Res. 34(Database issue), D436–D441 (2006)

24. Bader, J.S., Chaudhuri, A., Rothberg, J.M., Chant, J.: Gaining confidence in high-throughput protein interaction networks. Nat. Biotechnol. 22(1), 78–85 (2004)

25. Suthram, S., Shlomi, T., Ruppin, E., Sharan, R., Ideker, T.: A direct comparison of protein interaction confidence assignment schemes. BMC Bioinformatics 7, 360 (2006)

26. Roberts, C.J., Nelson, B., Marton, M.J., Stoughton, R., Meyer, M.R., Bennett, H.A., He, Y.D., Dai, H., Walker, W.L., Hughes, T.R., Tyers, M., Boone, C., Friend, S.H.: Signaling and circuitry of multiple MAPK pathways revealed by a matrix of global gene expression profiles. Science 287(5454), 873–880 (2000)

27. The Gene Ontology Consortium: Gene ontology: tool for the unification of biology. Nat. Genet. 25, 25–29 (2000)

28. Heiman, M.G., Walter, P.: Prm1p, a pheromone-regulated multispanning membrane protein, facilitates plasma membrane fusion during yeast mating. J. Cell Biol. 151(3), 719–730 (2000)

29. Aguilar, P.S., Engel, A., Walter, P.: The plasma membrane proteins prm1 and fig1 ascertain fidelity of membrane fusion during yeast mating. Mol. Biol. Cell 18(2), 547–556 (2007)

30. Despons, L., Wirth, B., Louis, V.L., Potier, S., Souciet, J.L.: An evolutionary scenario for one of the largest yeast gene families. Trends Genet. 22(1), 10–15 (2006)

31. Weigt, M., White, R.A., Szurmant, H., Hoch, J.A., Hwa, T.: Identification of direct residue contacts in protein-protein interaction by message passing. Proc. Natl. Acad. Sci. U S A 106(1), 67–72 (2009)

# Formal Analysis of the Genetic Toggle

Giampaolo Bella[1] and Pietro Liò[2]

[1] Dipartimento di Matematica e Informatica,
Università di Catania, Viale A. Doria 6, 95125 Catania, Italy
giamp@dmi.unict.it
[2] Computer Laboratory University of Cambridge
15 JJ Thomson Avenue, Cambridge CB3 0FD, UK
pl219@cam.ac.uk

**Abstract.** The formal analysis of the toggle switch, which is among the most common motifs of genetic networks, shows that along with the powerful development of mathematical modelling, formal methods can be of great help in investigating the properties of genetic networks. In particular, a general approach to modelling genetic networks through the language of higher-order logic is advanced and mechanised in the theorem prover Isabelle. An inductive definition provides a formal model for the genetic toggle as the set of all possible evolutions of such network. Gene polymerase and protein concentration are formalised as primitive recursive functions. The main properties of the genetic toggle are confirmed upon the model: it is possible that one protein exceeds a stated concentration threshold and the other protein does not; it is impossible that both proteins exceed their respective concentration thresholds at the same time. To the best of the authors' knowledge, this is the first contribution of theorem proving in the area of genetic network analysis, and as such may set the foundations for a new niche of research.

## 1 Introduction

Genes function in highly interconnected, hierarchical, and nonlinear chemical networks. Representing interactions between biological molecules as a network provides us with a conceptual framework that allows us to identify general principles that govern these complex systems. A network is best represented as a graph that is made up of nodes, which denote the components, and links, which denote the interaction between the components. In a genetic or transcriptional regulatory network, nodes represent DNA binding proteins or target genes and directed edges represent regulatory interaction where the protein regulates the expression of the target gene. A meaningful approach to investigate biological networks is the identification of network motifs. Motifs can be defined as a set of network nodes with specific molecular functions which are arranged together and perform some 'useful' processes. The behaviours of motifs are generally not separable from the rest of the system and they constitute only part of a recognizable systems level function. There are several known motifs in genetic networks for example toggle switches, amplitude filters, oscillators, frequency filters, noise

P. Degano and R. Gorrieri (Eds.): CMSB 2009, LNBI 5688, pp. 96–110, 2009.

filters and amplifiers, combinatorial logic, homeostats, rheostats, logic (e.g. see [1,2,3,4]). The first step of the analysis of a network motif is to model the process of gene expression, i.e. the process through which the RNA polymerase produces mRNA molecules from a DNA template, and the process of protein synthesis, which is the formation of proteins from an mRNA template. The mathematical description of the variation of biomolecular concentrations may be based on a set of differential equations (ODE) or on stochastic simulations by means of the Gillespie algorithm [5]. The differential equations approach is effective in describing the average behaviour of the reaction set and allows to incorporate non linearities; instead, the stochastic approach provides a more realistic approach because, in many cases, in the cellular environment, the number of molecules tend to be small and concentrations may fluctuate. For example, using the ODE approach, the rate of change of mRNA (R) and of protein (P) concentrations depend on the formation $\lambda$ and the degradation $\delta$ rates and can be written as:

$$\frac{dR}{dt} = \lambda_R - \delta_R R \; ; \qquad \frac{dP}{dt} = \lambda_P R - \delta_P P$$

The steady-state solutions of the equations are:

$$\tilde{R} = \frac{\lambda_R}{\delta_R} \; ; \qquad \tilde{P} = \frac{\lambda_P}{\delta_P R} = \frac{\lambda_R \lambda_P}{\delta_R \delta_P}$$

The most important and ubiquitous motif network is the toggle switch, which consists of two genes, each of which produces a protein that represses the transcription of the other gene. The toggle can be described by a system of coupled nonlinear differential equations, which are usually very difficult to solve exactly, leaving us to two choices: either solve the system numerically or find some kind of approximate solution. A stochastic implementation always results in intensive simulations. If we consider a network with a large number of genes, the mathematical analytical treatment becomes impossible and the stochastic simulation computationally very intensive. Therefore, methods that provide a different, complementary understanding of the behaviour of biological networks are greatly needed, and formal methods are good candidates. Some of them come with compelling computer support, which turns out particularly useful also when reasoning about biological problems. For example, this is the case with the use of model checking techniques in this area, a synergy that has produced significant findings in the last years. A non-exhaustive account on such findings would have to mention the seminal paper where a genetic network is simplified into a finite-state machine amenable to model checking [6]. Then, the account should cover the first complete application to a paradigmatic case study, the genetic toggle, where the language of linear-temporal logic is used to express the main properties of the network [7]. There are also more recent developments (omitted here due to space limitations).

Theorem proving is another popular computer-assisted technology to exceed the limitations of pen-and-paper reasoning. Its comparison to model checking, which lies beyond our focus, has notoriously been the subject of vast research.

In brief, model checking excels for the limited amount of human intervention, which makes it particularly appreciated in the industrial world, but can only handle systems of limited size [8]. By contrast, theorem proving typically reaches outstanding levels of detail in the expression of a system of unbounded size and of its properties, but requires more human intervention [9].

The main contribution of this paper is to lay the ground for the use of theorem proving in the field of biological problems such as genetic network analysis. To our knowledge, the potentialities of theorem proving in improving our understanding of genetic networks have never been explored before. Our findings are promising. A genetic network is formalised by mathematical induction as the set of all its possible evolutions, which in the real world are determined by stochastic factors. Such factors may safely be abstracted away because, whatever network evolution they may cause, that evolution is certain to belong to the inductive model of the genetic network. Invariant properties of the model therefore apply to the real network. Our inductive model of a genetic network is qualitative as it provides an operational semantics of the network. In particular, gene polymerase and protein concentration are defined by primitive recursion. All definitions are fed to the interactive theorem prover Isabelle [9], which supports the verification of properties of the model. This paper studies the genetic toggle to demonstrate the accuracy and expressiveness of the approach we propose.

The organisation of this manuscript is simple. A very brief outline of Isabelle (§2) precedes the description of our approach to formally analysing genetic networks (§3). Then, the genetic toggle is modelled and verified as a case study (§4). Some concluding remarks terminate the presentation (§5).

## 2   Isabelle

Isabelle is a generic, interactive theorem prover. *Generic* means that it can reason in a variety of formal systems. This paper refers to Isabelle/HOL [9], which supports the formal language *higher-order logic*, a typed formalism that allows quantification over functions, predicates and sets, but has no temporal operators. *Interactive* means that it is not entirely automatic and, rather, requires a good amount of human intervention. But Isabelle also provides much automation. Its *simplifier*, which can be invoked by the proof method `simp`, combines rewriting with arithmetic decision procedures. Its *automatic provers* can solve most simple proof scenarios. For example, the proof method `blast` implements a fast classical reasoner, and `auto` combines that with the simplifier.

Proofs are conducted interactively. In a typical proof, the user directs Isabelle to perform a certain induction and then to simplify the resulting subgoals. Any surviving subgoals may be given to an automatic prover or be reduced to other subgoals by means of some lemma. Failure to find a proof for a conjecture may simply mean that the user is not skilled enough; otherwise, it may exhibit what in the modelled system contradicts the conjecture and hence help in locating a system bug or an erroneous human intuition. The command list used to prove a theorem can be seen as a proof sketch. Confidence that the proof is sound comes from inspecting the line of reasoning adopted and the lemmas it requires.

# 3  Formal Analysis of Genetic Networks

Our approach to modelling genetic networks is inspired to an existing model of security protocols [10], and is mechanised in the theorem prover Isabelle. An inductive definition provides an operational semantics for the computer network underlying the distributed execution of a security protocol. The computer network is qualitatively seen as the interaction of agents who exchange the protocol messages, and hence our idea to view a genetic network as the interaction of genes and proteins who exchange some signals.

Our presentation proceeds by defining the basics of a general model of genetic networks, which can then be used to tackle any case study. Due to space limitations, the subsidiary lemmas about the general model, which, as is typical with theorem proving, assess its correctness, must be skipped. Likewise, a didactic example of how to model a simple network cannot be presented, but an example of realistic size is given in the next Section.

Following the observation that the main agents of a genetic network are genes and proteins, they are introduced as free types, that is as type declarations:

**typedecl** *gene*
**typedecl** *protein*

We model four main events that may take place in a genetic network. As a start, what a gene may do is to code a protein. A protein may in turn either trigger or inhibit a gene expression. Proteins may also diffuse in the cell volume and be degradated (misfolding and ubiquination) before they succeed in regulating a gene. These four events are respectively formalised by the four keywords *Produces*, *Triggers*, *Inhibits* and *Degrades*, which are introduced as type constructors of the datatype *event*. It can be seen how each type constructor takes two freetypes as parameters, except for the protein degradation event, which, as expected, only takes one:

**datatype**
```
  event = Produces gene protein |
          Triggers protein gene | Inhibits protein gene |
          Degrades protein
```

For example, the event *Triggers A b* is defined in our language, indicating the act of a protein *A* that induces a gene *b* to express its protein. The details or intermediate steps underlying each event, such as transcription to RNA, are hidden behind each event. An operational semantics only expresses what event takes place rather than specifying how each event is carried out. The datatype may of course be extended in the future to capture additional network events.

We continue by modelling the main condition for a gene to express a protein, that is the gene level of polymerase. It is assumed that all genes have some positive polymerase level since the very beginning of the observation, so that they can begin expressing their proteins. Such initial polymerase is introduced as constant for all genes for simplicity:

**consts** *initialpolymerase* :: *int*

Specifying different initial polymerases for the genes would be trivial as the constant just introduced would gain an extra parameter of type *gene*.

The requirement that the initial polymerase is strictly positive can be easily introduced as an axiom of our model:

**axioms** *initialpolymerase_value [iff]: initialpolymerase > 0*

The *[iff]* declaration tells both the simplifier and the classical reasoner of Isabelle that axiom named *initialpolymerase_value* can automatically be appealed to in solving the subgoals that may arise from subsequent proof attempts. For example, it may lead a set of facts to a desired absurd when the proof strategy proceeds from the falsification of the thesis.

The polymerase clearly is a dynamic value. It is influenced by the events that proteins carry out, which are triggers or inhibitions, as defined above. Therefore, such induced polymerase can be easily declared as:

**consts** *inducedpolymerase :: gene ⇒ event list ⇒ int*

The second parameter introduces the list of events that have taken place, which modify the induced polymerase. It may be currently obscure where such a list comes from, but the rest of the treatment will show how it can be effectively defined by mathematical induction. The function modelling the induced polymerase can be defined by primitive recursion on the structure of the list of events that is considered. It is useful to remind that Isabelle's lists are built from right to left, that *[]* denotes the empty list, and that # is the list cons operator:

**primrec**
```
    inducedpolymerase x [] = 0
    inducedpolymerase x (head#rest) = (case head of
        Produces y Y ⇒ inducedpolymerase x rest
      | Triggers Y y ⇒ if x=y then inducedpolymerase x rest + 1
                                else inducedpolymerase x rest
      | Inhibits Y y ⇒ if x=y then inducedpolymerase x rest - 1
                                else inducedpolymerase x rest
      | Degrades Y   ⇒ inducedpolymerase x rest
    )
```

The primitive recursive definition consists of two equations. The first, which is for the base case, is very simple as it states that the polymerase that an empty list of events induces upon a gene *x* is zero. Intuitively, because no events are available in this case, they induce no variation upon the polymerase. The second equation is the inductive step, as it considers a list of events of which the head event *head* is emphasized. The equation contains a case analysis deriving from each of the four possible events that may get instantiated as *head*. The first and last line of the case analysis respectively make sure that a gene expression of a protein and a protein degradation do not change the polymerase that the rest of the list *rest* induces upon the generic gene *x*. To describe it with the terminology of term rewriting, the expression *inducedpolymerase x (head#rest)* is evaluated as *inducedpolymerase x rest*, confirming that the *head* event just cancels itself out during the symbolic evaluation.

The two internal lines, respectively *Triggers Y y* and *Inhibits Y y*, are perhaps more interesting. Let us begin by studying the former. It resolves into an if-then-else expression whose condition is that the generic gene *y* that the event considers be exactly the gene *x* whose induced polymerase is being evaluated. Only if this check is true does the evaluation resolve into *inducedpolymerase x rest + 1*, which in our qualitative approach is to be interpreted as a generic increase of the polymerase that the rest of the list *rest* induces upon gene *x* (of course, a symbolic parameter might be used instead of the unit). Otherwise, the symbolic evaluation just skips *head*, returning *inducedpolymerase x rest*. The other line can be interpreted similarly, but an *Inhibits Y y* event should induce a decrease over the polymerase of the gene, and this is expressed by the – 1 (or minus some parameter instead).

Having defined the initial and the induced polymerase, the current polymerase can be introduced as a sum of the two. The declaration must account for two parameters due to the induced polymerase. The definition is simple, and can be introduced compactly by Isabelle's **constdefs** command as follows:

**constdefs**
```
currentpolymerase :: gene ⇒ event list ⇒ int
currentpolymerase x nt == initialpolymerase + inducedpolymerase x nt
```

It can be seen that the current polymerase has a static part, which was available initially, and a dynamic part induced by the events in the considered list.

The codomain of our functions for the polymerase is *int* rather than *nat*, which might have been more intuitive. Our choice is determined by the primitive recursive definition of the induced polymerase seen above, which decreases the current value when an appropriate inhibition event is evaluated. For example, the polymerase induced over a gene by a list of inhibition events on that gene is a negative value. It may be interpreted in the model as extra incapacity to express the protein. In consequence, also the other two polymerase functions must return an integer to avoid type clashes. Alternatively, the definition of the induced polymerase should have accommodated the extra case analysis that protein inhibition decreases the induced polymerase if and only if it is not already zero. We made the simpler modelling choice.

Turning our focus to proteins, the concentration appears to be the main prerequisite to a protein activity. A protein typically begins to inhibit or trigger genes depending on whether its concentration exceeds some threshold. The same mechanisms used above can be adopted here to define the concentration:

**consts**
```
concentration :: protein ⇒ event list ⇒ int
```
**primrec**
```
concentration X [] = 0
concentration X (head#rest) = (case head of
      Produces y Y ⇒ if X=Y then concentration X rest + 1
                            else concentration X rest
   | Triggers Y y ⇒ concentration X rest
   | Inhibits Y y ⇒ concentration X rest
```

```
| Degrades Y   ⇒ if X=Y then concentration X rest - 1
                        else concentration X rest
)
```

The definition insists that the concentration be initially zero. The inductive step provides the symbolic evaluation of the concentration of a generic protein X over a list of events whose head event is `head`. The first line of the case analysis states that when `head` exactly is a production event of protein X, then the concentration increases by being evaluated as `concentration X rest + 1`. The last line follows the same pattern to make sure that when the protein degrades its concentration decreases. The two central lines leave the symbolic evaluation unaltered by skipping the `head` event.

A first version of this definition decreased the concentration also in the cases of trigger or inhibition events to signify that a protein that binds to a promoter of a gene is no longer available. However, it was later preferred to simplify it as it now stands with a negligible overhead: making sure that a trigger or inhibition event is always followed by a degradation event. This proviso must be accounted for when defining the network, as we shall see below.

The remark made about the `int` codomain also applies here: it is a technical requirement of our formalisation. Although no negative concentration exists in the real world, it may be interpreted in the model as extra distance from a required threshold. Alternatively, it may be removed by adding a case study to the symbolic evaluation of the concentration in case of a degradation event.

Note that only one function suffices to define the concentration because we are assuming that the concentrations of all proteins be initially zero. Removing this assumption would be costless, resembling the pattern of the definitions introduced for the polymerase, that is to introduce an initial value, then an induced value and finally the current value.

## 4 A Case Study: The Genetic Toggle

The toggle switch is a cyclic digenic system (say genes $g_1$ and $g_2$) with negative feedbacks between the genes through the negative regulation of their proteins. A simple way to model the regulatory control of a protein on a gene is to modify R by considering the binding properties of the protein on a gene, $\sigma_1$ or $\sigma_2$. In a toggle switch network we would expect the system to exhibit two mutually exclusive behaviours - either $g_1$ is high, keeping expression of $g_2$ low, or conversely, $g_2$ is high, keeping expression of $g_1$ low. The toggle can switch from one stable state to the other by means of an external signal. Noteworthy, the toggle is at the core of the synthetic biology [11]. For simplicity, we assume that the switch is composed of symmetric elements so that the rate constants are identical for each half of the network and the mathematical model takes the form:

$$\frac{dR_1}{dt} = \lambda_R\sigma_1 - \delta_R R ; \qquad \frac{dR_2}{dt} = \lambda_R\sigma_2 - \delta_R R$$

$$\frac{dP_1}{dt} = \lambda_P R_1 - \delta_P P_1 ; \qquad \frac{dP_2}{dt} = \lambda_P R_2 - \delta_P P_2$$

Here, the λs and δs respectively are formation and degradation parameters. The numerical solution of this set of equations provides the determination of the two steady-state thresholds of proteins P1 and P2, which depend upon one another (see also [1]). The set of equations identifies curves that are called nullclines and their intersection is called equilibrium point. Although recent methodological developments have made it possible to estimate rate constants from experimental observations such as microarray data [12], from a qualitative perspective the identification and characterization of the presence of a toggle switch may come from the formal analysis of the behaviour of the system, as we shall see below.

### 4.1  Modelling the Genetic Toggle

The general model of genetic networks introduced above can be demonstrated upon the genetic toggle. Few additional definitions are necessary for this specific example. As our approach is qualitative, we abstract away the formation and degradation parameters seen above.

The pair of threshold concentrations of a protein can be introduced as a function parameterised over the integers and the proteins:

```
consts
  threshold :: nat ⇒ protein ⇒ int
axioms
  threshold_1A_less_2A [iff]: threshold 1 A < threshold 2 A
  threshold_1B_less_2B [iff]: threshold 1 B < threshold 2 B
  threshold_1A_pos    [iff]: threshold 1 A > 0
  threshold_1B_pos    [iff]: threshold 1 B > 0
```

By having the integers (rather than the booleans) as a parameter, our declaration also supports more than two thresholds. The axioms formalise the typical requirements for this case study.

We define the total concentration of proteins that can be produced as the sum of the two higher thresholds. Because we omit naming the function definition, Isabelle assigns the default name $totalproteins\_def$.

```
constdefs
  totalproteins :: int
  [iff]: totalproteins == threshold 2 A + threshold 2 B
```

The model for the genetic toggle can be now defined inductively as follows:

```
inductive_set network :: event list set where

  base:  [] ∈ network

| aprA:  ⟦ nt1 ∈ network; currentpolymerase a nt1 > 0;
             concentration A nt1 + concentration B nt1 < totalproteins ⟧
         ⟹ Produces a A # nt1 ∈ network

| bprB:  ⟦ nt2 ∈ network; currentpolymerase b nt2 > 0;
             concentration A nt2 + concentration B nt2 < totalproteins ⟧
         ⟹ Produces b B # nt2 ∈ network
```

```
| Ainb:  [ nt3 ∈ network; concentration A nt3 ≥ threshold 1 A ]
          ⟹ Degrades A # Inhibits A b # nt3 ∈ network

| Aina:  [ nt4 ∈ network; concentration A nt4 ≥ threshold 2 A ]
          ⟹ Degrades A # Inhibits A a # nt4 ∈ network

| Bina:  [ nt5 ∈ network; concentration B nt5 ≥ threshold 1 B ]
          ⟹ Degrades B # Inhibits B a # nt5 ∈ network

| Binb:  [ nt6 ∈ network; concentration B nt6 ≥ threshold 2 B ]
          ⟹ Degrades B # Inhibits B b # nt6 ∈ network

| dgrA:  [ nt7 ∈ network; Produces a A ∈ set nt7 ]
          ⟹ Degrades A # nt7 ∈ network

| dgrB:  [ nt8 ∈ network; Produces b B ∈ set nt8 ]
          ⟹ Degrades B # nt8 ∈ network
```

The base case sets the base of the induction. The last two inductive rules model a stochastic degradation of a protein, as they may fire any time provided that the protein was produced. The remaining rules model the distinctive features of the genetic toggle. Rules aprA and bprB produce proteins A and B respectively upon condition that the polymerase of the involved gene is positive and that the total concentration of proteins has not been reached. Then, four inhibition rules follow, all with the same structure enforcing the proviso that a protein degrades after it links to a gene. Precisely, rules Ainb and Bina state that a protein inhibits the other gene when the protein concentration exceeds its first threshold. Rules Aina and Binb state that a protein inhibits its own gene when the protein concentration exceeds its second threshold.

## 4.2 Verifying the Genetic Toggle

As a start, following the axioms defining the thresholds, we can prove that also the second thresholds are positive:

```
lemma threshold_2A_pos: threshold 2 A > 0
apply (blast intro: less_trans)
done

lemma threshold_2B_pos: threshold 2 B > 0
apply (blast intro: less_trans)
done
```

These lemmas are sometimes useful below. Each can be proved by a single invocation to the classical reasoner, the blast method, with an appeal to lemma less_trans, which confirms the transitivity of the less relation.

Having observed the model of the genetic toggle, our first conjecture was that whenever a gene produces a protein, the polymerase of the gene must be positive. This can be formalised as follows:

```
lemma aprA_cp: [ Produces a A ∈ set nt; nt ∈ network ] ⟹
                 currentpolymerase a nt > 0
```

Precisely, the statement insists upon two preconditions: one is that $nt$ is a network list of the model toggle $network$; the other one is that the event $Produces$ a $A$ appears in $nt$. The conclusion is that the current polymerase of gene a in the network list $nt$ is positive. Note Isabelle's meta implication "$\Longrightarrow$", not to be confused with object-level implication "$\longrightarrow$".

We begin proving this conjecture, which is the goal of our proof attempt, by bringing the main premise into the inductive formula. This can be done by Isabelle's method $erule$, which applies resolution of the goal with the stated theorem, reverse modus-ponens $rev\_mp$ in this case, and then eliminates the preconditions that were necessary to the application:

**apply** $(erule\ rev\_mp)$

As a result, $Produces$ a $A$ $\longrightarrow$ $currentpolymerase$ a $nt > 0$ becomes the conclusion of the proof goal. Structural induction can be applied now, which precisely is the resolution of the current proof goal with the $network.induct$ theorem that Isabelle builds automatically to reflect the inductive structure of constant $network$ (it is very long but can be viewed by interacting with Isabelle):

**apply** $(erule\ network.induct)$

Our goal is now split up into 9 subgoals, one per each rule defining $network$. We can apply the simplifier to all goals by the method $simp\_all$ but three subsidiary lemmas proved earlier (omitted here) must be invoked as rewrite rules for symbolic evaluation of the current polymerase function over network events. Note that $currentpolymerase\_Triggers$ is unnecessary because the genetic toggle features no trigger events.

**apply** $(simp\_all\ add:\ currentpolymerase\_Produces$
$\qquad\qquad\qquad\quad currentpolymerase\_Inhibits$
$\qquad\qquad\qquad\quad currentpolymerase\_Degrades)$
**oops**

By inspecting the resulting proof state, we realise that the proof cannot be terminated and hence the **oops** command tells the prover to leave it and proceed. The proof state, consisting of two subgoals, can be analysed from Figure 1.

In general, the inability to prove a conjecture may be due to insufficient skill of the human analyser, or to falseness of the conjecture. The proof state confirms that the latter applies here. It can be seen that the conclusion of each subgoal cannot be proved upon the available preconditions — the second one in each subgoal being the inductive hypothesis. The subgoals take our attention upon the rules that cause them, $Aina$ and $Bina$, which are the only rules that inhibit the very gene a mentioned in the conjecture and therefore decrease its polymerase. Hence, the subgoals highlight two counterexamples to the conjecture: one is a network list where the polymerase is positive and then a is inhibited by firing of the rule $Aina$; the other one is analogous but involves firing of the rule $Bina$.

So, we decided to weaken the conjecture by involving only network lists whose last event (i.e. the head event) is a gene production. Using the predefined function $hd$ to extract the head of a list, this can be formalised and proved as follows:

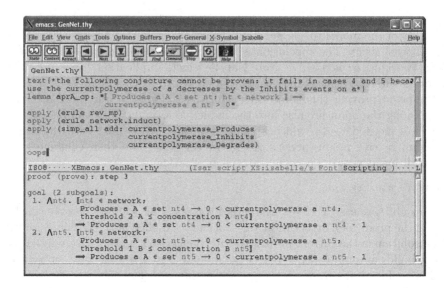

**Fig. 1.** The final proof state of a conjecture

**lemma** aprA_cp: ⟦ Produces a A = hd(nt); nt ∈ network ⟧ ⟹
                   currentpolymerase a nt > 0
**apply** (erule rev_mp)
**apply** (erule network.induct)
**apply** (simp_all add: currentpolymerase_Produces)
**apply** (simp add: currentpolymerase_def)
**done**

The proof terminates, so that the conjecture holds. The script needs an extra application of the simplifier to solve the subgoal deriving from rule base, which is Produces a A = hd [] ⟶ 0 < currentpolymerase a [], because the prover needs to evaluate the head of an empty list, unlike in the previous proof attempt. However, this subgoal can be solved straightforwardly by unfolding the definition of the current polymerase, which obtains the required postcondition by positiveness of the initial polymerase (axiom initialpolymerase_value, §3). Moreover, only the rewrite rule currentpolymerase_Produces is necessary in the main simplification step because the theorem precondition Produces a A = hd(nt) only requires symbolic evaluation of the current polymerase over production events, as it rules out all other events.

Our primary aim is to prove the main property of the genetic toggle, which is bistability. We address as *bistabilityA* the situation in which protein A exceeds its second threshold but protein B does not; as *bistabilityB* the opposite situation where protein B exceeds its second threshold but protein A does not; and as *instability* the situation in which both proteins exceed their respective second thresholds. Therefore, proving bistability evaluates to proving that:

the network may evolve to bistabilityA; the network may evolve to bistabilityB; the network may not evolve to instability.

Isabelle is best tailored to proving safety properties, that is invariants of all network lists. Hence, we set about the first two properties by attempting to prove that they fail to hold, that is that the network may not evolve to either bistabilityA or bistabilityB. The prover will exhibit the counterexamples we are looking for, if these exist.

Here is the conjecture that bistabilityA may not be reached and its (failed) proof attempt:

**theorem** `no_bistabilityA: nt` $\in$ `network` $\Longrightarrow$
     $\neg$ `(concentration A nt` $\geq$ `threshold 2 A` $\wedge$
          `concentration B nt < threshold 2 B)`
**apply** `(erule network.induct)`
**apply** `(insert threshold_2A_pos)`
**apply** `auto`
**oops**

After applying induction, lemma `threshold_2A_pos` is inserted in the first subgoal, which arises from rule `base` — simplifying that subgoal without inserting that lemma would leave `threshold 2 A` $\leq$ `0` $\longrightarrow$ $\neg$ `0 < threshold 2 B`. Then,

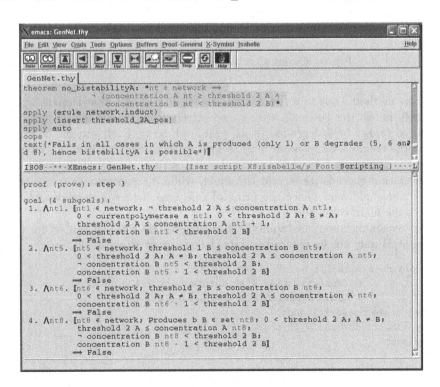

**Fig. 2.** Proving that bistabilityA may be reached

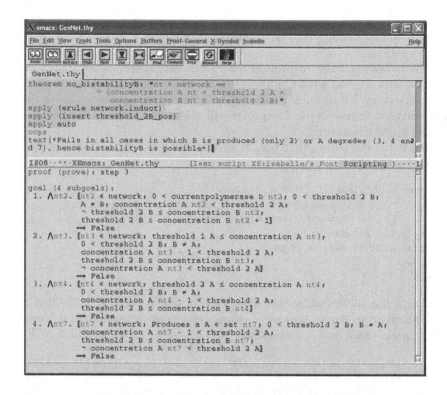

**Fig. 3.** Proving that bistabilityB may be reached

the `auto` method leads to a proof state, shown in Figure 2, that cannot be terminated. It features subgoals arising respectively from rules where protein `A` is produced (`aprA`) or `B` degraded (`Bina`, `Binb` and `dgrB`). Each subgoal describes one of the counterexamples we were seeking to establish that bistabilityA may be reached. For example, the second subgoal describes a network list where `A`'s concentration exceeds its second threshold, while `B`'s concentration equals its second threshold, and hence the firing of rule `Bina`, where `B` degrades, clearly falsifies the main conjecture.

The reasoning about bistabilityB is analogous. Here is the conjecture that bistabilityB may not be reached and its (failed) proof attempt:

**theorem** `no_bistabilityB: nt` ∈ `network` ⟹
        ¬ (`concentration A nt < threshold 2 A` ∧
            `concentration B nt` ≥ `threshold 2 B)`
**apply** (`erule network.induct`)
**apply** (`insert threshold_2B_pos`)
**apply** `auto`
**oops**

The final proof state, shown in Figure 3, cannot be terminated. It has subgoals arising respectively from rules where protein `B` is produced (`bprB`) or `A` degraded

(`Ainb`, `Aina` and `dgrA`). Each subgoal describes one of the counterexamples we were seeking to establish that bistabilityB may be reached. For example, the first subgoal describes a network list where `A`'s concentration is less than its second threshold, while `B`'s concentration equals its second threshold minus 1, and hence the firing of rule `bprB`, where `B` is produced, clearly falsifies the main conjecture.

The final step is to prove that instability is impossible as:

**theorem** `no_instability: nt` $\in$ `network` $\implies$
$\qquad\qquad$ $\neg$ `(concentration A nt > threshold 2 A` $\wedge$
$\qquad\qquad\qquad$ `concentration B nt > threshold 2 B)`
**apply** `(erule network.induct)`
**apply** `(insert threshold_2A_pos)`
**apply** `auto`
**done**

## 5  Conclusions

We have formally analysed the genetic toogle by theorem proving and established its main correctness property, bistability: proteins may prevail against each other, but never at the same time. The toggle had already been studied by model checking [7]. This method simplifies the toggle as a finite state machine so that the model checker can exhaustively search the state space, whose size is inversely proportional to the speed of the search. A temporal logic is used to express the property that a protein may prevail on some path in the state space. Our approach requires more intervention to the human analyser but scales up to systems of unbounded size. Not only is this claim supported by experience from other application areas [10], but also by the very nature of an inductive proof. The theorem prover only helps the human check a conjecture against each inductive rule defining the model. Also, the language of higher-order logic implemented in Isabelle/HOL is most expressive for the properties of the toggle.

Our contribution therefore is the first investigation about the use of theorem proving to formally analysing genetic networks. The motivation for taking this approach lies in the mathematical complexity of even small genetic networks. Noteworthy, besides genetic (i.e. transcriptional regulatory) networks, many different types of biological networks exist, such as the protein interaction networks where nodes represent proteins and links represent physical interaction between the proteins, and metabolic networks where nodes represent small molecules and links represent direct enzymatic conversion between the small molecules. The mathematical modeling of such interaction heterogeneity may scale up quickly with the number of components and of interaction types, but formal methods offer a complementary, complexity-reducing approach.

Moreover, formal methods have a dialectic relationship with mathematical methods. Highly automatic methods based on model checking may take the analysis of genetic networks well beyond the world of mathematicians, for example by reaching industrial worlds where such methods are widely adopted. Typically, the human launches the checker on a property and waits for its output, which is only either positive or negative. This paper makes the case for highly expressive theorem proving, which is expected to open the field of genetic network analysis to the world of pure logicians. Typically, the human develops the proof of a property with the help of the prover, and then anyone can inspect and self-evaluate that proof. The formal analysis of biological problems has only just begun.

**Acknowledgments.** The authors gratefully acknowledge funding for reciprocal visits at their respective Institutions from the British Council's British-Italian partnership programme for year 2008, grant entitled "Computer-assisted verification for safety properties of genetic networks underlying liver regeneration".

# References

1. Yeger-Lotem, E., et al.: Network motifs in integrated cellular networks of transcription-regulation and protein-protein interaction. Proc. Natl. Acad. Sci. U.S.A. 101, 5934–5939 (2004)
2. Tyson, J.J., Chen, K.C., Novak, B.: Sniffers, buzzers, toggles and blinkers: dynamics of regulatory and signaling pathways in the cell. Curr. Opin. Cell Biol. 15, 221–231 (2003)
3. Ninfa, A.J., Selinsky, S., Perry, N., Atkins, S., Xiu Song, Q., Mayo, A., Arps, D., Woolf, P., Atkinson, M.R.: Using two-component systems and other bacterial regulatory factors for the fabrication of synthetic genetic devices. Meth. Enzymol. 422, 488–512 (2007)
4. Atkinson, M.R., Savageau, M.A., Myers, J.T., Ninfa, A.J.: Development of genetic circuitry exhibiting toggle switch or oscillatory behavior in Escherichia coli. Cell 113, 597–607 (2003)
5. Gillespie, D.: Exact simulation of coupled chemical reactions. Journal of Chemical Physics 81, 2340–2345 (1997)
6. Batt, G., de Jong, H., Geiselmann, J., Page, M.: Analysis of genetic regulatory networks: a model-checking approach. In: Benerecetti, M., Pecheur, C. (eds.) Working Notes of the Second Workshop on Model Checking and Artificial Intelligence, MoChArt 2003, Acapulco, Mexico, pp. 51–58 (2003)
7. Batt, G., Bergamini, D., de Jong, H., Garavel, H., Mateescu, R.: Model checking genetic regulatory networks using GNA and CADP. In: Graf, S., Mounier, L. (eds.) SPIN 2004. LNCS, vol. 2989, pp. 158–163. Springer, Heidelberg (2004)
8. McMillan, K.: Symbolic Model Checking. Kluwer Academic Publisher, Dordrecht (1993)
9. Nipkow, T., Paulson, L.C., Wenzel, M.T.: Isabelle/HOL: A Proof Assistant for Higher-Order Logic. LNCS, vol. 2283. Springer, Heidelberg (2002)
10. Bella, G.: Formal Correctness of Security Protocols. Information Security and Cryptography. Springer, Heidelberg (2007)
11. Gardner, T.S., Cantor, C.R., Collins, J.J.: Construction of a genetic toggle switch in Escherichia coli. Nature 403, 339–342 (2000)
12. Tian, T., Xu, S., Gao, J., Burrage, K.: Simulated maximum likelihood method for estimating kinetic rates in gene expression. Bioinformatics 23, 84–91 (2007)

# Control Strategies for the Regulation of the Eukaryotic Heat Shock Response

Elena Czeizler[*], Eugen Czeizler[*], Ralph-Johan Back, and Ion Petre

Department of IT, Åbo Akademi University
Turku 20520, Finland
{elena.czeizler,eugen.czeizler,backrj,ipetre}@abo.fi

**Abstract.** Elevated temperatures cause proteins in living cells to mis-fold. They start forming larger and larger aggregates that can eventually lead to the cell's death. The heat shock response is an evolutionary well conserved cellular response to massive protein misfolding and it is driven by the need to keep the level of misfolded proteins under control. We consider in this paper a recently proposed new molecular model for the heat shock response in eukaryotes, consisting of a temperature-induced acti-vation mechanism, chaperoning of misfolded proteins and self-regulation of the chaperon synthesis. We take in this paper a control driven ap-proach to studying this regulatory network. We modularize the network by identifying its main functional modules. We distinguish three main feedback loops. The main question we are addressing is why is this level of complexity needed for implementing what could in principle also be achieved with an open-loop design. We answer the question by compar-ing the numerical behavior of various knockdown mutants where one or more feedback loops are missing. We also discuss a new approach for a biologically-unbiased model comparison.

## 1 Introduction

**Decomposing large biological networks.** Much experimental and theoret-ical effort is invested nowadays in compiling large, system-level models for bio-chemical processes, including regulatory networks, signaling pathways, metabolic pathways, etc. Models can encompass many thousands of reactants and reac-tions, see [2]. On this scale, understanding the details of the network, especially the interactions among its various parts, or even noticing a high-level functional separation in the network become considerable challenges. Recognizing that sim-ilar problems are also encountered in engineering (and elsewhere), see [3], one strategy towards a system-level understanding of a biological network is to adapt to systems biology specific methods coming from engineering sciences, in par-ticular from control theory, see [6], [7], [10], [14], [15], [16], [17]. Distinguishing among the main functional modules of a biological network and identifying their individual contribution to the overall behavior can provide great insight into the basis of its reactivity and efficiency. Using a control-driven approach, one

---

[*] Authors with equal contributions.

P. Degano and R. Gorrieri (Eds.): CMSB 2009, LNBI 5688, pp. 111–125, 2009.
© Springer-Verlag Berlin Heidelberg 2009

often aims to identify the main regulatory components, including feedback and feed-forward mechanisms. To disentangle their individual contribution to the network, knockdown mutants are often considered, see [13]. Such mutants, lacking one or more of the regulatory components, are then numerically compared in an effort to identify and quantify the exact contribution of each component.

**A new approach for model comparison.** Comparing alternative computational models for a biological process is in general a difficult problem, where one has to consider the differences in the underlying reaction network, biological assumptions, kinetics, and initial conditions. When comparing alternatives that are submodels of a larger model, e.g., as in the case of a functional analysis of the various modules in a large network, the problem is somewhat simpler: the underlying reaction networks are very similar (albeit not identical), the biological constraints are the same (given by those of the reference model), and the kinetics of the reactions are the same (given by the reference model). The main question is that of how to choose the initial numerical setup of each model (the initial values of all variables in the models) in such a way that the comparison is unbiased and identifies intrinsic differences in the structure of the models. One approach, related to a technique of mathematically controlled comparison, see [1], is to start from the initial values of the corresponding variables in the reference model, see [13] for a case study using this approach. While this approach is mathematically well-founded, it may in fact lead to biased comparisons in the case of biological processes. Especially in the case of regulatory networks, models are assumed to be at steady state in the absence of the trigger of the response. Initial values are often chosen in such a way that the reference model is in steady state in the absence of the stimulus. However, imposing the steady state condition to the initial values of the reference model does not imply in general that a submodel will also be at steady state using the same initial values. As such, the behavior of the submodel will in fact exhibit the combined behavior of two different efforts: that of migrating from a possibly unstable state and, perhaps simultaneously, that of responding to a specific stimulus. Questions related to the speed of the response or its effectiveness may thus get misleading answers. If the response to the stimulus of various different submodels is to be compared, we argue in this paper that a different approach yields biologically unbiased results. We propose an approach where the kinetics and the mass constants of the submodels are taken from the reference model. The initial distribution of reactants among their various forms is however different from model to model and it is chosen in such a way that the initial setup of each model forms a steady state of that model in the absence of the stimulus. In particular, this approach allows to consider each of the submodels as a viable alternative model to the biochemical process of interest.

**Our case study: the heat shock response.** We choose a recently proposed model for the eukaryotic heat shock response, see [11] and [12], as a case study for the model comparison method that we propose. The model is attractive for the purpose of this paper because on one hand, it is relatively small, with only ten reactants and twelve reactions, while on the other hand it contains a rather intricate control mechanism, with three different feedback mechanisms. We analyze all eight knockdown mutants obtained by combining the three feedbacks and identify their individual contribution to four performance indicators of the reference model.

## 2  The Heat Shock Response in Eukaryotes

**Molecular model.** The main role in the response is played by the *heat shock proteins* (hsp). They act as chaperons for the *misfolded proteins* (mfp), by forming hsp: mfp complexes and assisting them to refold. In the model recently introduced in [12], the heat shock response is controlled by regulating the transactivation of the hsp-encoding genes. The transcription process for these genes can start only after some transcription factors, called *heat shock factors* (hsf), trimerize ($hsf_3$) and then bind to some particular DNA sequences, called *heat shock elements* (hse), which are the promoters of the hsp-encoding genes. The binding of a trimer $hsf_3$ to a heat shock element is denoted by $hsf_3$: hse. As an intermediary step before trimerization, the heat shock factors go first through a dimerization stage, when they form $hsf_2$ complexes. Once the trimers $hsf_3$ are bound to the promoter sites, the transcription and translation of the hsp-encoding genes can start, leading eventually to the formation of new hsp molecules.

When the level of hsp is high enough, the transcription process is turned off through a cunning self-regulating mechanism. The hsp proteins sequestrate the heat shock factors hsf in three ways: by binding to free hsf, by breaking dimers and trimers, as well as by unbinding $hsf_3$ from the DNA promoter sites and, at the same time, breaking the trimer $hsf_3$. As a consequence, the heat shock factors are bound in hsp: hsf complexes and thus they are unable to form trimers anymore, turning off in this way the transcription of the hsp-encoding genes. As soon as the temperature is increased, proteins start misfolding, driving hsp away from hsf. Thus, hsf proteins are free and available to trimerize and bind to hse, promoting the synthesis of new hsp proteins. If the level of hsp becomes high enough, hsp unbinds $hsf_3$ from hse and again sequesters the majority of hsf into hsp: hsf, thus turning its own synthesis off. The molecular model for the heat shock response proposed in [12] is shown in Table 1.

**Model assumptions.** For the sake of simplicity, the molecular model of [12] includes several reductions. For instance, even though in a cell there exist several types of slightly different heat shock proteins, see [8], in this model they are all treated uniformly, with hsp 70 as base denominator. Similar conventiones are taken also for hse and hsf. The model does not differentiate either among different other types of proteins present in a cell. From the point of view of the heat shock response, the only relevant feature is whether they are correctly folded, in which case they are gathered under the name prot, or misfolded, in which case they are called mfp. Some of the cellular mechanisms, such as protein synthesis and protein degradation, are also greatly simplified in the model. We refer to [12] for a detailed discussion on this issue.

**Table 1.** The molecular model for the eukaryotic heat shock response proposed in [12]

| Reaction | Reaction name | |
|---|---|---|
| $2\,\mathsf{hsf} \leftrightarrows \mathsf{hsf}_2$ | (hsf dimerization) | (1) |
| $\mathsf{hsf} + \mathsf{hsf}_2 \leftrightarrows \mathsf{hsf}_3$ | (hsf trimerization) | (2) |
| $\mathsf{hsf}_3 + \mathsf{hse} \leftrightarrows \mathsf{hsf}_3\colon \mathsf{hse}$ | (DNA binding) | (3) |
| $\mathsf{hsf}_3\colon \mathsf{hse} \to \mathsf{hsf}_3\colon \mathsf{hse} + \mathsf{hsp}$ | (hsp synthesis) | (4) |
| $\mathsf{hsp} + \mathsf{hsf} \leftrightarrows \mathsf{hsp}\colon \mathsf{hsf}$ | (hsf sequestration) | (5) |
| $\mathsf{hsp} + \mathsf{hsf}_2 \to \mathsf{hsp}\colon \mathsf{hsf} + \mathsf{hsf}$ | (hsf$_2$ breaking) | (6) |
| $\mathsf{hsp} + \mathsf{hsf}_3 \to \mathsf{hsp}\colon \mathsf{hsf} + 2\,\mathsf{hsf}$ | (hsf$_3$ breaking) | (7) |
| $\mathsf{hsp} + \mathsf{hsf}_3\colon \mathsf{hse} \to \mathsf{hsp}\colon \mathsf{hsf} + 2\,\mathsf{hsf} + \mathsf{hse}$ | (hsp-forced hsf$_3$ unbinding) | (8) |
| $\mathsf{hsp} \to \emptyset$ | (hsp degradation) | (9) |
| $\mathsf{prot} \to \mathsf{mfp}$ | (protein misfolding) | (10) |
| $\mathsf{hsp} + \mathsf{mfp} \leftrightarrows \mathsf{hsp}\colon \mathsf{mfp}$ | (protein chaperoning) | (11) |
| $\mathsf{hsp}\colon \mathsf{mfp} \to \mathsf{hsp} + \mathsf{prot}$ | (protein refolding) | (12) |

The model assumes three conservation relations: for the total amount of heat shock factors, for the total amount of proteins (except heat shock proteins and heat shock factors), and for the total amount of the heat shock elements:

- $[\mathsf{hsf}] + 2 \times [\mathsf{hsf}_2] + 3 \times [\mathsf{hsf}_3] + 3 \times [\mathsf{hsf}_3\colon \mathsf{hse}] + [\mathsf{hsp}\colon \mathsf{hsf}] = C_1$,
- $[\mathsf{prot}] + [\mathsf{mfp}] + [\mathsf{hsp}\colon \mathsf{mfp}] = C_2$,
- $[\mathsf{hse}] + [\mathsf{hsf}_3\colon \mathsf{hse}] = C_3$,

for some constants $C_1, C_2, C_3$ called *mass constants*. The only variable not covered by conservation relations is $\mathsf{hsp}$, the main regulatory target of the network.

**Mathematical model.** We associate to the molecular model in Table 1 a mathematical model in terms of differential equations as follows. A time-dependant continuous variable is associated to each reactant and gives its concentration level. For a variable $x$, we denote by $[x](t)$ its concentration at time $t$. The dynamics of the system is then described through a system of differential equations. Its formulation is based on the principle of mass action, see [4], [5]. For each variable, its differential equation gives the cumulated consumption and production rates of the corresponding reactant as an effect of the reactions in Table 1. The rate of each reaction is, based on the principle of mass action, proportional with the concentration of the reactants. Due to lack of space we skip listing the resulting system of differential equations, referring instead to [11] and [12]. The kinetic rate constants and the initial values of all reactants were estimated in [12] so as to satisfy three conditions:

(i) For a temperature of $37°C$, the system is at steady state, i.e., the differentials of all variables are zero. This is a natural requirement since the model should exhibit no response in the absence of heat shock, i.e., at $37°C$.

(ii) For a temperature of $42°C$, the numerical prediction of the model for [hsf$_3$: hse]$(t)$ should be in agreement with experimental data of [9] on DNA binding of hsf$_3$.

(iii) For a temperature of $42°C$, the numerical prediction of the model for [hsp]$(t)$ should be correlated with experimental data of [12] on a *de-novo* fluorescent reporter-based experiment.

The numerical setup of the model is in Table 2. We refer to [12] for details on parameter estimation and on model validation.

The final model exhibits the following major (numerical) achievements, in-line with experimental evidence, see [12]:

- (A) *Makes economical use of the cellular resources:* the hsp-encoding gene is only transactivated for a short while when exposed to heat shock. The gene transcription is virtually non-existent in the absence of heat shock.
- (B) *It is fast to respond to a heat shock:* the hsp-encoding gene is quickly transactivated in response to heat shock.
- (C) *The response is effective:* the mfp concentration is kept low for mild heat shocks.

**Table 2.** The numerical values of the parameters and the initial values of the variables of the heat shock response model. A. The numerical values of the parameters. $k_i$ denotes the kinetic rate constant of the irreversible reaction (i). $k_i^+$ denotes the 'left-to-right' direction of reaction (i), while $k_i^-$ denotes its 'right-to-left' direction. For the expression of the temperature-dependant parameter $\phi_T$ we refer to [11] and [12]. B. The initial values of all variables.

A

| Param. | Value | Units |
|---|---|---|
| $k_1^+$ | 3.49 | $\frac{ml}{\#\cdot s}$ |
| $k_1^-$ | 0.19 | $s^{-1}$ |
| $k_2^+$ | 1.07 | $\frac{ml}{\#\cdot s}$ |
| $k_2^-$ | $10^{-9}$ | $s^{-1}$ |
| $k_3^+$ | 0.17 | $\frac{ml}{\#\cdot s}$ |
| $k_3^-$ | $1.21 \cdot 10^{-6}$ | $s^{-1}$ |
| $k_4$ | $8.3 \cdot 10^{-3}$ | $s^{-1}$ |
| $k_5^+$ | 9.74 | $\frac{ml}{\#\cdot s}$ |
| $k_5^-$ | 3.56 | $s^{-1}$ |
| $k_6$ | 2.33 | $\frac{ml}{\#\cdot s}$ |
| $k_7$ | $4.31 \cdot 10^{-5}$ | $\frac{ml}{\#\cdot s}$ |
| $k_8$ | $2.73 \cdot 10^{-7}$ | $\frac{ml}{\#\cdot s}$ |
| $k_9$ | $3.2 \cdot 10^{-5}$ | $s^{-1}$ |
| $k_{11}^+$ | $3.32 \cdot 10^{-3}$ | $\frac{ml}{\#\cdot s}$ |
| $k_{11}^-$ | 4.44 | $s^{-1}$ |
| $k_{12}$ | 13.94 | $s^{-1}$ |

B

| Variable | Initial conc. |
|---|---|
| [hsf] | 0.67 |
| [hsf$_2$] | $8.7 \cdot 10^{-4}$ |
| [hsf$_3$] | $1.2 \cdot 10^{-4}$ |
| [hse] | 29.73 |
| [hsf$_3$: hse] | 2.96 |
| [hsp] | 766.88 |
| [hsp: hsf] | 1403.13 |
| [mfp] | 517.352 |
| [hsp: mfp] | 71.65 |
| [prot] | $1.15 \times 10^8$ |

– (D) *Scalable:* higher response for higher temperature. The transactivation of the hsp-encoding gene reaches a higher peak and/or remains longer on the 100% level under higher temperature.

We disentangle in this paper the contribution of the various modules in the network to achieving these properties. We discuss the methodology in Section 3 and apply the method to the analysis of the heat shock response model in Section 4.

## 3    A Functional Decomposition of the Heat Shock Response Model

**A control-driven approach.** When designing the modular decomposition of a process, the first step is to identify and separate the process to be regulated, called the *plant*. The current state of that process is captured by specific *sensors* which send the information to a decision-making module, called the *controller*. The decision taken by the controller is implemented through a device, called the *actuator*, that modifies the state of the process, thus influencing the activity of the plant. A key concept of control systems is the existence of some *feedback mechanisms*, which are used to cope with the uncertainties of the system. A feedback sensor sends the current state of the process back to the controller, thus facilitating a dynamical compensation for disturbances of the system. In complex applications, the plant (and by effect, the controller, the sensors, and the actuators) can be chosen in several different ways depending on what is deemed to be the main target of the system, thus leading to several possibly different decompositions of the system.

An easy example illustrating these concepts and their interactions is given by the functioning principles of a motion activated spotlight. In this case, an electronic control unit, i.e., the controller of the system, reads the signal from the motion sensor in order to determine if there is any change in the environment. If the sensor's input falls outside of the parameters of the system's settings, then the control unit triggers a relay switch, i.e., the actuator, that operates the lighting system. As long as the sensor detects movement, it feeds this information to the control unit, keeping the switch on.

**Modularization of the heat shock response model.** In the case of the eukaryotic heat shock response we set the plant to focus on the main target, i.e., the misfolding and refolding of proteins. As protein refolding can be done only with the help of the chaperons hsp, we assign the actuator to model their synthesis and degradation. The controller's task is set to modulate the level of $hsf_3$: hse, since the DNA binding level determines the chaperon production. The controller's activity is influenced also by the information received from a sensor which measures the level of chaperons in the system. When this is high enough, the chaperons will sequestrate the heat shock factors by forming complexes hsp: hsf, leading in this way to a decrease in the level of $hsf_3$: hse. In particular, depending on the type of reactions which lead to the formation of the complex hsp: hsf, we identify three feedback mechanisms:

**Table 3.** The functional decomposition of the model in Table 1. We denote the 'left-to-right' direction of reaction (5) by $(5)^+$ and by $(5)^-$ its 'right-to-left' direction.

|  | Main Task | Reactions |
|---|---|---|
| *Plant* | Protein misfolding and refolding | (10), (11), (12) |
| *Actuator* | Regulate the level of hsp | (4), (9) |
| *Sensor* | Measure [hsp] | |
| *Controller* | Modulate the level of DNA binding | (1), (2), (3), $(5)^-$ |
| *Feedback $FB_1$* | Sequestration of hsf | $(5)^+$ |
| *Feedback $FB_2$* | Dimer and trimer breaking | (6), (7) |
| *Feedback $FB_3$* | hsp-forced DNA unbinding | (8) |

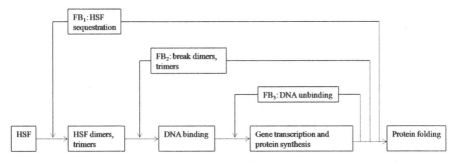

**Fig. 1.** The control structure of the heat shock response network

- $FB_1$: sequestration of hsf, i.e., reaction $(5)^+$ (i.e., the 'left-to-right' direction of reaction (5));
- $FB_2$: breaking hsf dimers and trimers, i.e., reactions (6) and (7);
- $FB_3$: unbinding $hsf_3$ from hse, i.e., reaction (8).

The functional decomposition of our model is summarized in Table 3, where the reaction numbers refer to the reactions in Table 1. A graphical illustration of the decomposition is in Fig. 1.

## 4   Disentangling the Functional Roles of the Control Feedbacks

**Approach.** We consider the eight knockdown mutants obtained through all combinations of feedbacks $FB_1$, $FB_2$, and $FB_3$. We analyze the dynamical behavior of the mutants at $42°C$. We choose this particular temperature since at $42°C$ the experimental data shows both a pronounced heat shock response in terms of increased levels of misfolded proteins, as well as an explicit response in terms of increased, transient DNA binding of $hsf_3$, see [12].

**Unbiased model comparison.** In order for our analysis to be biologically unbiased, we aim to eliminate all accidental differences between our models. In particular, they should satisfy some initial *biological constraints*. This way, we make sure that (i) the remaining dissimilarities in their dynamical behaviors reflect intrinsic differences in their architecture and (ii) they are all fairly considered as viable alternative models for the heat shock response. We impose the following two constraints.

(i) The reactions of all knockdown mutants assume kinetic rate constants identical to those of the corresponding reactions in the reference model. Also, the values of the mass constants $C_1$, $C_2$, $C_3$ of each of the mutants are identical to those of the reference model. Both these conditions are natural since each knockdown mutant is a submodel of the reference model. As such, they should all assume identical chemistry and mass constants.

(ii) The initial values of the variables of each knockdown mutant form a steady state of that model for a temperature of $37°C$. Note that the same condition has been applied in [12] when choosing the initial values of all variables for the reference model.

Note that our constraint (ii) is fundamentally different than that of [13]. Rather than taking an approach based on mathematically controlled comparison, see [1], where the submodels start from the same initial setup, we take a different view based on biologically meaningful constraints. We consider each submodel as a viable alternative model for the heat shock response and as such, we assume that before the heat shock is applied, they are in a steady state (at $37°C$). Consequently, we impose the condition that the initial values of each submodel form a steady state of that model at $37°C$. In particular, each model assumes a different initial numerical setup, determined by the kinetics of its underlying reaction network and by the mass constants.

Note also that we do not perform parameter estimation for each of the mutants. Indeed, this approach would yield different kinetics for each model and thus, the contribution of each feedback to the network would be wildly different from mutant to mutant.

Clearly, the results of our analysis remain heavily dependent on the numerical values of the mass constants and of the kinetic rate constants chosen in [12] for the reference model. We discuss this dependency in Section 5.

**The knockdown mutants.** Each of the knockdown mutant models will be denoted as $M_X$ where $X \subset \{1, 2, 3\}$ represents the indexes of the feedback mechanisms contained in $M_X$, as follows:

i) $M_0$ consists of reactions (1)-(4), (9)-(12). Using control-theory terminology, this model is referred as the *open-loop design*.
ii) $M_1$ consists of reactions (1)-(5), (9)-(12).
iii) $M_2$ consists of reactions (1)-(4), (6)-(7), (9)-(12).
iv) $M_3$ consists of reactions (1)-(4), (8)-(12).
v) $M_{1,2}$ consists of reactions (1)-(7), (9)-(12).
vi) $M_{1,3}$ consists of reactions (1)-(5), (8)-(12).
vii) $M_{2,3}$ consists of reactions (1)-(4), (6)-(12).

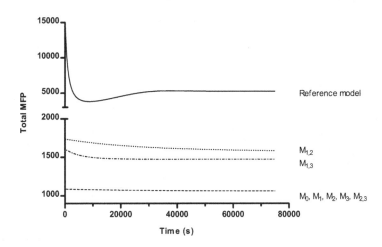

**Fig. 2.** The total amount of mfp in the seven knockdown mutants and in the reference model for a constant temperature of $42°C$

**Numerical comparison of the mutants.** The main task of the heat shock response is to keep the level of misfolded proteins under control. Focusing on this aspect alone, all knockdown mutants (even the open-loop design $M_0$), excel in achieving the goal. Perhaps surprisingly, they actually perform much better from this point of view than the reference model $M_{1,2,3}$, see Fig. 2: the total amount of mfp at a temperature of $42°C$ is considerably smaller in all mutants than in the reference model.

On the other hand, it is essential for the cell to be efficient in the use of materials and energy. In the case of the heat shock response, this translates in the amount of chaperons needed in each of the models to achieve the response. From this point of view, the tradeoff in the mutants is very high. Indeed, all mutants maintain much higher amounts of chaperons throughout the response than the reference model. For a comparison in relative terms we associate a time-dependant cost function describing the effort in each model with respect to its result in terms of chaperons per unit of misfolded protein:

$$\text{cost}(t) = [\text{total HSP}] \ / \ [\text{total MFP}]$$
$$= ([\text{hsp}](t) + [\text{hsp: hsf}](t) + [\text{hsp: mfp}](t))/([\text{mfp}](t) + [\text{hsp: mfp}](t)).$$

The evolution of the cost function for all models is illustrated in Fig. 3 and shows that all mutants are over-reacting, by over-producing hsp in an effort to maintain a very low level of mfp. (A similar plot can be obtained also by comparing the level of chaperons in the models in absolute terms.) The over-reaction to heat shock is also clearly seen in terms of DNA binding and gene transcription, see Fig. 4: DNA binding remains high throughout the heat shock response in all mutants, while it is only transient in the reference model.

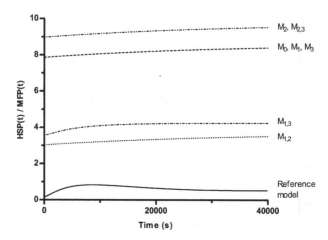

**Fig. 3.** The amount of chaperons per unit of misfolded protein in the seven knockdown mutants and in the reference model for a constant temperature of $42°C$

**The contribution of the three feedbacks.** We analyze in this section the functionality of each of the three feedbacks from the point of view of their contributions to the properties (A)-(D) of the heat shock response model, see Section 2. For this, we compare the eight knockdown mutants to capture the effects of adding or removing each of the feedbacks.

In line with biologically unbiased comparison discussed in this paper, we set the initial values for each of the knockdown mutants to obtain a steady state of the system for a temperature of $37°C$. As such, a large part of the differences among the various mutants can already be observed from their initial values, i.e., from their behavior in the absence of the heat shock. The initial numerical setup of all models is given in Table 4.

The mutants $M_0$, $M_1$, $M_2$, $M_3$, and $M_{2,3}$ exhibit DNA binding at the 100% level as their initial setup (equivalently, the initial value of [hse] is negligible in

**Table 4.** The initial numerical setup of all knockdown mutants. The initial values are chosen so that each mutant is at steady state for a temperature of $37°C$. The initial value of [prot] in all models is identical to its value in the reference model, see Table 2B.

|           | hsf    | hsf$_2$ | hsf$_3$ | hsf$_3$: hse | hse     | hsp     | hsp: hsf | hsp: mfp | mfp     |
|-----------|--------|---------|---------|--------------|---------|---------|----------|----------|---------|
| $M_0$     | 0.0028 | 0.0001  | 438.2   | 32.6895      | 0       | 8479.29 | 0        | 71.6476  | 46.79   |
| $M_1$     | 0.0028 | 0.0001  | 416.803 | 32.6895      | 0       | 8479.29 | 64.1905  | 71.6476  | 46.79   |
| $M_3$     | 0.2018 | 0.3516  | 437.893 | 32.6884      | 0.001   | 8479.03 | 0.0212   | 71.6476  | 46.7914 |
| $M_2$     | 35.168 | 0.2178  | 22.3266 | 32.6895      | 0       | 8479.29 | 1212.02  | 71.6476  | 46.79   |
| $M_{2,3}$ | 35.1759| 0.218   | 22.1625 | 32.6693      | 0.0202  | 8474.06 | 1212.56  | 71.6476  | 46.8189 |
| $M_{1,2}$ | 0.1316 | 0       | 0       | 14.6625      | 18.027  | 3803.28 | 1368.55  | 71.6476  | 104.317 |
| $M_{1,3}$ | 0.1172 | 0.1525  | 0.0069  | 16.405       | 16.2845 | 4255.28 | 1363.01  | 71.6476  | 93.2361 |
| $M_{1,2,3}$| 0.6692| 0.0009  | 0.0001  | 2.9565       | 29.733  | 766.875 | 1403.13  | 71.6473  | 517.352 |

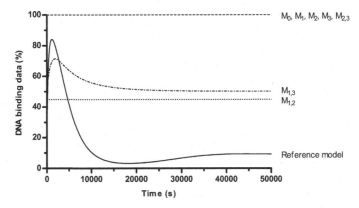

**Fig. 4.** The level of DNA binding in the seven knockdown mutants and in the reference model. The graph plots [hsf$_3$: hse]$(t)$ relative to the total number of heat shock elements (identical in all models) for a constant temperature of $42°C$.

all of these mutants). Thus, in these cases, the hsp-encoding gene is transactivated at the 100% level even in the absence of heat shock. This is not only a contradiction of experimental data of [8], but also a non-economical use of the cellular resources, see also Fig. 3 for a measure of the cost of the response. These models fail by over-reaching their target: the total amount of mfp in the system is kept at an unnecessarily low level, see Fig. 2. Indeed, the reference model allows for a higher [mfp] (that is still negligible relative to [prot]), for a dramatically lower price in terms of [hsp]. As such, we conclude already from their predicted relentless DNA transcription activity under physiological conditions that these mutants are non-viable models for the heat shock response.

In terms of the amount of chaperons in the system under a constant heat shock at $42°C$, $FB_1$ appears to be the main driver to minimize [hsp]. Indeed, removing $FB_1$ from the reference model yields model $M_{2,3}$, which is the only mutant containing two feedbacks that exhibits a behavior even worse than that of the open-loop model $M_0$ (that has no feedbacks at all), see Figs. 2-4.

Regarding the role of $FB_1$, we conclude that no viable heat shock response can be obtained in the absence of feedback $FB_1$. At the same time, $FB_1$ alone (see model $M_1$) cannot lead to a viable response either.

Regarding $FB_2$, we note that in models $M_2$ and $M_{2,3}$, the level of hsf$_3$ at $37°C$ is one degree of magnitude smaller than in the models $M_0$, $M_1$, and $M_3$. Even though the kinetic details of the reactions do not propagate this change towards a lower level of hsf$_3$: hse, this observation points to a critical role of $FB_2$ in lowering the level of hsf$_3$: hse. The same effect cannot be obtained by means of either $FB_1$ alone, or $FB_3$ alone.

Regarding the role of $FB_3$, consider now adding $FB_1$ to either of the other two feedbacks. We obtain models $M_{1,2}$ and $M_{1,3}$ that have drastically different behaviors under physiological conditions than the other knockdown mutants: the DNA binding level, [hsf$_3$: hse], is lowered from the 100% level in the absence of $FB_1$ to around the 50% level, see Fig. 4. As a direct consequence, [hsp] is lowered by about 50% and, as a tradeoff, [mfp] increases by about 50%. There is no big numerical difference between the two models at $37°C$. When a heat shock of $42°C$ is applied however, the behavior of the two models differs, especially in the speed of the response. While the response of $M_{1,2}$ in terms of DNA binding is static ([hsf$_3$: hse] remains roughly constant throughout the response, see Fig. 4), that of $M_{1,3}$ is adaptive. It is in fact the only knockdown mutant exhibiting an adaptive behavior in terms of DNA binding. Similarly as in the case of the reference model, [hsf$_3$: hse] is transiently increased in $M_{1,3}$ during the heat shock and eventually returns to the basal levels. Moreover, both the time to reach the peak of the response, as well as the time to return to the physiological level of [hsf$_3$: hse] in $M_{1,3}$, are very similar to those in the reference model. We conclude then that $FB_3$ (in addition to $FB_1$) has a main role in achieving a fast response to heat shock. Indeed, this conclusion is also supported by the plots in Fig. 2, showing that [mfp] is lowered very slowly in the absence of $FB_3$. Moreover, as seen in Fig. 4, in the absence of $FB_3$ the DNA binding levels are insensitive to a temperature upshift to $42°C$.

For a heat shock at $43°C$, the only mutant exhibiting a different response than at $42°C$ is $M_{1,3}$. Similarly to the reference model, [hsf$_3$: hse] reaches a higher peak in $M_{1,3}$, while it remains essentially unchanged for the other mutants.

Based on the numerical observations above, we can now summarize the contribution of the three feedbacks to the performance indicators (A)-(D) of the network as follows:

- (A) *Makes economical use of the cellular resources* (the hsp-encoding gene is only transactivated for a short while when exposed to heat shock): $FB_1$ plays the major role here, $FB_2$, $FB_3$ are also important. In the absence of $FB_1$, gene transcription is at the 100% level even without heat shock.
- (B) *It is fast to respond to a heat shock* (the hsp-encoding gene is quickly transactivated in response to heat shock): $FB_3$, $FB_1$ play the major role here. Model $M_{1,3}$ is as fast to react to heat shock as the reference model, albeit the scope of its response is lower than that of the reference model.
- (C) *The response is effective* (the mfp concentration is kept low for mild heat shocks): the underlying open-loop structure is enough to achieve this. No feedback mechanism is needed for maintaining a low [mfp] for a heat shock at $42°C$. Rather, the role of the feedbacks is to minimize the cost of the response.
- (D) *Scalable* (higher response for higher temperature): $FB_3$, $FB_1$ play the major role here. $M_{1,3}$ is the only mutant that scales its response to higher temperatures, similarly as the reference model. For higher temperatures, gene transactivation is faster to raise and it raises to a higher level.

In the list above, the role of $FB_2$ as a sole or a main contributor to one of the properties (A)-(D) is not clearly distinguished. $FB_2$ appears more as a modulator for $FB_1$ and $FB_3$. There is a clear numerical argument for this milder effect: $FB_2$ consists of a regulation mechanism acting on tow reactants, hsf$_2$ and hsf$_3$, having negligible concentrations in the reference model both at $37°C$ and throughout the response at $42°C$. In this context, it is most surprising however to see the dramatic effect that the removal of $F_2$ has on the reference model, see, e.g., Fig. 4.

## 5    Discussion

Numerical model comparison is a difficult problem even in the case of submodels of a larger network. We argued in this paper that the common practice of assuming for all submodels the same initial setup as that of the reference model is biologically biased. Indeed, a numerical setup chosen so that the reference model satisfies some biological constraints will not necessarily ensure the same for its submodels. As such, the submodels will already from the start be deemed as unviable alternative models for the biological process under study. We considered a different approach where the kinetics and the mass constants are taken from the reference model (so that the chemistry and the total amount of reactants are the same in all submodels), but the initial setup of each submodel is chosen in such a way that: (i) it satisfies all mass conservation relations and (ii) it satisfies all biological assumptions. We argued that in this way we obtain a biologically unbiased model comparison, where each submodel is evaluated as a real alternative model for the biological process.

Extending the idea of biologically unbiased model comparison to models that assume different underlying reaction networks is appealing. One possible way to do it is by combining numerical approaches such as those demonstrated in this paper with qualitative approaches of model refinement such as those developed in software engineering. Two different models to be compared could first be embedded into a larger model in such a way that they both become its submodels. Questions related to a systematic methodology for model embedding, dealing with conflicting model assumptions, and other issues appear to be very interesting.

Similarly to many other numerical techniques such as sensitivity analysis, the numerical comparisons demonstrated in this paper depend heavily on the numerical setup of the models, i.e., on the numerical values of the kinetic rate constants, mass constants, and on the initial values of the variables. We say then that our analysis is *local*. When repeating the analysis in a different numerical setup, the conclusions regarding the role of each of the feedbacks could be very different. This is most clear if we consider changes in the kinetic rate constants: the control in the regulatory network can easily be shifted elsewhere by drastically slowing down some reactions and speeding up others. The same is true however also when changing the mass constants. For example, we considered an experiment where we increased the total amount of hsf from around 1412 as in this paper to about 10000. We then changed all initial values of all variables in each of the models so that they form a steady state of the model for a temperature of $37°C$. When repeating the knockdown analysis in this setup, the conclusions were very different. It turned out that as far as the cost function is concerned, the open-loop design performs almost as well as the reference model, while the mutants

consisting of two feedbacks perform much worse. As far as gene transcription
activity is concerned under a constant heat shock at $42°C$, the reference model
exhibits a similar shape as in Fig. 4, while the mutants show a constant activity
at the 100% level. Taking a similar approach in the case when the total amount
of hsf is lowered instead to around 500, we obtained that the reference model has
by far the lowest cost, with $M_{1,2}$ and $M_{1,3}$ performing well, and the open-loop
model much worse. The gene transcription activity is similar to those in Fig. 4,
except for $M_{1,2}$ which shows a lower transactivation level. Two conclusions can
thus be formulated. On one hand, this is a local numerical analysis, in the same
category with approaches such as parameter fit and sensitivity analysis. They all
have to be taken in close relationship with the experimental data and available
biological knowledge and validated as such. On the other hand, repeating the
analysis in different, albeit less validated, numerical contexts can be very useful.
For example, modules whose functional role is hidden by other, more dominat-
ing modules, can be easier to analyze in numerical contexts where they assume
the dominant role. Clearly, projecting conclusions from one numerical context
to another is a challenge in itself that has to be addressed.

# References

1. Alves, R., Savageau, M.A.: Extending the method of mathematically controlled comparison to include numerical comparisons. Bioinformatics 16, 786–798 (2000)
2. Chen, W.W., Schoeberl, B., Jasper, P.J., Niepel, M., Nielsen, U.B., Lauffenburger, D.A., Sorger, P.K.: Input output behavior of ErbB signaling pathways as revealed by a mass action model trained against dynamic data. Molecular Systems Biology 5, 239
3. Csete, M.E., Doyle, J.C.: Reverse Engineering of Biological Complexity. Science 295, 1664–1669 (2002)
4. Guldberg, C.M., Waage, P.: Studies Concerning Affinity. C. M. Forhandlinger: Videnskabs-Selskabet i Christiana 35 (1864)
5. Guldberg, C.M., Waage, P.: Concerning Chemical Affinity. Erdmann's Journal für Practische Chemie 127, 69–114 (1879)
6. Hawkins, B.A., Cornell, H.V. (eds.): Theoretical Approaches to Biological Control. Cambridge University Press, Cambridge (1999)
7. Kitano, H.: Systems biology: A brief overview. Science 295, 1662–1664 (2002)
8. Holmberg, C.I., Tran, S.E., Eriksson, J.E., Sistonen, L.: Multisite phosphorylation provides sophisticated regulation of transcription factors. Trends Biochem. Sci. 27(12), 619–627 (2002)
9. Kline, M.P., Morimoto, R.I.: Repression of the heat shock factor 1 transcriptional activation domain is modulated by constitutive phosphorylation. Molecular and Cellular Biology 17(4), 2107–2115 (1997)
10. Lazebnik, Y.: Can a biologist fix a radio? - or what I learned while studying apoptosis. Cancer Cell 2, 179–182 (2002)
11. Petre, I., Mizera, A., Hyder, C.L., Mikhailov, A., Eriksson, J.E., Sistonen, L., Back, R.-J.: A new mathematical model for the heat shock response. In: Kok, J. (ed.) Algorithmic bioprocesses. Natural Computing. Springer, Heidelberg (2008)
12. Petre, I., Hyder, C.L., Mizera, A., Mikhailov, A., Eriksson, J.E., Sistonen, L., Back, R.-J.: A simple mathematical model for the eukaryotic heat shock response (submitted, 2009)

13. El Samad, H., Kurata, H., Doyle, J.C., Gross, C.A., Khammash, M.: Surviving heat shock: Control strategies for robustness and performance. Proc. Natl. Acad. Sci. 102(8), 2736–2741 (2005)
14. Sontag, E.: Some new directions in control theory inspired by systems biology. IEEE Systems Biology 1(1), 9–18 (2004)
15. Sontag, E.: Molecular systems biology and control. European Journal of Control 11, 396–435 (2005)
16. Stelling, J., Sauer, U., Szallasi, Z., Doyle, F.J., Doyle, J.: Robustness of cellular functions. Cell 118(6), 675–685 (2004)
17. Wolkenhauer, O.: Systems biology: the reincarnation of systems theory applied in biology? Brief Bioinform 2(3), 258–270 (2001)

# Computing Reachable States for Nonlinear Biological Models

Thao Dang, Colas Le Guernic, and Oded Maler*

CNRS-VERIMAG, 2, av. de Vignate, 38610 Gieres, France

**Abstract.** In this paper we describe reachability computation for continuous and hybrid systems and its potential contribution to the process
of building and debugging biological models. We then develop a novel algorithm for computing reachable states for *nonlinear* systems and report
experimental results obtained using a prototype implementation. We believe these results constitute a promising contribution to the analysis of
complex models of biological systems.

## 1 Introduction

The development of modeling formalisms and analysis techniques for the study
of biological systems is a central topic in systems biology. The formalisms proposed for representing biological processes are very diverse, differing at the levels
of abstraction, time scales and types of dynamics. The formalism chosen depends
naturally on the level of detail needed to answer the specific biological question
and on the granularity of available experiments. The contribution of this work
is at the level of abstraction of ordinary differential equations (ODEs), a widely
used modeling formalism. Biological systems, for instance metabolic networks
consisting of sets of reactions, can be viewed as continuous dynamical systems
with state variables denoting concentrations. The resulting differential equations
are derived, for example, from mass action rules and are, typically, polynomial.
Such equations can be numerically simulated from a given initial condition provided that the *exact* values of the parameters and the external environmental
conditions are known. In certain restricted cases it is possible to determine global
properties analytically.

Though widely used, ODEs suffer from several limitations. First, the passage
from a finite number of molecules to real-valued concentration is not always
justified, especially when the number of molecules is small [22]. Secondly, many
biological phenomena, for example gene activation, are more naturally modeled
as transitions between discrete states. Pure ODEs cannot easily accommodate
this mixture of continuous evolutions and discrete events. Alternatively, purely
discrete formalisms, based on transition systems expressed in various syntactic forms, suffer from a similar reciprocal limitation in the sense of not being
amenable to quantitative reasoning.

---

* Part of this work was done while the third author was a Weston visiting professor
at Weizmann Institute.

P. Degano and R. Gorrieri (Eds.): CMSB 2009, LNBI 5688, pp. 126–141, 2009.
© Springer-Verlag Berlin Heidelberg 2009

Second, the lack of quantitative information concerning molecular concentrations, reaction rates and other parameters is the rule, not the exception, in Biology. Consequently the value of predictions obtained using numerical ODEs models, where the values of the parameters are "guessed" or "tuned", is severely limited. Moreover, the validation of models based on ODEs with poorly-known parameters is difficult if not impossible because we are never sure to have covered all the qualitative behaviors compatible with a model by performing only a finite number of simulations, each with a different choice of parameters. This fact limits the applicability of such models for testing biological hypotheses.

To deal with this problem, qualitative approaches, notably based on qualitative versions of differential equations, have been proposed for representing genetic regulatory networks, molecular interaction networks or metabolic pathways [30, 40]. In these models only the *direction* of influence between variables is encoded (e.g. activation vs. inhibition) and much of the quantitative information is absent. As a consequence of such under-constrained descriptions, purely-qualitative approaches often lead to overly-conservative results in the sense of admitting many spurious behaviors. We propose a technique that can be used to analyze in a systematic manner quantitative models admitting this kind of uncertainty whose nature is *set-theoretic* rather than stochastic.

The analysis techniques that we use and extend originate from the study of *hybrid dynamical systems*, a domain situated in the intersection of control theory and computer science and are based on reachability analysis of hybrid automata. As their name suggests, hybrid automata are the result of marrying automata with differential equations. Each discrete state (mode) of the automaton is associated with one set of differential equations according to which the continuous variables evolve while being in that mode. When the variables satisfy certain conditions (transition guards) the automaton may switch to another mode where another set of equations will govern the evolution of the continuous variables. While hybrid automata allow us to express piecewise-continuous processes and can underlie numerical simulation, much of the analytic reasoning available for purely-continuous systems (especially for linear ones) is lost due to switching. In the last couple of years new techniques have been developed for the algorithmic analysis of hybrid systems, which open as well new opportunities for the analysis of purely-continuous systems subject to uncertainties. These techniques combine ideas from control theory, numerical analysis, graph algorithms and computational geometry in order to export algorithmic verification, also known as *model checking*, to the continuous and hybrid domains.

The principles of algorithmic verification can be summarized as follows. The system in question is modeled as an automaton whose transitions are labeled by input events. These inputs represent interactions of the automaton with its external environment (users, other systems). Each sequence of input events induces one behavior of the automaton, a trajectory over its state space. Simulation is the process of stimulating the automaton progressively with one input sequence and observing the behavior that this sequence induces starting from a given initial state. The problem is that the number of such sequences is prohibitively large.

Verification is based, instead, on computing with *sets* of states: starting from an initial set of states $P_0$, one computes *all* the one-step successors of $P_0$ (under all possible inputs) to obtain the set $P_1$, to which the same procedure is applied until all the states reachable from $P_0$ under any admissible input are computed.[1] Showing, for example, that some "bad" set of states is never reached (a "safety" property) amounts to checking whether the reachable set thus computed intersects the bad set. This computation replaces an infinite (or just huge) number of simulations. More complex properties that specify some temporal patterns of events can be specified and verified as well using similar methods.

The adaptation of this idea to continuous systems works as follows. Consider a differential equation of the form $\dot{x} = f(x, v)$ where $x$ is a vector of state variables and $v$ represents external disturbances and parameter uncertainties which are not known exactly but are always taken from a bounded convex set $V$. Given a subset $P_0$ of the state space (in a form of, say, a polytope) and a time step $r$, one can compute another polytope $P_1$, which contains all the points reachable from $P_0$ within the time interval $[0, r]$ under *any admissible value* of $v$ during that interval. Repeating this process we can obtain an over approximation of all the reachable states in any desired time horizon. To give a concrete example, one can compute all the possible evolutions of a reaction under all possible concentrations of a signalling molecule which are typically not precisely known, but which remain in a known interval. The principal contribution of this paper is in developing a new technique for conducting this type of analysis for *nonlinear* systems and in demonstrating its applicability on several biological models.

The rest of the paper is organized as follows. In Section 2 we give a brief introduction to the state-of-the-art in reachability computation for linear systems and explain why it cannot be applied in a straightforward manner to nonlinear systems. We then describe the *hybridization* approach [6] for handling nonlinear systems. Hybridization is based on over approximating a nonlinear system by a *piecewise-affine* system, a restricted type of a *hybrid automaton* without discontinuous jumps. Although, in principle, hybridization provides for the application of linear techniques to nonlinear systems, it suffers from inherent limitations that restrict its applicability to very low-dimensional systems. Section 3 describes our major contribution, a new *dynamic* hybridization scheme in which linearization is *not* based on a fixed partition of the state space and thus avoids much of the associated state explosion. For this algorithm we provide in Section 4 compelling experimental results, analyzing highly-nonlinear systems of 6 and 9 variables taken from systems biology. We conclude with a discussion of future work. Although we have tried to maintain the paper as self contained as possible, some readers might want to consult books like [42, 39, 28, 38] for some notions of geometry, linear algebra and dynamical systems or expository articles such as [34, 35] which discuss similarities and differences between transition systems and continuous dynamical systems.

---

[1] More precisely, the computation is guaranteed to converge for finite-state systems. In continuous domains we are currently satisfied with a bounded time horizon [34].

## 2    Reachability: Linear and Nonlinear Systems

Computing the states reachable by *all* trajectories of a dynamical system subject to disturbances and parameter variations emerged as a new research topic from the interaction between computer science and control. Reachability computation can be seen as a peculiar way to conduct *exhaustive* simulation which can be useful for the analysis of control systems, the verification of analog circuits, the debugging of biological models and, in fact, any other activity based on dynamical systems models. After a decade of intensive research, [2,25,11,15,26,4, 33,37,10,5,12,32,24] it is fair to say that a satisfactory solution has been provided for *time-invariant linear systems*. Existing algorithms manage to produce, within seconds, high-quality approximations of the reachable states of linear systems with *hundreds* of state variables, for time horizons of *thousands* of integration steps. Notwithstanding these achievements, the real challenge in almost any application domain, Biology included, is the treatment of *nonlinear* systems, a challenge that we address in the present paper.

Let us recall the rules of the game. Given a dynamical system $S$ defined by a differential equation $\dot{x} = f(x, v)$ with $v$ ranging over some bounded set $V$, a set $P$ of initial states and some time horizon $h$, we would like to compute the set of states reachable from points in $P$ by trajectories of $S$ within some $t \in [0, h]$. Fixing some time discretization step $r$, the reachable set is *approximated* by the union of the sets in a sequence $P_0, P_1, \ldots$ where $P_0$ contains all states reachable from $P$ within $t \in [0, r]$ and each $P_{i+1}$ includes states reachable from $P_i$ within $r$ time. Actual computations often work first in discrete time where $P_{i+1}$ is reachable from $P_i$ in one time step and then some error terms are added to bloat $P_{i+1}$ and compensate with respect to continuous time.

Reachability computation of linear systems is relatively easy. Consider first a discrete-time autonomous linear system defined by $x' = Ax$ and a set $P$ which admits a finite representation, for example, a polytope represented by its vertices or supporting halfspaces, an ellipsoid represented by its center and deformation matrix or a zonotope represented by its center and generators. Then the linear transformation "commutes" with the representation. For example, if $P = conv(\tilde{P})$, meaning a polytope $P$ being the convex hull of its finite set of vertices $\tilde{P}$, then

$$A \cdot P = A \cdot conv(\tilde{P}) = conv(A \cdot \tilde{P}), \tag{1}$$

that is, the vertices of the polytope obtained by applying $A$ to the whole set $P$ are the result of applying $A$ to the vertices of $P$.

The extension of this idea to systems with under-specified input, that is, $x' = Ax + v$ where $v$ ranges over a bounded convex set $V$, is more involved. The set of one-step successors of a set $P$ under such a dynamics is captured by the Minkowski sum $P' = AP \oplus V$, which yields a polytope $P'$ with more vertices than $P$. This repeated growth in the size of the representation of $P_i$ makes it impractical to iterate for a long time horizon because the number of points on which $A$ has to be evaluated becomes huge. Two approaches are commonly used to alleviate this problem:

1. For ellipsoids or for polytopes represented by their supporting halfspaces one can use techniques based on the maximum principle [41,13] to obtain an over approximation of $AP \oplus V$ whose representation size is not much larger than that of $P$;
2. The modified recurrence scheme of [27, 24] keeps the number of points to which the linear transformation is applied fixed. Its implementation using zonotopes [23,24], a subclass of polytopes which are closed under Minkowski sum, provides a very efficient solution which is, practically, exact for discrete time.

The technique that we present in this paper is invariant under the choice between these two approaches so we express it in terms of an abstract *successor* operator $\sigma$ which, given a set $P$, an affine differential inclusion (see below) of the form $\dot{x} \in Ax \oplus V$ and a time step $r$, it produces the set $\sigma(P, A, V, r)$ containing all points reachable from points in $P$ by trajectories of duration $r$ of the affine dynamics. The generic linear reachability algorithm can then be written as:

**Algorithm 1 (Linear Reachability)**
$P_0 := \tilde{R}_{[0,r]}(P)$
**repeat** $i = 1, 2, \ldots$
    $P_i := \sigma(P_{i-1}, A, V, r)$
**until** $i = k$

The set $\tilde{R}_{[0,r]}(P)$, the over approximation of the states reachable from $P$ within the time interval $[0, r]$, can be computed, for example, by bloating the convex hull of $P \cup \sigma(P, A, V, r)$ as in [4] or [6].

Moving to *nonlinear* systems of the form $x' = f(x)$ for arbitrary $f$ one observes that "convexity" properties such as (1) do not hold and new ideas are needed. In principle, it is possible to evaluate $f$ on some representative finite sample $\tilde{P} \subset P$ and then use the resulting points to construct a set which over approximates $f(P)$. However, the approximation can be very coarse and will require a costly optimization procedure to be refined, something that cannot be afforded as part of the inner loop of the reachability algorithm. The "hybridization" technique of [6] suggests a good tunable compromise between the quality of the approximation, the difficulty of the computation and the frequency in which it is invoked. Before explaining the idea, let us give some necessary definitions.

We consider a state space $X$, a bounded subset of $\mathbb{R}^n$ equipped with a metric $\rho$. Given two bounded closed subsets $Y$ and $Y'$ of $X$, the *Hausdorff distance* between them (the lifting of $\rho$ to sets) is

$$\rho(Y, Y') = \max\{\max_{y \in Y} \min_{y' \in Y'} \rho(y, y'), \max_{y' \in Y'} \min_{y \in Y} \rho(y, y')\}.$$

The trajectories of a dynamical system are viewed as *signals* over $X$.

**Definition 1 (Signals).** *A signal over $X$ is a partial continuous function $\xi$ from $T = [0, \infty)$ to $X$ whose domain of definition is $T$ or a prefix $[0, r]$ of it. In the latter case we say that $\xi$ is finite with duration $r$. The concatenation of a* *finite signal $\xi$ defined over $[0, r]$ and a signal $\xi'$ satisfying $\xi'(0) = \xi(r)$ is defined in the obvious way and is denoted by $\xi \cdot \xi'$.*

The continuous equivalent of a non-deterministic automaton is the relational vector field, also known as *differential inclusion* [7].

**Definition 2 (Relational Vector Fields).** *A relational vector field over $X$ is a function $f : X \to 2^X - \{\emptyset\}$ which is assumed to be $K$-Lipschitz, satisfying*

$$\rho(\{x\}, \{x'\}) < a \Rightarrow \rho(f(x), f(x')) < Ka.$$

When $f$ is a (deterministic) function we write $f(x) = y$ rather than $f(x) = \{y\}$.

**Definition 3 (Dynamical Systems, Trajectories and Reachable Sets).** *A (continuous) dynamical system is a pair $S = (X, f)$ where $X$ is a state space and $f$ is a vector field. A trajectory of $S$ starting from $x$ is a signal $\xi$ over $X$ with $\xi(0) = x$ and for every $t$ in the domain of definition of $\xi$, $\xi(t) \in X$ and $d\xi(t)/dt \in f(\xi(t))$. The set of all trajectories of $S$ starting from any $x \in P$ is denoted by $\mathcal{L}(S, P)$. The sets of states reachable from $P$ within a time interval $[h, h']$ is*

$$R_{[h,h']}(P) = \{\xi(t) : \xi \in \mathcal{L}(S, P) \wedge t \in [h, h']\}.$$

Hybridization takes a nonlinear system $S = (X, f)$ and produces another dynamical system $(S', f')$ which over approximates it, that is, $\mathcal{L}(S, P) \subseteq \mathcal{L}(S', P)$ for every $P$, and then computes the reachable states of $S'$. A formal definition of $S'$ as a hybrid automaton can be found in [6]. Since our algorithm does not use hybrid automata explicitly we only give an informal explanation.

Consider a partition of $X$ into hyper rectangles (we use the term *box* hereafter). For each box $X_q$ one can compute a linear function $A_q$ and an error polytope $V_q$ such that for every $x \in X_q$, $f(x) \in A_q x \oplus V_q$. In other words, $A_q$ is a local *linearization* of $f$ with error bounded in $V_q$. Thus the vector field $f'$ is defined as $f'(x) = A_q x \oplus V_q$ iff $x \in X_q$. To perform reachability computation on $S'$ one applies linear reachability using $A_q$ and $V_q$ as long as the reachable states remain within box $X_q$. Whenever some $P_i$ reaches the boundary between $X_q$ and $X_{q'}$ it is intersected with the switching surface (the transition guard, in the terminology of hybrid automata) and the obtained result is used as an initial set for reachability computation in $q'$ using $A_{q'}$ and $V_{q'}$, as illustrated in Figure 1-(a,b). The main advantage of hybridization is that the costly procedure of finding a good linear approximation is not invoked in every step, only in the passage between boxes. Although this scheme is clean and general, it suffers from some serious difficulties on the way to realization:

- Although the intersection of the actual set of reachable states inside a box with a facet of the box is typically a convex set, its computation can be inefficient and inaccurate. To see why, consider a subsequence of sets $P_j, \ldots, P_k$ computed using some linear technique, all of which intersect the boundary

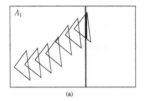
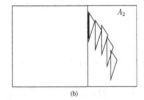

**Fig. 1.** Computing reachable states of the hybridization: (a) applying linear reachability using $A_1$ until intersection with the boundary; (b) taking the intersection as an initial set for linear reachability using $A_2$

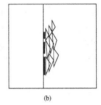
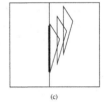

**Fig. 2.** (a) the intersection with the boundary spans over several iterations; (b) continuing with each intersection separately; (c) continuing with an approximation of the union of intersections

$G$ as illustrated in Figure 2-(a). In this case we have either to spawn several computations with the dynamics of the subsequent box, each starting with some $P_i \cap G$ (Figure 2-(b)) or to over approximate $\bigcup_i P_i \cap G$ by a convex set, an operation that may lead to a large over-approximation error (Figure 2-(c)).

– The size of the partition of the state space is, of course, exponential in the dimension, hence care should be taken in order to avoid state explosion. As suggested in [6], the partition can be generated *on-the-fly* as the reachability computation evolves, rather than being precomputed for the whole state space in advance. However, even on-the-fly generation cannot cope with the fact that in high dimension, a tube of reachable states will typically leave a box via *exponentially* many facets. This situation is illustrated in Figure 3-(a). Since each of these parts of the reachable set goes to a different box, they have to be handled separately (Figure 3-(b)) even though they continue to evolve close to each other.[2] Merging these sets when they converge to the same box is a tedious process and a source of further approximation errors. This problem is particularly severe because making the boxes smaller is the recommended recipe for improving accuracy.

---

[2] A similar phenomenon has been encountered in the analysis of timed automata [9].

## 3    Dynamic Hybridization

In this section we describe our novel nonlinear reachability algorithm which, unlike the scheme of [6], is not based on a fixed partitioning of the state space but rather generates overlapping linearization domains around the reachable states. An important ingredient of any hybridization methodology is the linearization procedure that we first define formally.

**Definition 4 (Linearization in a Domain).** *A linearization operator is a function $L$ which, for a given nonlinear function $f$ and a convex set $B$ (linearization domain), produces a matrix $A$ and a convex polytope $V$ such that for every $x \in B$, $f(x) \in Ax \oplus V$.*

We use the notation $L(f, B) = (A, V)$. In our current implementation the linearization domains are boxes, but other forms are possible. In addition to the linearization operator $L$ and the linear successor operator $\sigma$ we assume a procedure $\beta$ which takes as input a set $P$ and produces a linearization domain $B = \beta(P)$ which contains $P$. The form of $B$, the relation between its size and the size of $P$ as well as the position of $P$ inside $B$ are implementation details that may vary according to the system in question. We first present in general terms the algorithm for approximates the reachable states, prove its correctness and then discuss our implementation of $L$ and $\beta$.

**Algorithm 2 (Dynamic Hybridization) Input:** *A nonlinear dynamical system $S = (X, f)$ and an initial set $P$*
**Output:** *A sequence of sets $P_0, P_1, \ldots P_k$ whose union includes $R_{[0,h]}(P)$*

$B := \beta(P)$
$(A, V) := L(f, B)$
$P_0 := \tilde{R}_{[0,r]}(P)$
$i := 0$
**repeat**
    $P_{i+1} := \sigma(P_i, A, V, r)$
    **if** $P_{i+1} \subseteq B$
        $i := i + 1$
    **else**
        $B := \beta(P_i)$
        $(A, V) := L(f, B)$
**until** $i = k$

The algorithm performs linear reachability in a linearization domain $B$ as long as the computed sets remain inside $B$. Once a newly-computed set $P_{i+1}$ is not fully contained in $B$ we backtrack to $P_i$ and construct a new domain $B'$ around $P_i$ along with its corresponding linearization which is used for subsequent computations starting from $P_i$, as illustrated in Figure 4. The advantage of this approach is obvious: the linearization mesh is constructed *along the reachable set* and thus we avoid artificial splitting of sets due to the structure of the mesh. Needless to say, the intersection operation is altogether avoided.

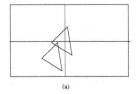
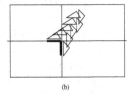

(a)                                    (b)

**Fig. 3.** (a) the reachable set leaves a box through several boundaries; (b) the computation is continued separately for each intersection although the computed sets remain close to each other and even go later to the same box

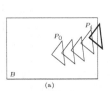
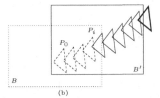

(a)                                    (b)

**Fig. 4.** Dynamic hybridization: (a) Computing in some box until intersection with the boundary; (b) Backtracking one step and computing in a new box

**Theorem 1 (Correctness of Algorithm 2).** *Let $P_0, P_1, \ldots$ be a sequence of sets produced by Algorithm 2. Then for every $k \leq k'$, we have*

$$R_{[kr,k'r]}(P) \subseteq \bigcup_{i=k}^{k'} P_i.$$

**Proof.** The proof is by induction on the number of switchings between linearization domains that the algorithm makes. The base case where no switching occurs follows from the correctness of the linear reachability algorithm and the fact that the linearized system over approximates $f$. For the inductive case, assume the claim holds for $s$ switchings and consider a run of the algorithm with $s + 1$ switchings, the last of which occurring after $P_j$, $k \leq j < k'$. By the inductive hypothesis $R_{[jr,jr]}(P) \subseteq P_j$ and since $P_j$ serves as the initial set for subsequent iterations inside a single linearization domain, the base case applies and $P_{j+1}, \ldots, P_{k'}$ includes $R_{[(j+1)r,k'r]}(P)$ which, together with $P_k, \ldots P_j$, include the states reachable within $[kr, k'r]$.  ∎

Algorithm 2 is implemented in C and uses the polytope-based algorithms of **d/dt** [13]. Below we explain the novel technical aspects, namely the dynamic construction of the linearization domain and its respective linearization.

The difference between the function $f$ and its linear approximation $A$ relative to a domain $B$ is $\Delta_B(f, A) = \{f(x) - Ax : x \in B\}$. To guarantee conservative approximation it is sufficient to find some $V$ such that $\Delta_B(f, A) \subseteq V$ and this can be done easily for *any* choice of a domain $B$ and a linearization $A$. However to obtain high-quality approximations, we need to choose $B$ and $A$ that minimize, roughly speaking, the diameter of $\Delta_B(f, A)$ which represents the error incurred by the linear over approximation. Clearly the smaller is $B$, the smaller is the error but then the linearization procedure has to be invoked more frequently. The problem of finding good $B$ and $A$ can be formulated, in principle, as some sort of a constrained *optimization* problem but this computation can be very costly and we use instead the following easy-to-compute heuristic which turns out to work in practice despite being non optimal. The first simplification that we do with respect to an optimized solution is to decouple the choice of the new domain $B = \beta(P)$ from the computation of the linearization $(A, V) = L(B, f)$.

The operator $\beta(P)$ which produces a box containing $P$ is realized as follows. Based on $f$ and $X$ we fix a standard rectangular frame $\mathcal{B}$ of size $d_1 \times, \cdots, \times d_n$. Given a polytope $P$ we define its centroid $c(P)$ to be the average of its vertices and let $\beta(P)$ be a copy of $\mathcal{B}$ whose center coincides with $c(P)$. The only problematic situation occurs when during reachability computation $P$ gets too large and cannot fit (either immediately or after few steps) within the frame $\mathcal{B}$. To prevent Algorithm 2 from getting stuck in the *else* branch, we split $P$ into two or more sets which are then treated separately. In principle, this splitting may lead to state explosion but, in this case, the explosion is due to *intrinsic properties* of the set of reachable states and not due to an arbitrary choice of the coordinate system underlying the mesh. This phenomenon will not occur too often while analyzing stable systems having a contracting dynamics.

To handle the splitting we first compute a tight bounding box $B(P)$ around $P$. This computation is performed by projecting the vertices on each of the dimensions and taking the minimum and maximum. Let us denote by $e_1 \times, \cdots, \times e_n$ the size of the obtained bounding box. If for every $i$, $m \cdot e_i < d_i$, where $m > 1$ is a fixed constant, then $P$ is sufficiently small and no splitting takes place. Otherwise we take the direction $i$ which maximizes the ratio $e_i/d_i$ and split $P$ into two parts along this direction by intersecting it with complementary halfspaces orthogonal to direction $i$ (see Figure 5). We repeat the process until the obtained sets are sufficiently small. We thus end up with one or more polytopes around each of which we put a properly-centered copy of $\mathcal{B}$.

Once the linearization domain $B$ is fixed we compute $A$ and $V$ as follows. Let $f = (f_1, \ldots, f_n)$, and let $y = c(B)$ be the *center* of $B$. The matrix $A$ is obtained by the evaluating (numerically) the Jacobian matrix of $f$ at $y$, that is,

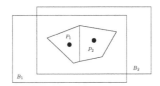

**Fig. 5.** A set $P$ and its bounding box $B(P)$. The set is too large and is split in the vertical dimension into $P_1$ and $P_2$, around which the respective linearization domains $B_1$ and $B_2$ are constructed.

$A = \frac{\partial f}{\partial x}(y)$ where $A_{ij} = \frac{\partial f_i}{\partial x_j}$. Then a box $V = V_1 \times V_2 \times \ldots \times V_n$, guaranteed to contain $\Delta_B(f, A)$, is computed as follows. For each dimension $i$ we let $V_i$ be the interval $[l_i, u_i]$ where $l_i = \min\{\pi_i(\Delta_B(f, A))\}$ and $u_i = \max\{\pi_i(\Delta_B(f, A))\}$ with $\pi_i$ denoting projection on $i$. These intervals are over approximated based on the Taylor expansion of $f(x) - Ax$.

## 4  Experimental Results

To test the feasibility of our algorithm we applied it to two nonlinear systems whose parameters and qualitative behaviors are documented in the literature.

The *Lac Operon* is a biochemical feedback mechanism through which the bacterium *E. Coli* adapts to the lack of Glucose in its environment by switching to a Lactose diet. We use the model appearing in [31] where the behavior of the system is described by the following system of differential equations:

$$\dot{R}_a = \tau - \mu * R_a - k_2 R_a O_f + k_{-2}(\chi - O_f) - k_3 R_a I_i^2 + k_8 R_i G^2$$
$$\dot{O}_f = -k_2 r_a O_f + k_{-2}(\chi - O_f)$$
$$\dot{E} = \nu k_4 O_f - k_7 E$$
$$\dot{M} = \nu k_4 O_f - k_6 M$$
$$\dot{I}_i = -2k_3 R_a I_i^2 + 2k_{-3} F_1 + k_5 I_r M - k_{-5} I_i M - k_9 I_i E$$
$$\dot{G} = -2k_8 R_i G^2 + 2k_{-8} R_a + k_9 I_i E$$

The differential variables denote the concentrations of different reactants, such as $R_a$ (active repressor) $O_f$ (free operator), $E$ (enzyme), $M$ (mRNA), $I_i$ (internal inducer), and $G$ (glucose). We studied the behavior of this 6-dimensional system around a quasi-steady state for the first 4 variables and the obtained results are consistent with the simulation results obtained on a simplified 2-dimensional model shown in [31], page 285. As a set of initial states we take a small box where

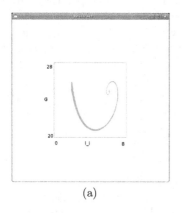

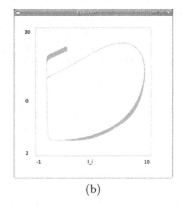

(a)                                    (b)

**Fig. 6.** Lac operon: (a) a stable focus, $k_{-1} = 2.0$; (b) a limit cycle, $k_{-1} = 0.008$

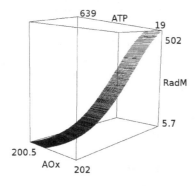

**Fig. 7.** Results obtained for the aging model

$I_i \in [1.9, 2.0]$ and $G \in [25.9, 26]$. When $k_{-1} = 2.0$ the system exhibits a stable focus and when $k_{-1} = 0.008$ the system exhibits a limit cycle (see Figure 6). Computation times are 3 and 5 minutes, respectively.

We conclude with a model of an *aging process*, based on the mitochondrial theory of aging. The highly-nonlinear differential equations, which include ratios between variables, can be found in [31], page 252. The model admits 9 variables and we show in Figure 7 the reachable set after 300 iterations projected on 3 variables, namely, the concentration of antioxidants (AOx), of radicals (RAdM) which suffer damages, and of ATP (adenosine triphosphate). After 1000 iterations, we observe the convergence towards a steady state. The computation time for 1000 iterations is 23.3 minutes.

## 5   Discussion

We made progress toward a very ambitious goal: automatic reachability analysis of nonlinear systems as a methodology for investigating under-specified biological models. Let us mention other attempts to solve this problem starting with methods that share with hybridization the idea of approximating the original systems by partitioning the continuous state space and producing a hybrid automaton with a simpler dynamics in each state. In the extreme case where no continuous dynamics is left the finite automaton is the sole responsible for approximating the dynamics. This approach is common in AI and qualitative physics and has been used extensively in Biology [21, 30, 19]. The technique of predicate abstraction applied to hybrid systems [3] is another elaboration of this idea where partition boundaries are based on predicates appearing in specifications. A more refined approach, incorporated into the tools HyTech [20] and PHAVer [17] over approximates the nonlinear system by hybrid automata where in each state the dynamics is defined by a *constant differential inclusion* of the form $A\dot{x} \leq c$. Since in each state the derivative does not depend on the real variables, it is easy to compute the reachable states exactly using linear algebra, however the over approximation with respect to the original system is large (zero-order compared to first-order approximation in the hybridization of [6]). The translation of continuous systems into timed automata [36] is another example.

Other, more direct, approaches perform reachability on the original nonlinear systems without relying on convexity properties. For example, the face lifting technique [25, 15, 26], which is based on computing the maximal projections of $f$ on all the normals of the facets of a polyhedron, may lead to large over-approximation errors. Other approaches use more complex classes of sets which are not necessarily convex. In [37] the evolution of the reachable states is transformed into a partial differential equation (PDE) where the boundary of the set is represented as the set of zeros of a function defined over the state space. The work of [14] uses Bezier simplices to represent reachable states for systems defined by polynomial differential equations. Finally in [18, 1] dynamic linearization and computation of error bounds is performed at every reachability step. None of these methods, to the best of our knowledge, can cope with systems of the size and complexity of the examples presented in this paper.

Let us also mention the whole domain of *interval analysis* [40], a branch of numerical analysis motivated by producing rigorous numerical answers to diverse mathematical questions despite round-off errors. As its name suggests, for the computation of a scalar function, the result is typically an interval guaranteed to contain the correct answer. The generalization to many dimensions leads naturally to bounding boxes. Although the motivation is different from ours as the uncertainty is due to the computation itself rather than the imperfection of the model, there are similarities between some of the techniques and we foresee more future cross fertilization between the domains.

Parameter uncertainty in biological models is a well-known problem that has been subject to extensive work using various techniques. We mention two recent attacks on the problem of *parameter synthesis*, namely, finding or approximating the range of model parameters for which some qualitative behavior is exhibited. The work of [8] takes a hybrid model (piecewise multi-affine dynamics) with parameter uncertainty and abstracts it into a finite automaton. When the property in question is violated by the automaton, the domain of parameter values is refined, a new abstraction is created and so on. A more direct and efficient way to explore the space of parameter values is described in [16] based on adaptive sampling of the parameter space and using ordinary numerical simulation. This technique uses numerical sensitivity information to guide the refinement of the parameter space.

To go further we need to combine dynamic hybridization with the algorithm of [24, 27] which can treat linear systems an order of magnitude larger than those treated in the present paper. To this end we need to develop a good splitting procedure for the sets computed by that algorithm. As the reader might have noticed, we have focused here on systems where the uncertainty is restricted to the initial set and we need to extend our linearization operator to nonlinear functions with input, something that can be done using similar principles.

To conclude, we have demonstrated the feasibility of our approach by computing reachable states for nonlinear systems of unpreceded size and complexity.

We intend to pursue this direction further and make reachability computation a useful tool for analyzing complex biological systems. A parallel effort should be invested in making modelers of biological systems aware of the potential of this analysis technology.

# References

1. Althoff, M., Stursberg, O., Buss, M.: Reachability Analysis of Nonlinear Systems with Uncertain Parameters using Conservative Linearization. In: CDC 2008 (2008)
2. Alur, R., Courcoubetis, C., Halbwachs, N., Henzinger, T.A., Ho, P.-H., Nicollin, X., Olivero, A., Sifakis, J., Yovine, S.: The Algorithmic Analysis of Hybrid Systems. Theoretical Computer Science 138, 3–34 (1995)
3. Alur, R., Dang, T., Ivancic, F.: Counterexample-guided Predicate Abstraction of Hybrid Systems. Theoretical Computer Science 354, 250–271 (2006)
4. Asarin, E., Bournez, O., Dang, T., Maler, O.: Approximate Reachability Analysis of Piecewise Linear Dynamical Systems. In: Lynch, N.A., Krogh, B.H. (eds.) HSCC 2000. LNCS, vol. 1790, pp. 21–31. Springer, Heidelberg (2000)
5. Asarin, E., Dang, T.: Abstraction by Projection and Application to Multi-affine Systems. In: Alur, R., Pappas, G.J. (eds.) HSCC 2004. LNCS, vol. 2993, pp. 32–47. Springer, Heidelberg (2004)
6. Asarin, E., Dang, T., Girard, A.: Hybridization Methods for the Analysis of Nonlinear Systems. Acta Informatica 43, 451–476 (2007)
7. Aubin, J.-P., Cellina, A.: Differential Inclusions. Springer, Heidelberg (1984)
8. Batt, G., Belta, C., Weiss, R.: Model Checking Genetic Regulatory Networks with Parameter Uncertainty. In: Bemporad, A., Bicchi, A., Buttazzo, G. (eds.) HSCC 2007. LNCS, vol. 4416, pp. 61–75. Springer, Heidelberg (2007)
9. Ben Salah, R., Bozga, M., Maler, O.: On Interleaving in Timed Automata. In: Baier, C., Hermanns, H. (eds.) CONCUR 2006. LNCS, vol. 4137, pp. 465–476. Springer, Heidelberg (2006)
10. Botchkarev, O., Tripakis, S.: Verification of hybrid systems with linear differential inclusions using ellipsoidal approximations. In: Lynch, N.A., Krogh, B.H. (eds.) HSCC 2000. LNCS, vol. 1790, pp. 73–88. Springer, Heidelberg (2000)
11. Chutinan, A., Krogh, B.H.: Verification of polyhedral-invariant hybrid automata using polygonal flow pipe approximations. In: Vaandrager, F.W., van Schuppen, J.H. (eds.) HSCC 1999. LNCS, vol. 1569, pp. 76–90. Springer, Heidelberg (1999)
12. Chutinan, A., Krogh, B.H.: Computational Techniques for Hybrid System Verification. IEEE Trans. on Automatic Control 48, 64–75 (2003)
13. Dang, T.: Verification and Synthesis of Hybrid Systems, PhD thesis, Institut National Polytechnique de Grenoble, Laboratoire Verimag (2000)
14. Dang, T.: Approximate Reachability Computation for Polynomial Systems. In: Hespanha, J.P., Tiwari, A. (eds.) HSCC 2006. LNCS, vol. 3927, pp. 138–152. Springer, Heidelberg (2006)
15. Dang, T., Maler, O.: Reachability Analysis via Face Lifting. In: Henzinger, T.A., Sastry, S.S. (eds.) HSCC 1998. LNCS, vol. 1386, pp. 96–109. Springer, Heidelberg (1998)
16. Donze, A., Clermont, G., Legay, A., Langmead, C.J.: Parameter Synthesis in Nonlinear Dynamical Systems: Application to Systems Biology. In: RECOMB 2009, pp. 155–169 (2009)

17. Frehse, G.: PHAVer: Algorithmic Verification of Hybrid Systems Past HyTech. In: Morari, M., Thiele, L. (eds.) HSCC 2005. LNCS, vol. 3414, pp. 258–273. Springer, Heidelberg (2005)
18. Han, Z., Krogh, B.H.: Reachability Analysis of Nonlinear Systems using Trajectory Piecewise Linearized Models. In: American Control Conference, pp. 1505–1510 (2006)
19. Halasz, A., Kumar, V., Imielinski, M., Belta, C., Sokolsky, O., Pathak, S.: Analysis of Lactose Metabolism in E.coli using Reachability Analysis of Hybrid Systems. IEE Proceedings - Systems Biology 21, 130–148 (2007)
20. Henzinger, T.A., Ho, P.-H., Wong-Toi, H.: Algorithmic Analysis of Nonlinear Hybrid Systems. IEEE Trans. on Automatic Control 43, 540–554 (1998)
21. de Jong, H., Page, M., Hernandez, C., Geiselmann, J.: Qualitative Simulation of Genetic Regulatory Networks: Method and Application. In: IJCAI 2001, pp. 67–73 (2001)
22. Gillespie, D.T.: Stochastic Simulation of Chemical Kinetics. Annual Review of Physical Chemistry 58, 35–55 (2007)
23. Girard, A.: Reachability of Uncertain Linear Systems using Zonotopes. In: Morari, M., Thiele, L. (eds.) HSCC 2005. LNCS, vol. 3414, pp. 291–305. Springer, Heidelberg (2005)
24. Girard, A., Le Guernic, C., Maler, O.: Efficient Computation of Reachable Sets of Linear Time-invariant Systems with Inputs. In: Hespanha, J.P., Tiwari, A. (eds.) HSCC 2006. LNCS, vol. 3927, pp. 257–271. Springer, Heidelberg (2006)
25. Greenstreet, M.R.: Verifying Safety Properties of Differential Equations. In: Alur, R., Henzinger, T.A. (eds.) CAV 1996. LNCS, vol. 1102, pp. 277–287. Springer, Heidelberg (1996)
26. Greenstreet, M.R., Mitchell, I.: Reachability Analysis Using Polygonal Projections. In: Vaandrager, F.W., van Schuppen, J.H. (eds.) HSCC 1999. LNCS, vol. 1569, pp. 103–116. Springer, Heidelberg (1999)
27. Le Guernic, C.: Calcul efficace de l'ensemble atteignable des systémes linaires avec incertitudes, Master's thesis, Université Paris 7 (2005)
28. Hirsch, M., Smale, S.: Differential Equations, Dynamical Systems and Linear Algebra. Academic Press, London (1974)
29. 29 Jaulin, L., Kieffer, M., Didrit, O., Walter, E.: Applied Interval Analysis. Springer, Heidelberg (2001)
30. de Jong, H., Page, M., Hernandez, C., Geiselmann, J.: Qualitative Simulation of Genetic Regulatory Networks: Method and Application. In: IJCAI 2001, pp. 67–73 (2001)
31. Klipp, E., Herwig, R., Kowald, A., Wierling, C., Lehrach, H.: Systems Biology in Practice: Concepts, Implementation and Application. Wiley, Chichester (2005)
32. Kurzhanskiy, A., Varaiya, P.: Ellipsoidal Techniques for Reachability Analysis of Discrete-time Linear Systems. IEEE Trans. Automatic Control 52, 26–38 (2007)
33. Kurzhanski, A., Varaiya, P.: Ellipsoidal tehcniques for reachability analysis. In: Lynch, N.A., Krogh, B.H. (eds.) HSCC 2000. LNCS, vol. 1790, p. 202. Springer, Heidelberg (2000)
34. Maler, O.: A Unified Approach for Studying Discrete and Continuous Dynamical Systems. In: CDC 1998, pp. 2083–2088 (1998)
35. Maler, O.: Control from Computer Science. Ann. Rev. in Control 26, 175–187 (2002)
36. Maler, O., Batt, G.: Approximating Continuous Systems by Timed Automata. In: Fisher, J. (ed.) FMSB 2008. LNCS (LNBI), vol. 5054, pp. 77–89. Springer, Heidelberg (2008)

37. Mitchell, I., Tomlin, C.J.: Level Set Methods for Computation in Hybrid Systems. In: Lynch, N.A., Krogh, B.H. (eds.) HSCC 2000. LNCS, vol. 1790, pp. 310–323. Springer, Heidelberg (2000)
38. Sastry, S.: Nonlinear systems. Analysis, Stability and Control. Springer, Heidelberg (1999)
39. Schrijver, A.: Theory of Linear and Integer Programming. Wiley, Chichester (1986)
40. Thomas, R., D'Ari, R.: Biological Feedback. CRC Press, Boca Raton (1990)
41. Varaiya, P.: Reach Set computation using Optimal Control. In: KIT Workshop, pp. 377–383. Verimag, Grenoble (1998)
42. Zeigler, G.M.: Lectures on Polytpoes. Springer, Heidelberg (1995)

# On Coupling Models Using Model-Checking: Effects of Irinotecan Injections on the Mammalian Cell Cycle

Elisabetta De Maria, François Fages, and Sylvain Soliman

Project-team Contraintes, INRIA Paris-Rocquencourt, France

**Abstract.** In systems biology, the number of models of cellular processes increases rapidly, but re-using models in different contexts or for different questions remains a challenging issue. In this paper, we show how the validation of a coupled model and the optimization of its parameters with respect to biological properties formalized in temporal logics, can be done automatically by model-checking. More specifically, we illustrate this approach with the coupling of existing models of the mammalian cell cycle, the p53-based DNA-damage repair network, and irinotecan metabolism, with respect to the biological properties of this anticancer drug.

## 1 Introduction

In systems biology, the number of models of cellular processes increases rapidly. To date, most of the effort has been devoted to building models and making them freely available, through the design of standard exchange formats, such as for instance the Systems Markup Language SBML [26], the making of model repositories, such as for instance http://biomodels.net/, the making of biological ontologies to establish the links between molecular synonyms, species, units, etc., and the development of modeling tools, such as Cell Designer, BIOCHAM [7], BioNetGen [4], Pathway Logic [15], Bio-ambients [35], etc. Despite these efforts however, re-using models in different contexts or for different questions remains a challenging issue. In practice, most of the models are developed, refined, simplified or coupled with respect to other models by hand with no direct support from the tools to re-use models in a systematic way using a specification of the expected behavior.

In this paper, we show how the validation of a coupled model and the optimization of its parameters with respect to biological properties formalized in temporal logics, can be done automatically by model-checking. More specifically, we illustrate this approach with the coupling of existing models of the mammalian cell cycle, the p53-based DNA-damage repair network, and irinotecan metabolism, with respect to the biological properties of the latter one.

Irinotecan is an anti-carcinogenic inhibitor of topoisomerase-1 which started to be used in clinical treatments approximately twenty years ago. It shows significant efficacy against a variety of solid tumors, including lung, colorectal,

P. Degano and R. Gorrieri (Eds.): CMSB 2009, LNBI 5688, pp. 142–157, 2009.

and cervical cancers. Scientists are currently trying to optimize the irinotecan therapy in order to understand how to limit its toxicity and to increase its efficacy. In this context, it is crucial to comprehend how the presence of this medicament influences cellular proliferation. In this work we present a model-checking approach to the problem. There are in the literature many models of the mammalian cell-cycle, a few ones of the cell's DNA-damage repair pathways, and recently some preliminary models of irinotecan intracellular pharmacodynamics. However these modules need to be assembled in a coherent way to provide meaningful answers. After reformulating the existing models in the rule-based language of the biochemical abstract machine Biocham [19,7], we assemble them into a coupled model under the constraint of satisfying some relevant biological properties, formalized in temporal logic and automatically checked by model-checking.

Model-checking is the process of algorithmically verifying whether a given structure is a model for a given temporal logic formula [13]. In literature, there are various applications of model-checking techniques to biology. In [8,15], temporal logic was introduced as a query language for biochemical networks and for validating boolean models of biological processes by model-checking techniques. Some experimental results were obtained on Kohn's map of the mammalian cell cycle control (800 reaction rules, 500 variables) using the symbolic model-checker NuSMV, and on a small ordinary differential equation (ODE) model using the constraint-based model checker DMC. This approach to verifying biological processes has pushed the development of model-checking techniques for quantitative properties, and continuous, stochastic or hybrid models.

For (non-linear) ODE models, numerical integration techniques provide numerical traces on which linear time temporal logic with numerical constraints can be evaluated by model-checking. Simpathica [2] and the biochemical abstract machine Biocham [6] are two examples of computational tools integrating such model-checkers for quantitative models. This approach has been generalized to temporal logic constraint solving in [18], allowing for efficient kinetic parameter optimization [37] and robustness analyses [36] w.r.t. quantitative temporal logic properties.

In [25], Heath et al. apply the probabilistic model-checker PRISM to the study of a complex biological system, namely, the Fibroblast Growth Factor (FGF) signalling pathway. In [12], Clarke et al. apply statistical model-checking on a stochastic model of a T-cell receptor.

In [3] Batt et al. develop a modeling framework based on differential equations to analyze genetic regulatory networks with parameter uncertainty. The values of uncertain parameters are given in terms of intervals and dynamical properties of the networks are expressed in temporal logic. Model-checking techniques are then exploited to prove that, for every possible parameter value, the modeled systems satisfy the expected properties and to find valid subsets of a given set of parameter values (such an approach is exploited in RoVerGeNe, a tool for robust verification of gene networks). In [33], Piazza et al. propose semi-algebraic hybrid systems as a natural framework for modeling biochemical networks, taking advantage of the decidability of the model-checking problem for TCTL (Timed Computation Tree Logic) over this large class of systems.

In this paper, we focus on the use of model-checking for integrating biological models. To compose the selected models, we assume a finite set of hypotheses concerning the structure of the "linking reaction rules", and we search for kinetic parameter values that will make the composite model interacts in a proper way. For this, the biological properties of the coupled model are formalized in temporal logic with numerical constraints, and model-checking with parameter optimization techniques are used to find parameter values for the new kinetic rules so that the expected properties are satisfied by the resulting model.

The paper is organized as follows. Section 2 provides the needed biological background on the mammalian cell cycle, on the tumor-suppressor protein p53, and on irinotecan. Section 3 describes Biocham models and explains how such models can be queried in temporal logic. Section 4 presents our coupled model on the effects of irinotecan and Section 5 describes the biological properties of the model that were automatically checked. Finally, in Section 6 we outline some ongoing developments of this work. All the models used are available in BIOCHAM format at http://contraintes.inria.fr/supplementary_material/CMSB09/.

## 2    Mammalian Cell Cycle, DNA-Damage Repair and Irinotecan

### 2.1    Mammalian Cell Division Cycle

Cells reproduce by duplicating their contents and then dividing in two. To produce a pair of genetically identical daughter cells, the DNA must be faithfully replicated, and the replicated chromosomes must be segregated into two separate cells. The duration of the cell cycle varies greatly from one cell type to another; in a mammalian cell it lasts about 24 hours. The cycle is traditionally divided into the following four distinct phases [1]: the **G1-phase**, that is the temporal gap between the completion of mitosis and the beginning of DNA synthesis, the **S-phase (synthesis)**, that is the period of DNA replication, the **G2-phase**, that is the temporal gap between the end of DNA synthesis and the beginning of mitosis, and the **M-phase (mitosis)**, when replicated DNA molecules are finally separated in two daughter cells.

The cell cycle is regulated by different checkpoints, that are moments when the cell progression is stopped to verify the state of the cell and, if needed, to repair it before damaged DNA is transmitted to progeny cells. DNA damaging agents trigger checkpoints that produce arrest in G1 and G2 stages of the cell cycle. Cells can also arrest in S, which amounts to a prolonged S phase with slowed DNA synthesis. Arrest in G1 allows repair before DNA replication, whereas arrest in G2 allows repair before chromosome separation in mitosis.

The proper alternation between synthesis and mitosis is coordinated by a complicated network that regulates the activity of a family of key proteins. These proteins are composed of two subunits: a catalytic subunit, the cyclin-dependent kinase, *cdk* for short, and a regulatory subunit, a *cyclin*. A cdk has to associate with a cyclin partner to form a dimer and has to be appropriately phosphorylated in order to be active. The progression through cell cycle is orchestrated by the rise and fall of the Cdk/cyclin dimers.

In this work we refer to the model of mammalian cell division proposed by Novák and Tyson in [30] where the authors present both a set of Ordinary Differential Equations (ODE) and a process diagram to represent the molecular network regulating the mammalian cell cycle. The model comprises 18 differential equations and 4 steady-state relations.

## 2.2  Protein p53

This subsection is devoted to the description of protein p53, a tumor suppressor protein which is activated in reply to DNA damage. In normal conditions, the concentration of p53 in the nucleus of a cell is feeble: its level is controlled by another protein, Mdm2. These two proteins present a loop of negative regulation. In fact, p53 activates the transcription of Mdm2 while the latter accelerates the degradation of the former. DNA damage increases the degradation rate of Mdm2 so that the control of this protein on p53 becomes weaker and p53 can exercise its functions. This protein is responsible for the activation of many mechanisms: in an indirect way, it stops the DNA synthesis process, it activates the production of proteins charged with DNA reparation, and it can lead to apoptosis, that is, cell death.

When DNA is damaged, Mdm2 loosens its influence on p53 and it is possible to observe some oscillations of p53 and Mdm2 concentrations. The answer to a stronger damage is a bigger number of oscillations. Oscillations have a very regular period. In literature, several models have been proposed to model the oscillatory behaviour of proteins p53 and Mdm2. The most interesting models are undoubtedly the ones proposed by Chickermane et al. [9], by Ciliberto et al. [10], and by Geva-Zatorsky et al. [20]. In this work we build upon the one described in [10], that consists of 6 differential eqiations.

## 2.3  Irinotecan

Camptothecins are substances that can be extracted from the Chinese tree "Camptotheca acuminata Decne" and are mainly used for the treatment of digestive cancers. Their anticancerogenic properties have been discovered at the end of the Fifties but the first clinical tests have been interrupted owing to heavy effects due to the toxicity of the substances. In the Eighties researchers discovered that camptothecins are inhibitors of topoisomerase-1(Top1 for short), essential enzyme for DNA synthesis. Afterwards, they started to focus on some semi-synthetic derivative of water-soluble camptothecins, such as irinotecan and topotecan. Irinotecan is pro-medicine and must be transformed in its active metabolite, SN38, to be effectively cytotoxic. In fact the anticancerogenic activity of irinotecan (CPT11) is approximately 100 times less effective than the one of SN38. The activation is due to carboxylesterase, an enzyme mainly located in the liver, in the intestine, and in the tumoral tissues. SN38 is then detoxified through glucorono-conjugation: this realizes uridine diphosphate glucoronosyl transferase 1A1.

Mechanisms through which irinotecan damages the cell are very complex and have not completely been explained yet. It is sure that DNA lesions appear after the inhibition of Top1 by SN38. Top1 is a protein which is present in all living organisms and which checks DNA replication and transcription. It intervenes to modify the DNA winding degree, acting on one strand. More precisely, Top1 links itself to extremity 3' of DNA forming a transitory cleavage complex and cuts a DNA strand, that in such a way is able to unroll. Then such a complex dissociates and a new ligature comes up. In normal conditions, the connection process is favored with respect to the cleavage one. The target of irinotecan, and above all of its active metabolite SN38, is the complex Top1-DNA. SN38 links to the complex through a covalent bond, preventing in such a way from the ligature of the DNA strand. As clearly written in the title of [34], SN38 acts like a "foot in the door": it keeps opened the DNA strand to which Top1 is linked as to prevent a door from closing. These complexes are still reversible and do not cause DNA lesions. However, they favor them: some lesions can rise as a consequence of the possible collisions with the transcription complexes or with the replication fork. This induces the arrest of the cell cycle. In this case we speak of irreversible complexes. Lesions due to the inhibition of Top1 are therefore consecutive to the stages of the cell metabolism. It means that irinotecan injections must be repeated and abundant in order to be effective. Besides irinotecan is more effective during the DNA replication phase [31,38]. Furthermore, the inhibition of the DNA synthesis takes rapidly place (in a few minutes) and lasts several hours.

Defence answers of cells subjected to irinotecan injections are multiple and vary according to the drug dose. The administration of a very light dose suffices to slow down the S phase of the cell cycle and to delay the G2-M transition. If the dose is more substantial, the lag time in the S phase is much more significant and the cell cycle arrest in the G2-M transition can last more than sixty hours or even be permanent. In this latter case, some genes responsible for the cell cycle arrest (as an example, p21) and involved in the aptototic pathway are over-expressed. These genes are activated by p53, and this suggests the intervention of the protein in reply to a DNA damage due to the dissociation of Top1 from DNA [38].

In this work we refer to a pharmacokinetics/pharmacodinamics (PK/PD) model of irinotecan developed by Dimitrio [14], that takes aim at representing the action of the drug on the body (pharmacodinamic) and the action of the body on the drug (pharmacokinetic), and thus the drug metabolism and its transformations. Such a model is made up of 8 differential equations.

# 3   The BIOCHAM Abstract Machine

In the last years, one of the main challenges of computational system biology became the creation of powerful simulation, analysis, and reasoning tools for biologists to decipher existing data, devise new experiments, and thus understand functional properties of genomes, proteomes, cells, organs, and organisms. Many of the goals of this emerging discipline have been investigated in the logical setting of temporal logics and have been partly achieved in this apprroacch in the Biochemical Abstract Machine *(BIOCHAM)* [19,7], a formal modeling environment for network biology developed at INRIA-Rocquencourt since 2002.

In this section we briefly recall BIOCHAM syntax, its continuous differential semantics and show how a temporal logic with numerical constraints is used to formalize the biological properties of the models and automatically check their satisfaction.

### 3.1  Syntax and Differential Semantics of BIOCHAM Reaction Models

Following SBML and BIOCHAM conventions, a model of a biochemical system is formally a set of reaction rules of the form $e$ for $S \Rightarrow S'$, where $S$ is a *solution*, that is, a set of molecules given with their stoichiometric coefficient, $S'$ is the transformed solution, and $e$ is a kinetic expression involving the concentrations of molecules. The reaction rules represent biomolecular interactions between chemical or biochemical compounds, ranging from small molecules to proteins and genes.

Reaction rules transform one formal solution into another one. The following abbreviations are used: A =[C]=> B for the catalyzed reaction A+C => C+B, and A <=> B for the reversible reaction equivalent to the two symmetrical reactions A => B and B => A. The constant _ represents the empty solution. It is used for instance in protein *degradation rules*, such us A =>_ , and in *synthesis rules*, such us _=[G]=> A for the synthesis of A by (activated gene) catalyst G. The other main rule schemas are *(de)complexation rules*, such us A + B => A-B for the complexation of A and B, *(de)phosphorylation rules*, such us A =[B]=> A~{p} for the phosphorylation of A catalyzed by kinase B, and *transport rules*, such us A::nucleus => A::cytoplasm for the transport of A from the nucleus to the cytoplasm.

Reactions can be given kinetic expressions. For instance, k*[A]*[B] for A=[B]=>A~{p} specifies a mass action law kinetics with parameter $k$ for the reaction. Classical kinetics expressions are the following ones:

- the **mass action law kinetics** $k * \prod_{i=1}^{n} x_i^{l_i}$, which refers to a reaction with $n$ reactants $x_i$, where $l_i$ is the stoichiometric coefficient of $x_i$ as a reactant;
- the **Michaelis-Menten kinetics** $V_m * x_s/(K_m + x_s)$, which states for an enzymatic reaction of the form $x_s = [x_e] \Rightarrow x_p$, where[1] $V_m = k * (x_e + x_e * x_s/K_m)$;
- the **Hill's kinetics** $V_m * x_s^n/(K_m^n + x_s^n)$, of which the Michaelis-Menten kinetics is a special case for $n = 1$.

---

[1] $x_e * x_s/K_m$ is the concentration of the enzyme-substrate complex, supposed constant in the Michaelian approximation, and $x_e + x_e * x_s/K_m$ is thus the total amount of enzyme.

Kinetic expressions can be written either explicitly, allowing any kinetics, or using shortcuts such us `MA(k)` for a Mass Action law with parameter $k$, or `MM(Vm,Km)` for a Michaelian kinetics.

A set of reaction rules $\{e_i \text{ for } S_i => S'_i\}_{i=1,...,n}$ over molecular concentration variables $\{x_1, \ldots, x_m\}$ can be interpreted under different semantics. In this paper we refer to the traditional *differential semantics*, that interprets the rules by the following system of Ordinary Differential Equations (ODE):

$$dx_k/dt = \sum_{i=1}^{n} r_i(x_k) * e_i - \sum_{j=1}^{n} l_j(x_k) * e_j,$$

where $r_i(x_k)$ (resp., $l_i$) is the stoichiometric coefficient of $x_k$ in the right (resp., left) member of rule $i$. Given a set of reaction rules, BIOCHAM allows to obtain a simulation of the model by solving the set of corresponding differential equations.

## 3.2 Querying BIOCHAM Models in Temporal Logic

Temporal logics and model-checking algorithms have been proved to be useful to respectively express biological properties of complex biochemical systems and automatically verify their satisfaction [17]. Having a formal language not only for describing models, (i.e., transition systems based on process calculi [35,32], rules [15], Petri nets [22], ODEs [39], etc.), but also for formalizing the biological properties of the system known from biological experiments under various conditions, opens a whole avenue of research for designing automated reasoning tools inspired from circuit and program verification to help the modeler.

In this paper, we use a version of *linear time* logic LTL with numerical constraints, named *Constraint-LTL* [6]. Constraint-LTL formulae are formed over first-order atomic formulae with equality, inequality and arithmetic operators ranging over real values of concentrations and of their derivatives, using the logical connectives and the usual *temporal operators* of LTL: in particular the "always in the future" operator $\mathbf{G}$, and the "sometimes in the future" operator $\mathbf{F}$), the next time operator $\mathbf{X}$, and the binary operator until $\mathbf{U}$.

For instance, $F([A] > 10)$ expresses that the concentration of $A$ eventually gets above the threshold value 10 and $G([A] + [B] < [C])$ states that the concentration of $C$ is always greater than the sum of the concentrations of $A$ and $B$. Oscillation properties, abbreviated as $oscil(M, K)$, are defined as a change of sign of the derivative of $M$ at least $K$ times: $F((d[M]/dt > 0)$ & $F((d[M]/dt < 0)$ & $F((d[M]/dt > 0) \ldots )))$. The abbreviated formula $oscil(M, K, V)$ adds the constraint that the maximum concentration of $M$ must be above the threshold $V$ in at least $K$ oscillations.

LTL formulae are interpreted in linear Kripke structures which represent either an experimental data time series or a simulation trace, both completed with loops on terminal states. Given the system of ordinary differential equations (ODE) corresponding to a reaction model, under the hypothesis that the initial state is completely defined, a discrete simulation trace is easily obtained by means of numerical integration methods (such as Runge-Kutta or Rosenbrock method for stiff systems). Since constraints refer not only to concentrations, but also to their derivatives, traces of the form

$$(< t_0, x_0, dx_0/dt, d^2x_0/dt^2 >, < t_1, x_1, dx_1/dt, d^2x_1/dt^2 >, \dots)$$

are considered, where at each time point $t_i$, the trace associates the concentration values $x_i$ to the variables, and the values of their first and second derivatives $dx_i/dt$ and $d^2x_i/dt^2$. It is worth noting that in adaptive step size integration methods of ODE systems, the step size $t_{i+1}$ - $t_i$ is not constant and is determined through an estimation of the error made by the discretization.

Observe that the notion of *next state* refers to the state of the following time point in a discretized trace, and thus does not necessarily imply a real time neighborhood. The rationale is that the numerical trace contains enough relevant points, and in particular those where the derivatives change abruptly, to correctly evaluate temporal logic formulae. An innovative feature of BIOCHAM is that it places at the user's disposal a procedure `learn_parameters` for finding parameter values such that a given LTL formula is satisfied [37]. This search procedure actually replicates and automates part of what the modeler currently does by hand: trying different parameter values, between bounds that are thought reasonable, or computed by other methods such as bifurcation diagrams, in order to obtain behaviors in accordance with the experimental knowledge.

## 4  Coupling the Models of the Mammalian Cell Cycle, p53-Based DNA-Damage Repair System, and Irinotecan Pharmacodynamics

### 4.1  Three Different Modules

The first step of our approach to the investigation of the influence of irinotecan on the mammalian cell cycle consists in the encoding of the selected models of irinotecan [14], p53/Mdm2 [10], and mammalian cell cycle [30] in the BIOCHAM rule-based language.

SBML versions of the irinotecan and p53/Mdm2 modules being available, they were imported in BIOCHAM. The renaming of the variable representing DNA-damage was the only modification necessary in this precise case. More generally it would be necessary to rely on existing databases and ontologies to match corresponding entities in different models.

For the other model, we looked in parallel at the corresponding set of ordinary differential equations and at the available diagrammatic notation to obtain a set of BIOCHAM reaction rules. Since ODEs can be automatically extracted back from the reactions, and displayed with the BIOCHAM command `show_kinetics`, it was possible to check that the obtained model was indeed coherent with the original one.

### 4.2  A Diagrammatic Coupled Model

In order to assemble the sub-models to get the coupled model, we reviewed the literature about known links between the different building blocks.

The expected behaviour of the cell is graphically depicted in Figure 1 and can be described as follows. Injections of irinotecan (CPT11) induce DNA damage. In reply to this, the cell reacts by activating protein p53, which blocks the cell cycle at a checkpoint. This arrest aims at repairing critical damage before DNA replications occurs, thereby avoiding the propagation of genetic lesions to progeny cells. Thus, while the cell cycle is arrested, the protein p53 will activate the DNA-damage repair mechanisms. If it is possible for the cell to recover, the cell cycle will be restarted; otherwise, if the damage is too extensive, the cell will undergo apoptosis.

The strategy we chose to draw the three models together is illustrated in Figure 2. As remarked in Section 2, in literature we found evidence of the fact that, if irinotecan is injected in a cell during the S phase of the cell cycle, then more DNA damage will be caused with respect to the other phases of the cell cycle [31,38]. Keeping this fact in mind, and considering that the dimer characterizing S phase is CycA/Cdk2 (CycA for short), we inserted in cell cycle model a rule stating that a high concentration of CycA determines a high concentration of Top1, an enzyme that, as previously explained, contributes to cause DNA single-strand breaks in presence of irinotecan. In this way we linked the cell cycle model to the irinotecan one.

The link between the irinotecan model and the p53/Mdm2 one is given by DNA damage. In fact, irinotecan injections cause DNA damage, which in turn triggers the activity of protein p53. In order to link the p53/Mdm2 and cell

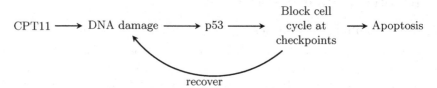

**Fig. 1.** Expected behaviour of the coupled model

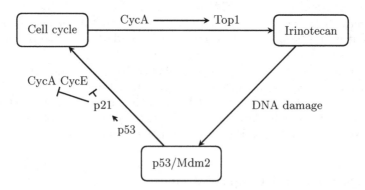

**Fig. 2.** Our coupled model

cycle models, we inserted in the p53/Mdm2 model a rule which fixes that p53 activates p21, and two further rules imposing that p21 inhibits CycA and CycE, respectively.

It is worth noting that we also investigated the possibility to abstract the previous *expanded* rules by letting p53 directly inhibit CycA and CycE. In the following, we will refer to this last version of the link as to the *contracted* one.

Finally, in order to define precisely the kinetics underlying the links and to validate the coupled model, some biological knowledge was formalized as temporal logic specifications allowing us to use model-checking and parameter optimization techniques.

## 5  Specifying and Validating the Links through Model-Checking

In this section, we will show how the integration of model-checking and parameter learning techniques allowed us to both specify and validate our linking rules. In both cases, we took advantage of constraint-LTL formulae to query numerical simulations in a much more flexible way than by doing curve fitting. Our Kripke structures are constituted of simulation traces over a time window of 100 hours, obtained by numerical integration (using Rosenbrock's implicit method for stiff systems), extended with the first and second derivatives of the system's variables.

As a matter of fact, both at specification and validation time, first we considered two models at a time, and finally we dealt with the complete model made up by the three ones. In the following, we provide examples of biological properties that helped us in defining the kinetics underlying the links: for each property, first of all we express it through the natural language, then we formalize it in constraint-LTL, and finally we present the results of the corresponding model-checking query.

We encoded the link between the cell cycle and irinotecan models (see Figure 2) by means of the following kinetics rule:

`MA(top1bis) for _=[CycA]=>TOP1.`

Then, we used the procedure `learn_parameters` to find out the minimum value for `top1bis` such that property F1, that correlates the concentration values of CycA and Top1, holds.

**F1:** Whenever CycA gets over 1, there exists a future state where Top1 is greater than 1.
**LTL:** $G((([CycA] > 1) \rightarrow F([TOP1] > 1))$.
**Results:** The minimum value for `top1bis` such that F1 is satisfied turned out to be 0.45.

The following Biocham rules encode the link between the p53/Mdm2 and cell cycle models (see Figure 2):

`MA(k5321) for _=[p53]=>p21.`
`MA(kA21) for CycA=[p21]=>_.`

`MA(kA21) for CycE=[p21]=>_.`

As for the contracted version, the encoding is the following one:

`MA(kA53) for CycA=[p53]=>_.`
`MA(kA53) for CycE=[p53]=>_.`

Again, we took advantage of the procedure **learn_parameters** to find suitable parameter values for $k5321$ and $kA21$ ($kA53$ in the second case) so that property F2, that expresses the CycA oscillating behaviour exhibited in the cell cycle model, is conserved in the coupled model in case there are no irinotecan injection.

**F2:** Within a time interval of 100 time units, CycA is greater than 2 in at least 8 oscillations.

**LTL:** $oscil([CycA], 8, 2)$.

**Results:** We found out that suitable parameter values are $k5321 = 0.18$, $kA21 = 0.08$, and $kA53 = 0.25$. Property F2 also turned out to be true when there are injections but the p53/Mdm2 model is not taken into account. In fact, as expected, even if DNA damage occurs, when protein p53 does not act, the cell cycle is not affected, and thus CycA exhibits a regular oscillating behaviour.

On the other hand, when the p53/Mdm2 model is added, in case of repeated injections (and thus of sustained DNA-damage) the oscillations of CycA should be affected.

**F3:** When there is sustained DNA damage (after an initial period), the amplitude of CycA decreases before 70 time units and then stays low.

**LTL:** $F((Time < 13) \land G([DNAdam] > 0.5)) \rightarrow F((Time < 70) \land G([CycA] < 1.2))$.

**Results:** With the expanded version of the links the amplitude of oscillations gradually decreases, satisfying the property. With the contracted one oscillations are very irregular, as graphically depicted in Figure 3.

The next property regards the DNA repairing power of the cell.

**F4:** After an irinotecan injection is performed, DNA damage is able to go under the threshold of 0.1 before the next injection is done.

**LTL:** $G(([CPT11] > 9.45) \lor (([CPT11] \leq 9.45)U([DNAdam] < 0.1)))$.

**Results:** Before testing the property, we decided to parameterize the lapse of time between consecutive irinotecan injections. Then we took advantage of the procedure **learn_parameters** to find the minimum $k$ such that, if one injection is performed every $k$ hours, then property F4 is true. We found out that the minimum $k$ multiple of 12 which makes F4 true is 36. Thus, one injection every 36 hours should be performed in order to allow DNA damage to be recovered before the next injection. Then we tried to see what it happens if, at each injection, we double the irinotecan dose. In this case, one injection every 48 hours should be done.

The next property requires the oscillating trend of proteins p53 and Mdm2 to stop before a new injection is performed.

**F5:** When an injection is performed, p53 and Mdm2 are in a steady state, that is, their derivatives approach 0.

**LTL:** $G(([CPT11] > 9.45) \rightarrow ((d[p53] \leq 0.05) \land (d[p53] \geq -0.05) \land (d[Mdm2 :: n] \leq 0.05) \land (d[Mdm2 :: n] \geq -0.05)))$.

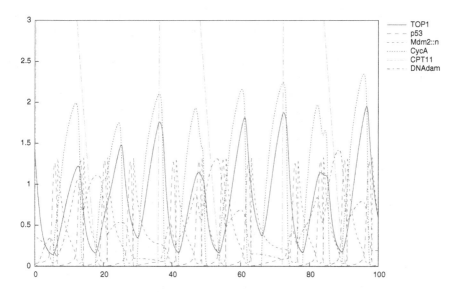

**Fig. 3.** Zoomed simulation plot of the resulting model with injections every 36 hours

**Results:** As for the previous specification, we parameterized the lapse of time between consecutive irinotecan injections and we used the procedure `learn_parameters`. The the minimum $k$ multiple of 12 which makes F5 true is 48.

The next property deals with DNA damage.

**F6:** If p53 is not functional, DNA damage is an increasing function.

**LTL:** $G([p53] \leq 0.1) \rightarrow G(d([DNAdam])/dt \geq 0)$.

**Results:** As matter of fact, the premise of this property only holds when the p53/Mdm2 model is not taken into account, thus DNA damage is not recovered and continuously increases until apoptosis is reached.

Finally, the last two properties concern the oscillating trend of proteins p53 and Mdm2 caused by irinotecan injections.

**F7:** An irinotecan injection causes at least one oscillation of proteins p53 and Mdm2.

**LTL:** $G(([CPT11] > 9.45) \rightarrow F(oscil([p53], 1) \wedge F(oscil([Mdm2], 1))))$.

**F8:** p53 oscillations are alternated by Mdm2 ones.

**LTL:** $G(oscil([p53], 1) \rightarrow X((\neg oscil([p53], 1))U(oscil([Mdm2 :: n], 1))))$.

**Results:** Both F7 and F8 are satisfied in the coupled model.

Such a model-checking approach turned out to be very efficient: the execution times were lower than one second in almost all cases[2]. Furthermore, it proved to be effective, allowing us to express relevant biological properties of the model (and concentration values that make specifications true) that could not be easily encoded as curve fitting problems for instance. There is a common recognition of

---

[2] All the experiments have been run on a Centrino Duo 2.00 GHz Windows machine.

the potentialities of model-checking in bio-informatics and our contribution goes in this direction, showing its utility to compose and validate biological models.

In Figure 3 there is a 100 hours-simulation of the complete model, where the contracted version of the link between the p53/Mdm2 and cell cycle models is considered and one injection every 36 hours is performed. The plot puts in evidence how DNA damage increases after every injection. The oscillating trend of proteins p53 and Mdm2 is well highlighted. Furthermore, it is possible to notice the irregular behaviour assumed by CycA after irinotecan injections and the dependence of Top1 from CycA.

## 6    Conclusion and Perspectives

In this paper we presented a model-checking approach to the investigation of the influence of irinotecan on the mammalian cell cycle. We coupled in the rule-based language of BIOCHAM three models of respectively the mammalian cell cycle, p53-based DNA-damage repair, and irinotecan intracellular PK/PD, using BIOCHAM's procedure for optimizing parameters with respect to biological properties formalized in temporal logic. Model-checking techniques proved to be particularly suitable for studying the irinotecan metabolic pathways. It let us get a better understanding of the drug influence on the mammalian cell cycle and infer some properties to be exploited in the drug therapy, such us optimal injection times and doses.

In order to study how irinotecan interferes with tumor cell proliferation, a further component should be taken into account, that is, the *circadian clock*, which regulates the synchronous progression of cells through each stage of the cell cycle, determining the daily time windows during which cells can traverse certain phases of the cell cycle [28]. The circadian synchronization of cell cycle progression, which characterizes healthy normal tissues, is often altered in cellular tissues affected by malignant tumors [29]. Roughly speaking, it means that, unlike safe cells, different tumor cells at the same time can be in different phases of the cell cycle.

As a matter of fact, the behaviour of a single cell is not sensibly affected by malignant tumors, but the presence of a disease is generally detected by observing groups of at least one hundred cells. At the level of single cells, a slowing down of biochemical reactions is the only observed phenomenon in presence of tumor. An interesting extension of the work so far presented would consist in adding to our coupled model a new "circadian clock module" to synchronize cells and simulating the resulting model on a group of about one hundred cells. The three already present modules would remain almost unchanged, apart from a speed reduction of the reactions inserted in Figure 2 to link them.

## Acknowledgements

This work is partly supported by EU FP6 STREP project TEMPO on cancer chronotherapies. We acknowledge fruitful discussions with the partners of this project.

# References

1. Alberts, B., Alexander, J., Julian, L., Martin, R., Keith, R.: In: Garland (ed.) Molecular biology of the cell (2008)
2. Antoniotti, M., Policriti, A., Ugel, N., Mishra, B.: Model building and model checking for biochemical processes. Cell Biochem. Biophys. 38(3), 271–286 (2003)
3. Batt, G., Belta, C., Weiss, R.: Temporal Logic Analysis of Gene Networks under Parameter Uncertainty. IEEE Transactions of Automatic Control 53, 215–229 (2008)
4. Blinov, M.L., Faeder, J.R., Goldstein, B., Hlavacek, W.S.: BioNetGen: software for rule-based modeling of signal transduction based on the interactions of molecular domains. Bioinformatics 20(17), 3289–3291 (2004)
5. Calder, M., Vyshemirsky, V., Gilbert, D., Orton, R.: Analysis of signalling pathways using the continuous time Markov chains. In: Priami, C., Plotkin, G. (eds.) Transactions on Computational Systems Biology VI. LNCS (LNBI), vol. 4220, pp. 44–67. Springer, Heidelberg (2006)
6. Calzone, L., Chabrier-Rivier, N., Fages, F., Soliman, S.: Machine learning biochemical networks from temporal logic properties. In: Priami, C., Plotkin, G. (eds.) Transactions on Computational Systems Biology VI. LNCS (LNBI), vol. 4220, pp. 68–94. Springer, Heidelberg (2006); CMSB 2005 Special Issue
7. Calzone, L., Fages, F., Soliman, S.: BIOCHAM: An Environment for Modeling Biological Systems and Formalizing Experimental Knowledge. Bioinformatics 22, 1805–1807 (2006)
8. Chabrier, N., Fages, F.: Symbolic model checking of biochemical networks. In: Priami, C. (ed.) CMSB 2003. LNCS, vol. 2602, pp. 149–162. Springer, Heidelberg (2003)
9. Chickermane, V., Ray, A., Sauro, H.M., Nadim, A.: A model for p53 Dynamics Triggered by DNA damage. Siam Journal on Applied Dynamical Systems 6(1), 61–78 (2007)
10. Ciliberto, A., Novak, B., Tyson, J.J.: Steady States on Oscillations in the p53/Mdm2 Network. Cell Cycle 4(3), 488–493 (2005)
11. Cimatti, A., Clarke, E., Giunchiglia, E., Giunchiglia, F., Pistore, M., Roveri, M., Sebastiani, R., Tacchella, A.: NuSMV 2: An openSource tool for symbolic model checking. In: Brinksma, E., Larsen, K.G. (eds.) CAV 2002. LNCS, vol. 2404, p. 359. Springer, Heidelberg (2002)
12. Clarke, E.M., Faeder, J.R., Langmead, C.J., Harris, L., Jha, S.K., Legay, A.: Statistical Model Checking in *Biolab*: Applications to the Automated Analysis of T-Cell Receptor Signaling Pathway. In: Heiner, M., Uhrmacher, A.M. (eds.) CMSB 2008. LNCS (LNBI), vol. 5307, pp. 231–250. Springer, Heidelberg (2008)
13. Clarke, E.M., Grumberg, O., Peled, D.A.: Model Checking. MIT Press, Cambridge (1999)
14. Dimitrio, L.: Irinotecan: Modelling intracellular pharmacokinetics and pharmacodynamics. M2 master thesis (in French, English summary), University Pierre-et-Marie-Curie and INRIA internal report (June 2007)
15. Eker, S., Knapp, M., Laderoute, K., Lincoln, P., Meseguer, J., Sönmez, M.K.: Pathway logic: Symbolic analysis of biological signaling. In: Proc. of the seventh Pacific Symposium on Biocomputing, pp. 400–412 (2002)
16. Emerson, E.A.: Temporal and modal logic. In: van Leeuwen, J. (ed.) The Handbook of Theoretical Computer Science, vol. B (ch. 16), pp. 995–1072. Elsevier Science Publisher, Amsterdam (1990)

17. Fages, F.: Temporal logic constraints in the biochemical abstract machine biocham (invited talk). In: Hill, P.M. (ed.) LOPSTR 2005. LNCS, vol. 3901, pp. 1–5. Springer, Heidelberg (2006)

18. Fages, F., Rizk, A.: From Model-Checking to Temporal Logic Constraint Solving. In: CP 2009: Proc. of the fifteenth International Conference on Principles and Practice of Constraint Programming. LNCS. Springer, Heidelberg (to appear, 2009)

19. Fages, F., Soliman, S., Chabrier-Rivier, N.: Modelling and querying interaction networks in the biochemical abstract machine BIOCHAM. Journal of Biological Physics and Chemistry 4(2), 64–73 (2004)

20. Geva-Zatorsky, N., Rosenfeld, N., Itzkovitz, S., Milo, R., Sigal, A., Dekel, E., Yarnitzky, T., Liton, Y., Polak, P., Lahav, G., Alon, U.: Oscillations and variability in the p53 system. Molecular System Biology, 2006.0033 (2006)

21. Gibson, M.A., Bruck, J.: Efficient exact stochastic simulation of chemical systems with many species and many channels. Journal of Physical Chemistry 104, 1876–1889 (2000)

22. Gilbert, D., Heiner, M., Lehrack, S.: A unifying framework for modelling and analysing biochemical pathways using Petri nets. In: Calder, M., Gilmore, S. (eds.) CMSB 2007. LNCS (LNBI), vol. 4695, pp. 200–216. Springer, Heidelberg (2007)

23. Gillespie, D.T.: General method for numerically simulating stochastic time evolution of coupled chemical-reactions. Journal of Computational Physics 22, 403–434 (1976)

24. Hansson, H., Jonsson, B.: A logic for reasoning about time and reliability. Formal Aspects of Computing 6, 512–535 (1994)

25. Heath, J.K., Kwiatkowska, M., Norman, G., Parker, D., Tymchyshyn, O.: Probabilistic model checking of complex biological pathways. In: Priami, C. (ed.) CMSB 2006. LNCS (LNBI), vol. 4210, pp. 32–47. Springer, Heidelberg (2006)

26. Hucka, M., et al.: The systems biology markup language (SBML): A medium for representation and exchange of biochemical network models. Bioinformatics 19, 524–531 (2003)

27. Kwiatkowska, M.Z., Norman, G., Parker, D.: Prism 2.0: A tool for probabilistic model checking. In: First International Conference on Quantitative Evaluation of Systems (QEST 2004), pp. 322–323. IEEE Computer Society, Los Alamitos (2004)

28. Matsuo, T., Yamaguchi, S., Mitsui, S., Emi, A., Shimoda, F., Okamura, H.: Control mechanism of the circadian clock for timing of cell division in vivo. Science 302, 255–259 (2003)

29. Mormont, M.C., Levi, F.: Circadian system alterations during cancer processes. International Journal of Cancer 70, 241–247 (1997)

30. Novák, B., Tyson, J.J.: A model for restriction point control of the mammalian cell cycle. Journal of Theoretical Biology 230, 563–579 (2004)

31. Ohdo, S., Makinosumi, T., Ishizaki, T., Yukawa, E., Higuchi, S., Nakano, S., Ogawa, N.: Cell Cycle-Dependent Chronotoxicity of Irinotecan Hydrochloride in Mice. Journal of Pharmacology and Experimental Terapeutics 283(3), 1383–1388 (1997)

32. Phillips, A., Cardelli, L.: A correct abstract machine for the stochastic pi-calculus. In: Proc. of Concurrent Models in Molecular Biology (Bioconcur 2004), affiliated with CONCUR 2004 (2004)

33. Piazza, C., Antoniotti, M., Mysore, V., Policriti, A., Winkler, F., Mishra, B.: Algorithmic Algebraic Model Checking I: Challenges from Systems Biology. In: Etessami, K., Rajamani, S.K. (eds.) CAV 2005. LNCS, vol. 3576, pp. 5–19. Springer, Heidelberg (2005)

34. Pommier, Y.: Camptothecins and Topoisomerase I: A Foot in the Door. Targeting the genome beyond Topoisomerase I with camptothecins and Novel Anticancer drugs: importance of DNA Replication, Repair and Cell Cycle Checkpoints. Preprint NIH, NCI (2004), http://discover.nci.nih.gov/pommier/Pommier.Top1.pdf

35. Regev, A., Silverman, W., Shapiro, E.Y.: Representation and simulation of biochemical processes using the pi-calculus process algebra. In: Proc. of the sixth Pacific Symposium of Biocomputing, pp. 459–470 (2001)

36. Rizk, A., Batt, G., Fages, F., Soliman, S.: A general computational method for robustness analysis with applications to synthetic gene networks. BioInformatics (July 2009) (to appear)

37. Rizk, A., Batt, G., Fages, F., Soliman, S.: On a Continuous Degree of Satisfaction of Temporal Logic Formulae with Applications to Systems Biology. In: Heiner, M., Uhrmacher, A.M. (eds.) CMSB 2008. LNCS (LNBI), vol. 5307, pp. 251–268. Springer, Heidelberg (2008)

38. Zhou, Y., Gwadry, F.G., Reinhold, W.C., Miller, L.D., Smith, L.H., Scherf, U., Liu, E.T., kohn, K.W., Pommier, Y., Weinstein, J.N.: Transcriptional Regulation of Mitotic Genes by Camptothecin-induced DNA Damage: Microarray Analysis of Doseand Time-dependent Effects. Cancer Research 62, 1668–1695 (2002)

39. Zwillinger, D.: Handbook of Differential Equations, 3rd edn. Academic Press, Boston (1997)

# The $\kappa$-Lattice: Decidability Boundaries for Qualitative Analysis in Biological Languages

Giorgio Delzanno[1], Cinzia Di Giusto[2], Maurizio Gabbrielli[2], Cosimo Laneve[2],
and Gianluigi Zavattaro[2]

[1] Dipartimento di Informatica e Scienze dell'Informazione,
Università di Genova, Italia
[2] Dipartimento di Scienze dell'Informazione,
Università di Bologna, Italia

**Abstract.** The $\kappa$-calculus is a formalism for modelling molecular biology where molecules are terms with internal state and sites, bonds are represented by shared names labelling sites, and reactions are represented by rewriting rules. Depending on the shape of the rewriting rules, a lattice of dialects of $\kappa$ can be obtained. We analyze the expressive power of some of these dialects by focusing on the thin boundary between decidability and undecidability for problems like reachability and coverability.

## 1 Introduction

For this reason, as in other applications of concurrency, an important foundational issue is the study of dialects for which qualitative analysis is computable in an effective way and the isolation of minimal fragments in which it is proved to be impossible. $\kappa$ is a formalism for modelling molecular biology where molecules are terms with internal state and with sites, bonds are represented by names that label sites, and reactions are represented by rewriting rules. For example, $EGFR[tk^0](1^z)$ represents a molecule of species $EGFR$ that is not phosphorilated – the internal state $tk$ is 0 – and that is bond to another molecule – its site 1 is labelled with a name $z$. The reaction in Fig. 1 defines the first step of the *Receptor Tyrosine Kinase* (RTK) growth factor $EGF$ (a dimeric form of $EGF$ binds two receptors $EGFR$, thus phosphorylating the tyrosine kinase site – $tk$ switches from 0 to 1). This reaction is rendered by the following $\kappa$ rule:

$$EGF(1^x + 2^y), EGF(1^x + 2^z), EGFR(1^y), EGFR[tk^0](1^z) \\ \rightarrowtail EGF(1^x + 2^y), EGF(1^x + 2^z), EGFR(1^y), EGFR[tk^1](1^z) \tag{1}$$

Qualitative problems, such as reachability of a given solution, turn out to be undecidable in $\kappa$. Therefore one is either compelled to design approximated analyses or to study these properties in dialects of $\kappa$. We choose the second direction, thus yielding a number of precise analyses that do not abstract away either from the multiplicity of molecules or from the exact structure of complexes.

P. Degano and R. Gorrieri (Eds.): CMSB 2009, LNBI 5688, pp. 158–172, 2009.

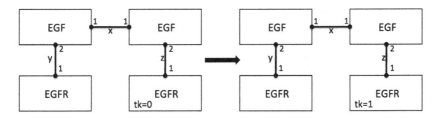

**Fig. 1.** Representation of the $\kappa$-rule (1)

To this aim, we consider a number of $\kappa$ dialects that, as we discuss in the following, take inspiration from biological phenomena such as the *molecular self-assembly* [1] or the *DNA branch migration* [2]. These dialects are ordered into a lattice by the sublanguage relation – see Figure 2 disregarding the ovals. Let us unravel the lattice with the restrictions imposed on $\kappa$ to obtain the sublanguages $\kappa^{-n}$, $\kappa^{-d}$, and $\kappa^{-d-u}$. The calculus $\kappa^{-n}$ follows by removing any form of destruction of molecules (the number of molecules never decrease). This fragment naturally models those systems where molecules always keep their "identity" even when they are part of a complex because, for example, they can subsequently dissociate from the complex. This is the case of polymers, that is chemical structures obtained by joining monomers that react on complementary surfaces. A simple polymerisation – the linear bidirectional one, where the complementary surfaces of monomers are two (that we respectively call $l$ and $r$ in the following) – is modelled by the following $\kappa^{-n}$ rules:

$$A(r), A(l) \;\rightarrowtail\; A(r^x), A(l^x) \tag{2}$$
$$A(r^x), A(l^x) \;\rightarrowtail\; A(r), A(l) \tag{3}$$

The reaction (2) defines polymerization (the creation of a bond between two monomers with free complementary surfaces); (3) defines depolymerization (the destruction of the bond, but not of the monomers).

The additional restriction yielding $\kappa^{-d}$ is the one that disallows the removal of bonds (depolymerizations are forbidden). This restriction is inspired by molecular self-assembly, which is a process where molecules, initially unbound, adopt a defined arrangement. The DNA-origami method is a popular example of self-assembly that allows to create arbitrary two-dimensional shapes, such as Borromean rings [3], using DNA. In $\kappa^{-d}$ self-assembly is directly enforced because bonds cannot be broken. The last dialect along this axis, called $\kappa^{-d-u}$, is obtained by considering molecules without internal states. In several cases such states are not useful. An example is the DNA self-assembly governed by the Watson-Crick complementary base pairing [4]. We also consider two other subcalculi that forbid destructions of molecules and bonds: $\kappa^{-d-i}$ and $\kappa^{-d-u-i}$. These dialects are obtained from $\kappa^{-d}$ and $\kappa^{-d-u}$, respectively, by restricting reductions to those that never verify the connectedness of reactants. For example, the polymerization (2) is a reaction of this type. It turns out that the Watson-Crick complementary base pairing may be defined in $\kappa^{-d-u-i}$.

Our analysis also takes into account a different axis. In [5] a new reaction rule has been introduced, called *exchange*. According to this reaction, the interaction between two molecules may flip a bond from one to the other. For example, the reader may consider the case where a thief molecule $T$ may connect to a third site of the monomer $A$ and steals the polymer connected to the site $l$ of $A$:

$$T(t+s), A(h) \rightarrowtail T(t^x + s), A(h^x) \qquad (4)$$

$$T(t^x + s), A(h^x + l^y) \rightarrowtail T(t^x + s^y), A(h^x + l) \qquad (5)$$

(reaction 5 is an example of bond flipping). Bond flipping allows us to model other interesting DNA systems, such as those based on branch migration used to create, for instance, a nanoscale biped walking along a DNA strand [6]. The calculi including bond flipping are made evident with the superscript $+bf$. Finally, we consider also a more liberal form of flipping, called free flipping (see Figure 3), in which flipping can occur also between two unbound molecules. With free flipping, the thief molecule $T$ can steal the polymer to a monomer without previously connecting to it:

$$T(s), A(l^y) \rightarrowtail T(s^y), A(l) \qquad (6)$$

For all of the 14 dialects of $\kappa$ we investigate three problems: the *Reachability Problem* (RP), the *Simple Coverability Problem* (SCP) and the *Coverability Problem* (CP). The RP is the decision problem associated to the existence of a derivation (simulation) from an initial solution to a target. As shown in [7,8,9], this problem is of high relevance for validation of formal models of biological systems. The SCP is the decision problem associated to the existence of a derivation from an initial solution to a target with given components, regardless of their multiplicity. In other words the shape of complexes in the target solution is taken from a set fixed a priori. SCP is a generalization of the decision problem associated to the static analysis considered in [10]. Finally, CP is the decision problem associated to the existence of a derivation from an initial solution to a target that contains given components.

Our results about the (un)decidability of RP, SCP, and CP in the $\kappa$ lattice are illustrated in Figure 2.

The undecidability results are proved by modelling Turing complete formalisms in the calculi, while the decidability results are proved by reduction to decidable properties in finite state systems or Petri-nets. As far as the undecidability results are concerned, the most surprising one is the undecidability of CP in $\kappa^{-d-u}$. We prove that this very poor fragment of $\kappa$ – in which molecules have no state and bonds can be neither destroyed nor flipped – is powerful enough to encode Two Counter Machines [11], a Turing complete formalism. It is also interesting to observe that this result about $\kappa^{-d-u}$ relies on the possibility to test at least the presence of bonds. In fact, $\kappa^{-d-u-i}$ is no longer Turing complete because CP is decidable for this fragment (CP allows one to test whether a certain complex, for instance representing the termination of a computation, can be produced). While the dialects that include $\kappa^{-d-u}$ are Turing complete, many of them retain decidable SCP and/or RP properties. These

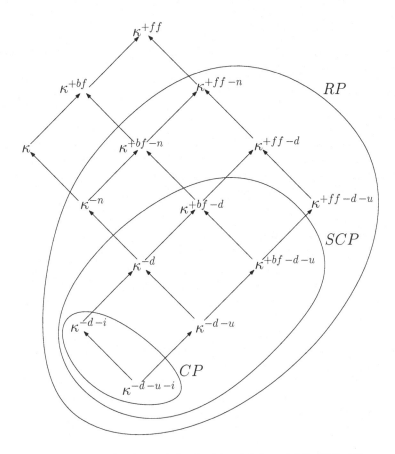

**Fig. 2.** The $\kappa$ lattice and the (un)decidability of RP, SCP, CP

facts, apparently contrasting with Turing universality of the calculi, are consequences of the following monotonic properties: reactions cannot decrease either (i) the total number of molecules in the solution or (ii) the size of the complexes in the solution. In the calculi satisfying the form of monotonicity (i) we show that it is possible to compute an upper-bound to the number of molecules in the solutions of interest for the analysis of RP. In this way, we reduce our analysis to a finite state system. For the calculi satisfying the form of monotonicity (ii) we show that it is possible to compute an upper-bound to the size of the complexes in the solutions of interest for the analysis of SCP. In this case, even if it is not possible to reduce to a finite state system (because there is no upper-bound to the number of instances of the complexes in the solutions of interest), we can reduce to Petri-nets in which reachability and coverability are decidable.

The paper is organised as follows: Section 2 recalls $\kappa$, its fragments and the needed terminology. Section 3 discusses the separation results between the fragments of $\kappa$. Section 4 discusses related contributions in literature. Section 5 concludes with few final remarks.

## 2    Preliminaries

This section introduces $\kappa$ and its dialects, together with the terminology that is necessary in the sequel.

**$\kappa$-calculi.** Two countable sets of *species* $A, B, C, \ldots$, and of *bonds* $x, y, z, \ldots$ are assumed. Species are sorted according to the number of sites, , and fields $h, i, j, \ldots$ they possess.

Sites may be either bound to other sites or unbound, i.e. not connected to other sites. The configuration of sites are defined by partial maps, called *interfaces* and ranged over by $\sigma, \rho, \ldots$. The interfaces associate to sites either a bond or a special empty value $\varepsilon$, which models the fact that the site is unbound.

For instance, if $A$ is a species with three sites, $(2 \mapsto x; 3 \mapsto \varepsilon)$ is one of its interfaces. This map is written $2^x + 3$ (the $\varepsilon$ is always omitted). We notice that this $\sigma$ does not define the state of the site 1, which may be bound or not. Such (proper) partial maps are used in reaction rules in order to abstract from sites that do not play any role in the reactions (similar for evaluations, see below). In the following, when we write $\sigma + \sigma'$ we assume that the domains of $\sigma$ and $\sigma'$ are disjoint. The functions $dom(\cdot)$ and $ran(\cdot)$ return the domain and the range of a function.

Fields represent the internal state of a species. The values of fields are also defined by partial maps, called *evaluations*, ranged over by $u, v, \ldots$. For instance, if $A$ is a species with three fields, $\{1 \mapsto 5; 2 \mapsto 0; 3 \mapsto 4\}$, shortened into $1^5 + 2^0 + 3^4$, is a possible evaluation. We assume there are finitely many internal states, that is every field is mapped into a finite set of values. As for interfaces, $u + v$, we implicitly assume that the domains of $u$ and $v$ are disjoint.

**Definition 1.** *A molecule $A[u](\sigma)$ is a term where $u$ and $\sigma$ are a total evaluation and a total interface of $A$.*
*Solutions, ranged over by $S$, $T$, $\ldots$, are defined by: $S ::= A[u](\sigma) \mid S, S$. Bonds in solutions occur at most twice; in case bonds occur exactly twice the solution is proper. A pre-solution is a sequence of terms $A[u](\sigma)$ where $u$ and $\sigma$ are partial functions and bonds occur at most twice. A pre-solution is proper if (similarly as before) bonds occur exactly twice. The set of bonds in $S$ is denoted $\mathbf{bonds}(S)$.*

In the rest of the paper the composition operator "," is assumed to be associative, so $(S, S'), S''$ is equal to $S, (S', S'')$ (therefore parentheses will be always omitted).

Let $\sigma \le \sigma'$ if $dom(\sigma) = dom(\sigma')$ and, for every $i$, if $\sigma(i) \ne \varepsilon$ then $\sigma(i) = \sigma'(i)$ (the two interfaces may differ on sites mapped to the empty value $\varepsilon$ by $\sigma$ as $\sigma'$ may map such sites to bonds).

Reactions have the shape $\mathsf{L} \rightarrowtail \mathsf{R}$, where $\mathsf{L}$ and $\mathsf{R}$ are pre-solutions called reactants and products, respectively. The general shape of reactions is defined in the next definition. Following [5], we extend the definition of [12] with exchange reactions, thus the calculus is an extension of the $\kappa$-calculus.[1]

---

[1] Another difference with [12] is that we allow newly produced molecules unbound from existing ones.

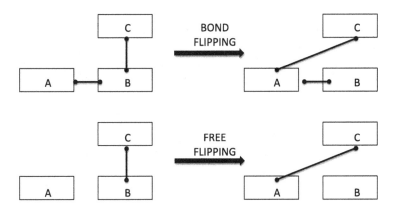

**Fig. 3.** Bond flipping and free flipping

**Definition 2.** *Reactions of the $\kappa^{+ff}$ calculus – the $\kappa$ calculus with free flipping rules – are either creations $C$, or destructions $D$, or exchanges $E$.*
*The format of creations is*

$$A_1[u_1](\sigma_1), \ldots, A_n[u_n](\sigma_n)$$
$$\rightarrowtail A_1[u_1'](\sigma_1'), \ldots, A_n[u_n'](\sigma_n'), B_1[v_1](\phi_1), \ldots, B_k[v_k](\phi_k)$$

*where, for every $i$, $dom(u_i) = dom(u_i')$, $\sigma_i \leq \sigma_i'$, and $v_i$ and $\phi_i$ are total. Reactants and products are proper.*
*The format of destructions is*

$$A_1[u_1](\sigma_1), \ldots, A_n[u_n](\sigma_n) \rightarrowtail A_{i_1}[u_{i_1}'](\sigma_{i_1}'), \ldots, A_{i_m}[u_{i_m}'](\sigma_{i_m}')$$

*where $i_1, \ldots, i_m$ is an ordered sequence in $[1 \ldots n]$, for every $i_j$, $dom(u_{i_j}) = dom(u_{i_j}')$, $\sigma_{i_j} \geq \sigma_{i_j}'$, and if $i_j \notin \{i_1, \ldots, i_m\}$ then $\sigma_{i_j}$ is total. Reactants and products are proper.*
*The format of exchanges is*

$$A[u](a^x + \sigma), B[v](b + \rho) \rightarrowtail A[u'](a + \sigma), B[v'](b^x + \rho)$$

*where $ran(\sigma) = ran(\rho)$.*

Creations may change state, produce new bonds between two unbound sites, or synthesise new molecules. Destructions behave the other way around. Exchanges are reminiscent of the $\pi$ calculus because they define a migration of a bond from one reactant to the other. We distinguish two types of exchanges: the one occurring between connected molecules, called *(connected) bond flipping*, and the one occurring between disconnected molecules, called *free (bond) flipping*. These are illustrated below:

The operational semantics of $\kappa^{+ff}$ calculus uses the following two definitions:

– the *structural equivalence* between solutions, denoted $\equiv$, is the least one satisfying (we remind that solutions are already quotiented by associativity of ","):

- $S, T \equiv T, S$;
- $S \equiv T$ if there exists a renaming $\imath$ on bonds such that $S = \imath(T)$.
  - $A_1[u_1 + u_1'](\sigma_1 \circ \imath + \sigma_1'), \ldots, A_n[u_n + u_n'](\sigma_n \circ \imath + \sigma_n')$ is an $(\imath, u_1', \cdots, u_n', \sigma_1', \cdots, \sigma_n')$ instance of $A_1[u_1](\sigma_1), \ldots, A_n[u_n](\sigma_n)$ if $\imath$ is a renaming on bonds and the maps $u_j + u_j'$ and $\sigma_j \circ \imath + \sigma_j'$ are total with respect to the species $A_j$.

**Definition 3.** *The reduction relation of the $\kappa^{+ff}$ calculus, written $\to$, is the least one satisfying the rules:*

- *let* $\mathsf{L} \rightarrowtail \mathsf{R}$ *be a reaction of* $\kappa^{+ff}$, $S$ *be an* $(\imath, \tilde{u}, \tilde{\sigma})$-*instance of* $\mathsf{L}$, *and* $T$ *be an* $(\imath, \tilde{u}', \tilde{\sigma}')$-*instance of* $\mathsf{R}$. *Then* $S \to T$;
- *let* $S \to T$ *and* $(\textbf{bonds}(T) \setminus \textbf{bonds}(S)) \cap \textbf{bonds}(R) = \emptyset$, *then* $S, R \to T, R$;
- *let* $S \equiv S'$, $S' \to T'$, *and* $T' \equiv T$, *then* $S \to T$.

The $\kappa^{+ff}$ calculus groups several sub-calculi that have in turn simpler formats of rules. We have already depicted in Figure 2 the fragments we study. We move from $\kappa^{+ff}$ along two different axes:

1. along one axis we restrict the shape of destructions rules:
   - the superscript $-n$: we restrict destructions by letting $i_m = n$ (i.e. forbidding cancellations of molecules),
   - the superscript $-d$: we remove destructions,
   - the superscript $-d-u$: we remove destructions and consider species with no fields,
   - the superscript $-d-u-i$: we remove destructions, fields and we restrict creations and exchanges such that no bond occurs in the left-hand side except the flipping one,
2. along the other axis we restrict the capabilities of the exchange rule:
   - the superscript $+bf$: we restrict exchanges by allowing bond-flipping only,
   - no superscript: we remove exchanges.

Some of the combinations are empty. For example, a calculus without checks of bonds and with cancellation of bonds is meaningless as, in order to remove one bond, it is necessary to test its presence first.

The reader may refer to the introduction for formalisations of relevant biological systems written in these calculi.

**Decision problems for qualitative analysis.** A first basic qualitative property is whether a solution eventually produces "something relevant" or not. Clearly this "something relevant" can be defined in a variety of ways. In this paper we consider its formalisation in terms of reachability and coverability, two standard properties which have been extensively investigated in many concurrent formalisms. Few preliminary notions are required.

**Definition 4 (Complex).** *Given a proper solution, a complex is a sub-solution that is connected (there is a path of bonds connecting every pair of molecules therein) and proper. Two complexes in a solution are equal if they are structurally equivalent.*

Let $S(S)$ be the set of different complexes in $S$; let also $\to^*$ be the transitive and reflexive closure of $\to$.

**Definition 5. RP:** *the* reachability *problem of $T$ from a proper solution $S$ checks the existence of $R$ such that $S \to^* R$ and $R \equiv T$;*
**SCP:** *the* simple coverability *problem of $T$ from a proper solution $S$ checks the existence of $R$ such that $S \to^* R$ and $S(R) = S(T)$ and $R \equiv T, T'$, for some $T'$;*
**CP:** *the* coverability *problem of $T$ from a proper solution $S$ checks the existence of $R$ such that $S \to^* R$ and $R \equiv T, T'$, for some $T'$.*

# 3 (Un)Decidability Results for $\kappa$ Dialects

In this section we study the (un)decidability of RP, SCP, and CP in the $\kappa$ lattice of Figure 2. The overall results represented in that figure are the consequences of theorems that we detail in the remainder of this section. For each decidability region – one for RP, one for SCP, and one for CP – we prove that the corresponding property is decidable in the top language of the region and undecidable in the bottom language(s) among those not included in the region.

We separate the presentation of our results in two subsections, the first one is devoted to decidability, the latter to undecidability.

## 3.1 Decidability Results

The proofs of decidability follow by reduction to decidable problems in either finite state systems or Place/Transition Petri nets (P/T nets). These nets are an interesting infinite state model for the representation and analysis of parallel processes because they retain several decidability problems, such as reachability or coverability [13]. We recall here the basic notation, for a full description of this computational model see [14].

**Definition 6.** *A P/T net is a tuple $N = (S, T, F, m_0)$, where $S$ and $T$ are finite sets, called* places *and* transitions, *respectively, such that $S \cap T = \emptyset$. A finite multiset over the set $S$ of places is called a* marking, *and $m_0$ is the initial marking. $F$ is the transition function associating to each transition $t$ two markings called the* pre-set *and the* post-set *of $t$.*

*The marking $m$ of a P/T net can be modified by means of transitions firing: a transition with pre-set $m'$ and post-set $m''$ can fire if $m' \subseteq m$; upon transition firing the new marking of the net becomes $(m \setminus m') \cup m''$ where $\setminus$ and $\cup$ are the difference and union operators for multisets, respectively.*

Our first positive result is for the $\kappa^{+f\!f -n}$ fragment.

**Theorem 1.** *RP is decidable in $\kappa^{+f\!f -n}$.*

*Proof.* We reduce RP to the reachability problem in a finite state system. Let $\mathcal{R}$ be a set of $\kappa^{+f\!f -n}$ reactions and let $S$ and $T$ be two proper solutions. We notice that, in order for $S \to^* T$, all intermediary solutions traversed by the computation must have a number of molecules which is less or equal to the number $n_T$ of molecules in $T$. This is because in $\kappa^{+ff - n}$ it is not possible to delete molecules.

Let $\mathcal{A}$ be the set of species occurring either in $S$ or in a rule of $\mathcal{R}$. Let also $\mathsf{set}^T(\mathcal{A})$ be the set of (proper) solutions with a number of molecules less than $n_T$. This set is finite *up-to structural equivalence* because the number of sites and fields of species is finite, the values of fields is finite, and the possible combinations of bonds is finite, as well. By mapping every solution $R$ to its canonical representative in the structural equivalence class, called $[R]$, we can build a *finite state system* $\mathsf{FSS}_T$ such that, by Definition 3, given two solutions in $\mathsf{set}^T(\mathcal{A})$, $R \to R'$ if and only if $[R] \to [R']$. We conclude the proof by observing that $S \to^* T$ if and only if $[S] \to^* [T]$, and this latter property is decidable in $\mathsf{FSS}_T$.                                                                    □

The proof technique adopted above cannot be used to prove the decidability of the $SCP$ problem for a given target $T$ in $\kappa^{+bf - d}$. As a matter of fact, $SCP$ allows one to specify only lower bounds, and no upper bounds, to the number of instances of complexes (thus also of the molecules) in the target solution. For this reason finite state systems are not sufficiently expressive to model the computations of interest. Nevertheless, we can move to P/T nets because it is possible to compute a finite set $\mathsf{SET}^T(\mathcal{A})$ containing the kinds of complexes to be considered in the reachability analysis. This set turns out to be finite since in $\kappa^{+bf - d}$ the size of one complex can never decrease and the size of the biggest complex in $T$ fixes an upper bound to the size of the complexes in $\mathsf{SET}^T(\mathcal{A})$. The idea is then to map each complex in $\mathsf{SET}^T(\mathcal{A})$ into one place, and define transitions according to the considered reactions. Hence we have the following theorem:

**Theorem 2.** *SCP is decidable for $\kappa^{+bf - d}$.*

The P/T net described above cannot be used to prove the decidability of the $CP$ problem for a given target $T$ in $\kappa^{+bf - d}$. In fact, according to $CP$, the target $T$ indicates only part of the complexes to be reached. Thus, the reached solution that contains the target complexes, could also contain other complexes of size greater than the size $m_T$ of the biggest complex in $T$. Nevertheless, since in $\kappa^{-d-i}$ bond names cannot be tested in the reactants of a reaction, we can remove from the P/T net representation of those complexes the structure of their bonds, and thus consider only the states and the free sites of their molecules. More precisely, the P/T net described above is now extended with places $\widehat{A}[u](\sigma)$ (for every species $A$, every evaluation $u$, and with partial functions $\sigma$ mapping every site to $\varepsilon$) used to represent the molecules in complexes of size greater than $m_T$. Due to the finiteness of species, evaluations, and sites we have that this additional set of places is finite. Moreover the set of transitions is straightforwardly extended to cope with the new places. Hence it is possible to prove the following:

**Theorem 3.** *CP is decidable in $\kappa^{-d-i}$.*

## 3.2  Undecidability Results

Our undecidability results follow by reduction to undecidable problems such as the halting problem for Two Counter Machines (2CMs), which is a Turing equivalent formalism. A 2CM [11] is a machine with *two registers* $R_1$ and $R_2$ holding arbitrary large natural numbers and a *program* $P$ consisting of a finite sequence of numbered instructions of the following type:

- $j : \mathsf{Succ}(R_i)$: increments $R_i$ and goes to the instruction $j + 1$;
- $j : \mathsf{DecJump}(R_i, l)$: if the content of $R_i$ is not zero, then decreases it by 1 and goes to the instruction $j + 1$, otherwise jumps to the instruction $l$;
- $j : \mathsf{Halt}$: stops the computation and returns the value in the register $R_1$.

A state of the machine is given by a tuple $(j, v_1, v_2)$ where $i$ indicates the next instruction to execute (the program counter) and $v_1$ and $v_2$ are the contents of the two registers. The user has to provide the initial state of the machine. In the rest of the paper, we consider 2CMs in which registers are initially set to zero and where the instruction 0 is $\mathsf{Halt}$. Our first negative result is for reachability of a solution in $\kappa$.

**Theorem 4.** *RP is undecidable in $\kappa$.*

*Proof.* We reduce the termination problem for 2CMs to RP. Let $M$ be a 2CM with $n$ instructions. To encode it in $\kappa$ we use five species:

1. $P$ is the program counter; it retains one field with values in $[0, \dots, n]$ and no site;
2. $Z_1$ and $Z_2$, both with one site, represent the value 0;
3. $R_1$ and $R_2$, both with two sites, represent the unity to be added to or removed from registries.

Let $j, l \in [0..n]$ and let $i \in \{1, 2\}$. The encoding $[\![\cdot]\!]_\kappa$ is defined as follows:

$$[\![j : \mathsf{Succ}(R_i)]\!]_\kappa = \begin{cases} P[1^j], Z_i(1) \rightarrowtail P[1^{j+1}], Z_i(1^x), R_i(1^x + 2) \\ P[1^j], R_i(2) \rightarrowtail P[1^{j+1}], R_i(2^x), R_i(1^x + 2) \end{cases}$$

$$[\![j : \mathsf{DecJump}(R_i, l)]\!]_\kappa = \begin{cases} P[1^j], Z_i(1) \rightarrowtail P[1^l], Z_i(1) \\ P[1^j], Z_i(1^x), R_i(1^x + 2) \rightarrowtail P[1^{j+1}], Z_i(1) \\ P[1^j], R_i(2^x), R_i(1^x + 2) \rightarrowtail P[1^{j+1}], R_i(2) \end{cases}$$

$$[\![j : \mathsf{Halt}]\!]_\kappa = \begin{cases} P[1^j], Z_1(1), Z_2(1) \rightarrowtail P[1^0], Z_1(1), Z_2(1) \\ P[1^j], Z_i(1^x), R_i(1^x + 2) \rightarrowtail P[1^j], Z_i(1) \\ P[1^j], R_i(2^x), R_i(1^x + 2) \rightarrowtail P[1^j], R_i(2) \end{cases}$$

It turns out that the 2CM halts if and only if the solution $P[1^0], Z_1(1), Z_2(1)$

□

The encoding of 2CMs described above does not apply to $\kappa^{-n}$ because in this dialect molecules cannot be removed. Nevertheless, we can rephrase the decrement operation of the encoding above by breaking the link between the two last molecules $R_i$ (or the molecules $Z_i$ and $R_i$ in case the register holds 1). Hence we have the following theorem:

**Theorem 5.** *SCP is undecidable in $\kappa^{-n}$.*

We observe that, without using fields and destructions, as in $\kappa^{-d-u}$, it is not possible to reuse the encoding scheme above. Nevertheless, using only creations we can model registers with grids containing two classes of molecules: the molecules of the first class represent units in the register, while those of the second class are used to replace units during decrement instructions. Given the register $R_i$ holding $n$, the corresponding grid contains in the topmost row $n$ molecules of the first class. More precisely, the encoding of the increment increases the topmost row of the grid with a molecule of the first class. The encoding of the DecJump instruction is more complex: The idea is to copy the topmost row of the grid replacing, if possible, one molecule of the first class with one of the second class. If this replacement occurs the subsequent instruction is activated, otherwise a jump is performed. Finally, the encoding of the Halt instruction simply produces the *Halt* molecule. Given this construction it follows that:

**Theorem 6.** *CP is undecidable in $\kappa^{-d-u}$.*

The previous encoding of 2CMs does not allow us to prove the undecidability of $SCP$ in $\kappa^{+\!f\!f-d-u}$ because the exact structure of the grids representing the two registers at the end of the computation is unknown as it depends on the number of increment and decrement instructions that are executed. Nevertheless, in $\kappa^{+\!f\!f-d-u}$ we can use free flipping to "destruct", at the end of the computation, the grids obtaining an unknown amount of complexes with a known structure. More precisely, we extend the previous construction in such a way that the molecule *Halt* triggers the following computation: one molecule is produced for each end of each bond in the grids, and all those ends are then passed to such new molecule. Thus we can state the following theorem:

**Theorem 7.** *SCP is undecidable in $\kappa^{+\!f\!f-d-u}$.*

## 4   Related Work

In this section we discuss some related works by first focusing on formal models specifically proposed for describing biological systems and then considering more generally the fields of term/graph rewriting and process calculi.

As we said in the Introduction, the closest work to this contribution is [10] where a syntactic restriction entailing a form of SCP is proposed. This restriction – $\kappa$ with local rule sets – is orthogonal to the ones proposed in this paper. It does not cover the reachability analysis of finite structures with recurrent patterns, such as finite polymers. In these cases, the analysis in [10] yields an over-approximation of the reachable complexes.

Apart from $\kappa$, the literature reports several proposals for describing (and reasoning on) biological systems, which use a variety of formal tools, including process calculi, term/graph rewriting, (temporal) logic, and rule based languages. However, the expressive power of most of these formalisms is the one of Petri nets. Therefore, the decidability of reachability and coverability problems is an immediate consequence of the corresponding results on Petri nets. Formalisms whose expressive power is similar to $\kappa$, miss results analogous to those contained in this paper. For example, the biochemical abstract machine Biocham [15,16] is a rule-based model similar to $\kappa$. However reactions are constrained to specify completely the reagent solution, unlike $\kappa$ where reactions partially specify reactants and products. It is worth noticing that the Biocham constraint do not allow finite descriptions of rules creating polymers of arbitrary length. As a consequence, when considering purely qualitative aspects, i.e. removing kinetic quantities, the Biocham can be reduced to a classical Petri net [15].

Another rule-based model for describing and analysing biological processes is Pathway Logic [17,18]. This model is based on rewrite logic, which allows to describe biological entities and their relations at different levels of abstractions and granularity by using elements of an algebraic data type (to describe states) and rewrite rules (to describe transitions between states and therefore behaviours). Even though Pathway Logic models of biological processes are developed in Maude system, which is Turing complete, yet the analysis of biological systems uses the, so called, Pathway Logic Assistant for representing models in terms of Petri Nets [18]. Therefore, also in this case, the relevant decidability results derive from the analogous results on Petri nets. This is the case also for the model used in [19]. A different model, based on graph transformation has been proposed by Blinov et al. [20]. However, in this case, the relevant properties (e.g. membership of a given species in a reaction network) are semi-decidable and we are not aware of suitable restrictions on the general model that ensure decidability for some of them.

We have not find results regarding the fields of term/graph rewriting and process calculi, from which we can immediately derive the ones obtained here. In particular, for term rewriting systems, the reductions to Petri net reachability can be applied to decide reachability for associative-commutative ground term rewriting (AC) [21] and for Process Rewrite Systems (PRS) [22]. However, AC and PRS are more expressive than Petri nets, but strictly less expressive than Turing machines [22]. On the other hand our positive results are given for fragments of $\kappa$ that are Turing-complete. As such, the set of derivatives of a $\kappa$ solution may not be a regular set of terms. Thus, decision procedures based on tree automata like those proposed in fragments of non-ground term rewriting [23,24,25,26] cannot be applied to the $\kappa$-lattice.

Decidability results for reachability in process calculi like Mobile Ambients, Boxed Ambients, and Bio-ambients are given in [27,28,29,30,31]. These results are obtained for fragments (or for weak semantics) that ensure the monotonicity of the generated ambient structures. In addition they consider process calculi (Mobile/Boxed/Bio Ambients) which operate on tree-like structures and without fresh name generation. This contrasts with the dialect of $\kappa$ of Figure 2, that operate on (possibly cyclic) graph-structures and admit dynamic creation of new names (bonds).

Concerning Graph Rewriting Systems (GRS) there exist folk theorems about reachability that state its undecidability in full-fledged GRS and its decidability for GRS in which rules do not add new nodes. We are not aware of (un)decidability results for decision problems like reachability and coverability in graph rewriting systems with features similar to those considered in our $\kappa$-lattice. The only specific results we are aware of are those given for reachability in context-free graph grammars [32] and for coverability in GRS that are *well-structured* with respect to the graph minor relation [33]. However, we consider here more general rules than those of context-free graph grammars. Furthermore, we do not see how to apply the decision procedure proposed in [33] to languages in the $\kappa$-lattice that, in general, do not enjoy strict compatibility with respect to the graph minor ordering.

## 5   Conclusions

We have investigated three decidability problems for several $\kappa$ dialects. These problems allow one to check whether, starting from a given initial solution, a sequence of reactions described in the $\kappa$ formalism produces a solution having some specific features. Hence our results, summarized in Figure 2, can be seen as a first step in the direction of qualitative analysis of $\kappa$ calculus.

Besides presenting techniques for qualitative analysis, we also characterise the computational power of $\kappa$-like biologically inspired models. In this respect, the main result is that we can remove bond and molecule destruction and the internal state of molecules from $\kappa$ without losing Turing completeness On the contrary, if we remove the possibility to test the presence of one bond in a reaction, the calculus is no longer Turing universal.

Our work can be extended along at least two lines. First, several other fragments of $\kappa$ can be considered for a similar investigation. Notably **nano**$\kappa$ that admits at most two reactants. In particular, our encoding of a 2CM into $\kappa^{-d-i}$ uses ternary (at the left hand side) rules and we conjecture that a 2CM cannot be encoded faithfully into $\kappa^{-d-i}$ with binary rules only.

Second, there are several other interesting properties to investigate, for example a form of coverability where one admits complexes strictly larger than the original ones. In this perspective, we plan to exploit the theory of well structured transition systems [34] as done in [33] to prove decidability of coverability w.r.t. the graph minor relation in classes of graph rewriting systems.

## References

1. Whitesides, G., Mathias, J., Seto, C.: Molecular self-assembly and nanochemistry – a chemical strategy for the synthesis of nanostructures. Science 254, 1312–1319 (1991)
2. Panyutin, I., Hsieh, P.: The kinetics of spontaneous DNA branch migration. Proc. National Academy of Science USA 91(6), 2021–2025 (1994)
3. Mao, C., Sun, W., Seeman, N.: Assembly of Borromean rings from DNA. Nature 386(6621), 137–138 (1997)

4. Watson, J., Crick, F.: A Structure for Deoxyribose Nucleic Acid. Nature 171, 737–738 (1953)
5. Credi, A., Garavelli, M., Laneve, C., Pradalier, S., Silvi, S., Zavattaro, G.: Modelizations and simulations of nano devices in nanok calculus. In: Calder, M., Gilmore, S. (eds.) CMSB 2007. LNCS (LNBI), vol. 4695, pp. 168–183. Springer, Heidelberg (2007)
6. Yin, P., Harry, M., Choi, M., Colby, R., Calvert, R., Pierce, N.: Programming biomolecular self-assembly pathways. Nature 451, 318–322 (2008)
7. Dill, D.L., Knapp, M., Gage, P., Talcott, C.L., Laderoute, K., Lincoln, P.: The pathalyzer: A tool for analysis of signal transduction pathways. In: Eskin, E., Ideker, T., Raphael, B., Workman, C. (eds.) RECOMB 2005. LNCS (LNBI), vol. 4023, pp. 11–22. Springer, Heidelberg (2007)
8. Fages, F.: Symbolic model-checking for biochemical systems. In: Palamidessi, C. (ed.) ICLP 2003. LNCS, vol. 2916, pp. 102–102. Springer, Heidelberg (2003)
9. Peleg, M., Yeh, I., Altman, R.B.: Modelling biological processes using workflow and petri net models. Bioinformatics 18(6), 825–837 (2002)
10. Danos, V., Feret, J., Fontana, W., Krivine, J.: Abstract interpretation of cellular signalling networks. In: Logozzo, F., Peled, D.A., Zuck, L.D. (eds.) VMCAI 2008. LNCS, vol. 4905, pp. 83–97. Springer, Heidelberg (2008)
11. Minsky, M.: Computation: finite and infinite machines. Prentice-Hall, Englewood Cliffs (1967)
12. Danos, V., Laneve, C.: Formal molecular biology. TCS 325(1), 69–110 (2004)
13. Esparza, J., Nielsen, M.: Decidability Issues for Petri Nets-a Survey. Bulletin of the European Association for TCS 52, 245–262 (1994)
14. Reisig, W.: Petri nets: an introduction. Springer, Heidelberg (1985)
15. Fages, F., Soliman, S.: Formal cell biology in biocham. In: Bernardo, M., Degano, P., Zavattaro, G. (eds.) SFM 2008. LNCS, vol. 5016, pp. 54–80. Springer, Heidelberg (2008)
16. Fages, F., Sollman, S., Chabrier-Rivier, N.: Modelling and querying interaction networks in the biochemical abstract machine biocham. Journal of Biological Physics and Chemistry 4, 64–73 (2004)
17. Eker, S., Knapp, M., Laderoute, K., Lincoln, P., Meseguer, J., Sönmez, M.K.: Pathway logic: Symbolic analysis of biological signaling. In: Pacific Symposium on Biocomputing, pp. 400–412 (2002)
18. Talcott, C.L.: Pathway logic. In: Bernardo, M., Degano, P., Zavattaro, G. (eds.) SFM 2008. LNCS, vol. 5016, pp. 21–53. Springer, Heidelberg (2008)
19. Heiner, M., Gilbert, D., Donaldson, R.: Petri nets for systems and synthetic biology. In: Bernardo, M., Degano, P., Zavattaro, G. (eds.) SFM 2008. LNCS, vol. 5016, pp. 215–264. Springer, Heidelberg (2008)
20. Blinov, M.L., Yang, J., Faeder, J.R., Hlavacek, W.S.: Graph theory for rule-based modeling of biochemical networks. In: Priami, C., Ingólfsdóttir, A., Mishra, B., Riis Nielson, H. (eds.) Transactions on Computational Systems Biology VII. LNCS (LNBI), vol. 4230, pp. 89–106. Springer, Heidelberg (2006)
21. Mayr, R., Rusinowitch, M.: Reachability is decidable for ground ac rewrite systems. In: Infinity 1998, pp. 53–64 (1998)
22. Mayr, R.: Process rewrite systems. Inf. Comput. 156(1-2), 264–286 (2000)
23. Coquidé, J.L., Dauchet, M., Gilleron, R., Vágvölgyi, S.: Bottom-up tree pushdown automata and rewrite systems. In: Book, R.V. (ed.) RTA 1991. LNCS, vol. 488, pp. 287–298. Springer, Heidelberg (1991)
24. Dauchet, M., Tison, S.: The theory of ground rewrite systems is decidable. In: LICS, pp. 242–248 (1990)

25. Jacquemard, F.: Decidable approximations of term rewriting systems. In: Ganzinger, H. (ed.) RTA 1996. LNCS, vol. 1103, pp. 362–376. Springer, Heidelberg (1996)

26. Salomaa, K.: Deterministic tree pushdown automata and monadic tree rewriting systems. J. Comput. Syst. Sci. 37(3), 367–394 (1988)

27. Busi, N., Zavattaro, G.: Reachability analysis in boxed ambients. In: Coppo, M., Lodi, E., Pinna, G.M. (eds.) ICTCS 2005. LNCS, vol. 3701, pp. 143–159. Springer, Heidelberg (2005)

28. Boneva, I., Talbot, J.M.: When ambients cannot be opened. Theor. Comput. Sci. 333(1-2), 127–169 (2005)

29. Busi, N., Zavattaro, G.: Deciding reachability in mobile ambients. In: Sagiv, M. (ed.) ESOP 2005. LNCS, vol. 3444, pp. 248–262. Springer, Heidelberg (2005)

30. Delzanno, G., Montagna, R.: On reachability and spatial reachability in fragments of bioambients. Electr. Notes Theor. Comput. Sci. 171(2), 69–79 (2007)

31. Zavattaro, G.: Reachability analysis in bioambients. ENTCS 227, 179–193 (2009)

32. Drewes, F., Kreowski, H.J., Habel, A.: Hyperedge replacement, graph grammars. In: Handbook of Graph Grammars, pp. 95–162. World Scientific, Singapore (1997)

33. Joshi, S., König, B.: Applying the graph minor theorem to the verification of graph transformation systems. In: Gupta, A., Malik, S. (eds.) CAV 2008. LNCS, vol. 5123, pp. 214–226. Springer, Heidelberg (2008)

34. Finkel, A., Schnoebelen, P.: Well-structured transition systems everywhere! TCS 256(1-2), 63–92 (2001)

# Approximation of Event Probabilities in Noisy Cellular Processes*

Frédéric Didier[1], Thomas A. Henzinger[1], Maria Mateescu[1], and Verena Wolf[1,2]

[1] EPFL, Switzerland
[2] Saarland University, Germany

**Abstract.** Molecular noise, which arises from the randomness of the discrete events in the cell, significantly influences fundamental biological processes. Discrete-state continuous-time stochastic models (CTMC) can be used to describe such effects, but the calculation of the probabilities of certain events is computationally expensive.

We present a comparison of two analysis approaches for CTMC. On one hand, we estimate the probabilities of interest using repeated Gillespie simulation and determine the statistical accuracy that we obtain. On the other hand, we apply a numerical reachability analysis that approximates the probability distributions of the system at several time instances. We use examples of cellular processes to demonstrate the superiority of the reachability analysis if accurate results are required.

## 1  Introduction

The traditional approach for a dynamical model of cellular reaction networks is based on the assumption that the concentrations of the chemical species change continuously and deterministically in time. During the last decade, however, stochastic models with discrete state spaces have seen growing interest [25,30,36,7,29,38,40,43]. The reason is that they take into account the effects of molecular noise in the cell. Molecular noise has a significant influence on important processes such as gene expression [19,3,27,24,6,39], decisions of the cell fate [1,23,22], and circadian oscillations [11,2,12].

An appropriate modeling approach for systems that are subject to molecular noise is a discrete-state continuous-time Markov process, also called *continuous-time Markov chain* (CTMC). This is particularly evident in the presence of *intrinsic noise* arising from random microscopic events in the cell, such as the location of molecules or the order of the reactions. As opposed to continuous models, the discrete-state stochastic model is able to capture the discreteness of the random events in the cell.

The evolution of such a CTMC is given by a master equation that is derived according to Gillespie's theory of stochastic chemical kinetics [10]. Since the state space grows exponentially in the number of involved chemical species, the state space of the CTMC is large, which renders its analysis difficult. Moreover, the discrete structure becomes

* This research was supported in part by the Swiss National Science Foundation under grant 205321-111840 and by the Excellence Cluster on Multimodal Computing and Interaction.

even larger when the number of molecules in the system grows. If the populations of certain chemical species are large, their effect on the system's variance is small and they can be approximated assuming a continuous deterministic change. For species with small populations, however, a continuous approximation is not appropriate and other approximation techniques are necessary to reduce the computational effort of the analysis.

Besides the computation of cumulative measures such as expectations and variances of the populations of certain chemical species, the computation of event probabilities is important for several reasons. First, cellular process may decide probabilistically between several possibilities, e.g., in the case of developmental switches [14,1,30]. In order to verify, falsify, or refine the mathematical model based on experimental data, the likelihood for each of these possibilities has to be calculated. But also full distributions are of interest, such as the distribution of switching delays [24], the distribution of the time of DNA replication initiation at different origins [28], and the distribution of gene expression products [42]. Finally, many parameter estimation methods require the computation of the posterior distribution because means and variances do not provide enough information to calibrate parameters [16].

Two different families of computational approaches have been proposed and used to estimate event probabilities and approximate probability distributions. The first kind of approach is based on numerical simulation, i.e., the generation of many sample trajectories (or *simulation runs*) of the system. The second kind of approach is based on numerical reachability analysis, i.e., the propagation of the probability mass through the state space. The former approach is known as *Gillespie simulation* [9], in which pseudo-random numbers are used to simulate molecular noise. Measures of interest are obtained via statistical output analysis. The main advantage of simulation is that it is easy to implement and the generation of trajectories is not limited by the size of the state space. Moreover, the precision level of the method can be easily adjusted by performing more or fewer simulation runs. For the computation of the probability of certain events, however, simulative approaches become computationally expensive, because a large number of runs have to be carried out to bound the statistical error appropriately. For estimating event probabilities, a higher precision level is necessary than for estimating cumulative measures such as expectations, and simulation becomes expensive because doubling the precision requires four times more simulation runs.

In contrast, approaches based on a numerical reachability analysis approximate probability distributions of the CTMC. As opposed to a statistical estimation of probabilities, which yields an indirect solution, the master equation is numerically solved by integrating the system's behavior over time. Standard numerical techniques are impractical for many systems because of the enormous size of the state space. Recently, however, more sophisticated numerical approximation methods have been proposed, which solve the system in an iterative fashion and consider only subsets of the state space during any given time interval [17,26,5]. They are significantly more efficient than global analysis because they use localization optimizations (such as "sliding windows") and dynamic adaptation ("on-the-fly" generation of windows). These methods efficiently compute the probability distribution of large CTMC at several time instances up to a small approximation error. They can also be used for infinite-state systems.

In this paper, we evaluate and compare the performance of the two different approaches for the computation of probabilities of certain events, i.e., the statistical estimation using simulation and the approximation using a numerical reachability analysis. For the latter we use a particular algorithm as a representative of the whole family of numerical analysis algorithms, because we have found it to perform best. Similar to the sliding-window method [17], our algorithm performs a sequence of local analysis steps on dynamically constructed abstractions of the system. The main improvement over the sliding-window method is that our algorithm is based on adaptive uniformization [41], which allows us to consider arbitrary sets of significant states, i.e., they may be located at different parts of the state space and are not restricted to a specific window shape. Moreover, adaptive uniformization is more robust if the system under study is stiff, i.e., if the chemical reactions occur at time scales that differ by several orders of magnitude. In contrast to [17], here, for the first time, we perform a systematic experimental performance comparison of a numerical reachability analysis with simulation.

The first example that we consider is the transcription regulation of a repressor protein in bacteriophage $\lambda$ [13], where we approximate the probability distribution at several time instances. In the second example, which is a gene expression network [39], we compute the distribution of the time until the number of produced proteins exceeds a certain threshold. In both examples the number of states reachable from the initial state is infinite. The number of chemical species is 6 and 2, and the number of chemical reactions is 10 and 4, respectively. We compare the running time of our numerical reachability analysis to that of the simulative approach for both examples, for different precision levels. Our results show that numerical approximation based on reachability analysis is superior to statistical estimation based on repeated simulation, especially if we increase the desired precision level. For instance, the numerical approximation of the first example needs 39 minutes for a total approximation error of $2 \times 10^{-5}$, which distributes among all states. Simulation requires more than six hours if the statistical error of a single event is to be bounded by $10^{-5}$ and more than sixty hours for $10^{-6}$.

# 2  Stochastic Model

According to the theory of stochastic chemical reaction kinetics, a continuous-time Markov chain (CTMC) can be derived from a set of biochemical reactions [18,10]. This discrete-state model has a regular structure, which gives rise to a functional description in terms of *transition class models* (TCMs) [33]. TCMs naturally represent coupled chemical reactions as each chemical reaction corresponds to a transition class. They provide, however, a more general description than a set of chemical reactions.

## 2.1  Transition Class Models

Consider a dynamical system with a finite number of discrete state variables such as the number of instances of some chemical species in a reaction volume. Assume that these variables change at discrete points in time. A *transition class* provides a rule for these changes and a function for the calculation of the state-dependent *transition rate* at which a state change occurs. Let $S$ be a countable set of states.

**Definition 1.** *A transition class $C$ is a triple $(G, u, \alpha)$ such that (i) the guard $G \subset S$ is a subset of $S$, (ii) $u : G \to S$ is an injective update function with $u(x) \neq x$ for all $x \in G$, (iii) $\alpha : G \to \mathbb{R}_{>0}$ is a rate function. A transition class model (TCM) $M = (y, \{C_1, \ldots, C_k\})$ consists of an initial state $y \in S$ and a finite set of transition classes $C_1, \ldots, C_k$.*

The set $G$ contains all states $x$ in which a transition of type $C$ is possible and $u(x)$ is the target state of the transition. The probability of the $C$-transition depends on the transition rate $\alpha(x)$ in the way explained below.

In practice, we can usually express $G$ by a finite number of constraints on the state variables, and $u$ and $\alpha$ by elementary arithmetic functions. Thus, a TCM provides a finite description of a (possibly infinite-state) system. Before we show how a CTMC is derived from a TCM, we present some examples of TCMs that describe biochemical reaction networks.

**Biochemical Reaction Networks.** We consider a fixed reaction volume with $n$ different chemical species that is spatially homogeneous and in thermal equilibrium. Then, the state space of the system is given by $S = \mathbb{N}_0^n$. We assume that molecules collide randomly and that collisions may lead to chemical reactions. For a given set of chemical reactions, we construct a TCM such that each transition class corresponds to a reaction and the associated propensity function is given by the rate function $\alpha$.

*Example 1.* We consider a simple transition class model for transcription of a gene into messenger RNA (mRNA), and subsequent translation of the latter into proteins [39]. This reaction network involves three chemical species, namely, gene, mRNA, and protein. As only a single copy of the gene exists, a state of the system is uniquely determined by the number of mRNA and protein molecules. Therefore, $S = \mathbb{N}_0^2$ and a state is a pair $(x_R, x_P) \in S$. We assume that initially there are no mRNA molecules and no proteins in the system, i.e., $y = (0, 0)$. The following four types of reactions occur in the system, namely $\emptyset \to mRNA$, $mRNA \to mRNA + P$, $mRNA \to \emptyset$, and $P \to \emptyset$. Let $i \in \{1, \ldots, 4\}$ and let $c_i > 0$ be a constant. Transition class $C_i = (G_i, u_i, \alpha_i)$ describes the $i$-th reaction type.

- We describe gene transcription by transition class $C_1$, which increases the number of mRNA molecules by 1. Thus, $u_1(x_R, x_P) = (x_R + 1, x_P)$. This transition class is possible in all states, i.e., $G_1 = S$. Transcription happens at the constant rate $\alpha_1(x_R, x_P) = c_1$, as only one reactant molecule (the gene) is available.
- We represent the translation of mRNA into protein by $C_2$. A $C_2$-transition is only possible if there is at least one mRNA molecule in the system. We set $G_2 = \{(x_R, x_P) \in S \mid x_R > 0\}$ and $u_2(x_R, x_P) = (x_R, x_P + 1)$. Note that in this case mRNA is a reactant that is not consumed. The translation rate depends linearly on the number of mRNA molecules. Therefore, $\alpha_2(x_R, x_P) = c_2 \cdot x_R$.
- Degradation is modeled by $C_3$ and $C_4$. Hence, $G_3 = G_2$, $G_4 = \{(x_R, x_P) \in S \mid x_P > 0\}$, $u_3(x_R, x_P) = (x_R - 1, x_P)$, and $u_4(x_R, x_P) = (x_R, x_P - 1)$. We set $\alpha_3(x_R, x_P) = c_3 \cdot x_R$ and $\alpha_4(x_R, x_P) = c_4 \cdot x_P$.

## 2.2 Chemical Master Equation

A transition class model $M = (y, \{C_1, \ldots, C_k\})$ represents a time-homogeneous, discrete-state Markov process $\{X(t)\}_{t \geq 0}$, that is, a CTMC with state space $S$. The $j$-th entry of the random vector $X(t) = (X_1(t), \ldots, X_n(t))$ represents the value of the $j$-th state variable. Let $C_m = (G_m, u_m, \alpha_m)$, $1 \leq m \leq k$, and assume that at time $t \geq 0$ the process is in state $x \in G_m$.

The probability of a transition of type $C_m$ occurring in the next infinitesimal time interval $[t, t + \tau)$, $\tau > 0$ is given by

$$Pr(X(t + \tau) = u_m(x) \mid X(t) = x) = \alpha_m(x) \cdot \tau.$$

Since $y$ is the initial state of $M$ we have $Pr(X(0) = y) = 1$, and for $x \in S$ we define the probability that $X$ is in state $x$ at time $t$ by $p^{(t)}(x) = Pr(X(t) = x \mid X(0) = y)$. Recall that $u_m$ is injective. To simplify our presentation, we define the set $H_m$ as the set of all states $x$ for which $u_m^{-1}(x)$ is defined, that is, that can be reached by a transition of type $C_m$. The *chemical master equation* describes the behavior of $X$ by the differential equation [18]

$$\frac{\partial p^{(t)}(x)}{\partial t} = \sum_{m: x \in H_m} \alpha_m(u_m^{-1}(x)) \cdot p^{(t)}(u_m^{-1}(x)) - \sum_{m: x \in G_m} \alpha_m(x) \cdot p^{(t)}(x). \quad (1)$$

**Unbounded Range.** For realistic systems, the state space of the Markov chain is extremely large, because its size grows exponentially in the number of involved chemical species. Moreover, if upper bounds on the state variables cannot derived from certain conservation laws, their range is assumed to be infinite although in practice the number of molecules is bounded. Then from the infinite structure, we can compute bounds that are kept with a very high probability. Even though every state in the infinite state space has a non-zero probability, certain attracting regions force most of the probability mass to remain within a finite range.

*Example 2.* In Ex. 1, the degradation rates $\alpha_3(x)$ and $\alpha_4(x)$ grow linearly in the state variables. Thus, the higher the number of mRNA or protein molecules the more likely is their degradation. Depending on the rate constants $c_1, \ldots, c_4$, the system becomes "stable" in different regions. As time approaches infinity, the main part of the probability mass will be close to a region where production and degradation of molecules cancel each other out. Below, we discuss in general under which conditions the system approaches such a stable distribution.

**Holding Times and Jump Probabilities.** A Markov chain $\{X(t)\}_{t \geq 0}$ defined in the way above is a *stable and conservative jump process* [4]. Thus, there exists a sequence of jump times $\{\tau(n)\}_{n \geq 0}$ and a sequence $\{\hat{X}(n)\}_{n \geq 0}$ of visited states such that

$$\tau(0) = 0 < \tau(1) < \tau(2) < \ldots \text{ and } X(t) = \hat{X}(n) \text{ if } \tau(n) \leq t < \tau(n + 1).$$

The distribution of the $n$-th holding time $\tau(n + 1) - \tau(n)$ under the condition $\hat{X}(n) = x$ is negative exponentially distributed with parameter $\lambda(x) = \sum_{m: x \in G_m} \alpha_m(x)$, also called *exit rate* of state $x$.

If the sum of all holding times is finite with positive probability, the Markov chain is said to *explode* and the limiting distribution does not exist. Explosive Markov chains are not of interest for the application area of this work since in this case the system "gets lost at infinity". It is possible to check if the Markov chain does not explode by using *Reuter's Criterion* [4]. For the remainder of our presentation we assume that the rate functions $\alpha_m$ are such that the Markov chain does not explode.

Assume that the $n$-th state of the Markov chain is $x$, that is, $\hat{X}(n) = x$. If at least one transition class is enabled in $x$, the successor state is $u_m(x)$ for some $m$ with $x \in G_m$. The probability of successor $u_m(x)$ is given by

$$Pr(\hat{X}(n+1) = u_m(x) \mid \hat{X}(n) = x) = \frac{\alpha_m(x)}{\lambda(x)}.$$

The holding times and the jump probabilities play an important role for the simulation of the Markov chain, which is used to estimate the probability of a certain events.

## 3  Statistical Estimation of Probabilities

In this section we shortly review the basic steps that have to be carried out to estimate the probability of a certain measurable event using stochastic simulation. Throughout this section, we will denote this event by $A$ and its probability by $\gamma$. For the analysis of biological systems, the events of interest may be the marginal distributions or even the joint distributions of certain chemical species. For instance, $A$ may have the form $X_j(t) = k$, that is, the number of type $j$ molecules is $k$.

Estimates are obtained in two steps. In the first step, a certain number of simulation runs of the Markov chain have to be generated, and in the second step, the results of the simulation runs are analyzed.

### 3.1  Trajectory Generation

A realization of the Markov chain, also called *trajectory* or *run*, is the random sequence of states visited by the process. If trajectories are produced by a computer, *pseudo-random numbers* are used to artificially generate randomness [20]. The basic steps of producing a single trajectory that starts in the initial state $y$ at time 0 are as follows:

1. Initialize time $t = 0$ and state $x = y$.
2. Generate the holding time $h$, i.e., a sample of a random variable being exponentially distributed with parameter $-\lambda(x)$.
3. Generate the successor state, i.e., a sample $m$ of a discrete random variable $Z$ that has probability distribution $P(Z = m) = \alpha_m(x)/\lambda(x)$.
4. Set $t = t + h$, $x = u_m(x)$ and go to Step 2 if $t < T$.

In Step 2, we generate the holding time of the current state $x$. Pseudo-random number generators usually draw from a uniform distribution. Thus, for a given random sample $r_1$ that is uniformly distributed on $(0, 1)$, we calculate an exponentially distributed sample by using the inverse transform method. More precisely, we compute the inverse $-\frac{\ln r_1}{\lambda(x)}$ of the cumulative distribution function of the exponential distribution. In Step 3, the same idea is used to decide, which reaction occurs next. The inverse of the cumulative distribution function of $Z$ is given by $m = \min\{i : \sum_{j=1}^{i} \alpha_j(x) > r_2 \cdot \lambda(x)\}$, where $r_2$ is again a random sample that is uniformly distributed on $(0, 1)$. In the final step, the current time and the current state are updated. The simulation is terminated if the time horizon $T$ of interest is reached and continued otherwise.

### 3.2   Output Analysis

The problem of estimating the probability $\gamma$ of the event $A$ can be reformulated as estimating the expectation of the random variable $\chi_A$ with

$$\chi_A(\omega) = \begin{cases} 1 \text{ if } \omega \in A, \\ 0 \text{ if } \omega \notin A, \end{cases}$$

where $\omega$ is a trajectory. The expectation $E[\chi_A]$ equals $\gamma$, since $E[\chi_A] = 1 \cdot Pr(\chi_A = 1) + 0 \cdot Pr(\chi_A = 0) = \gamma$. Therefore, we can resort to the standard estimation procedure for expectations. Assume that $N$ is the number of runs that have been carried out and $Y_1, \ldots, Y_N$ are independent and identically distributed as $\chi_A$. Thus, from the $i$-th run we get a realization of $Y_i$ by checking if $A$ has occurred or not. It is important to point out that we have to guarantee the independence of the $Y_i$'s. This implies that we generate $N$ independent trajectories of the Markov chain, each time with a different initial seed[1] for the pseudo-random number generator. The sample mean $\bar{Y} = \frac{1}{N} \sum_{i=1}^{N} Y_i$ is then an *unbiased* and *consistent estimator* [20] for $E[\chi_A]$. The former means that $E[\bar{Y}] = E[\chi_A]$ and the latter refers to the fact that as $N$ increases the estimator $\bar{Y}$ becomes closer to $\gamma$. Note that $\bar{Y}$ is equal to the relative frequency of the event $A$. Let $\sigma^2 = VAR[\chi_A]$ be the variance of $\chi_A$. We evaluate the quality of the estimator $\bar{Y}$ by applying the central limit theorem, which states that $\bar{Y}$ will approximately have a Normal distribution with mean $E[\chi_A] = \gamma$ and variance $\sigma^2/N$. Hence, for large $N$ the random variable

$$Z = \frac{\bar{Y} - \gamma}{\sqrt{\sigma^2/N}}$$

has a standard Normal distribution, that is, the mean is zero and the variance is one. Knowing the distribution of $Z$ enables us reason about the difference $|\bar{Y} - \gamma|$. Let $\beta \in [0, 1]$ be the *confidence level* and $z \in \mathbb{R}^+$ such that $\beta = Pr(|Z| \le z)$. Then

$$\beta = Pr(|Z| \le z) = Pr\left(\frac{|\bar{Y}-\gamma|}{\sqrt{\sigma^2/N}} \le z\right) = Pr\left(|\bar{Y} - \gamma| \le z\sqrt{\sigma^2/N}\right).$$

We estimate $\sigma^2$ with the sample covariance $S^2 = \frac{1}{N-1} \sum_{i=1}^{N}(Y_i - \bar{Y})^2$, which is an unbiased estimator for $\sigma^2$. Then, for large $N$ and a large number of realizations of the *confidence interval*

$$\left[\bar{Y} - z\sqrt{S^2/N}, \bar{Y} + z\sqrt{S^2/N}\right], \tag{2}$$

$\beta$ is the fraction of intervals that cover $\gamma$. It therefore measures the quality of the estimator $\bar{Y}$.

---

[1] The seed of a pseudo-random number generator is an initial value, on which the sequence of generated numbers depend [20].

For a practical application, two further remarks are important. Firstly, we usually choose $\beta \in \{0.95, 0.99\}$ and the corresponding value of $z$ can be found in the table of the standard Normal distribution. Let $\Phi$ be the cumulative distribution function of the standard Normal distribution. Then, using that the Normal distribution is symmetric,

$$\Phi(z) = Pr(Z \leq z) = 1 - \tfrac{1-\beta}{2} = \tfrac{1+\beta}{2} \iff z = \Phi^{-1}\left(\tfrac{1+\beta}{2}\right).$$

Secondly, both, $\bar{Y}$ and $S^2$ can be computed efficiently if during the trajectory generation the realizations of the two sums $\sum_{i=1}^{N} Y_i$ and $\sum_{i=1}^{N} Y_i^2$ are calculated, since it can be easily shown that

$$S^2 = \tfrac{\sum_{i=1}^{N} Y_i^2}{N-1} - \tfrac{\left(\sum_{i=1}^{N} Y_i\right)^2}{(N-1)N}.$$

Thus, if $r \in \{0, \dots, N\}$ is the number of times event $A$ occurred during the $N$ simulation runs, $\bar{Y} = r/N$ and $S^2 = \tfrac{r(N-r)}{N(N-1)}$.

If the interval in Eq. 2 is large relative to $\bar{Y}$ the quality of the estimator is poor and more simulation runs have to be carried out. For our experimental results in Section 5, we fixed the relative width of the interval to be 0.2 (which means that we have a relative error of at most 0.1) and chose confidence level $\beta = 0.95$. Thus, $z \approx 1.96$ and we can determine the number of necessary runs by bounding the relative width

$$2 \cdot \tfrac{z \cdot \sqrt{S^2/N}}{\gamma} \leq 0.2 \implies \tfrac{z^2}{0.01} \tfrac{S^2}{\gamma^2} \leq N \implies 384 \cdot \tfrac{S^2}{\gamma^2} \leq N$$

Assume now that we want to estimate the probability of events that occur at least with probability $\gamma$. Using the fact that $\sigma^2 = VAR[\chi_A] = \gamma(1 - \gamma)$ and replacing $S^2$ by $\sigma^2$ yields $N \geq 384 \cdot \tfrac{1-\gamma}{\gamma}$ [32]. For instance, the sufficient number of runs to guarantee that probabilities, having at least the order of magnitude of $10^{-5}$, are estimated with a relative error of at most 0.1 and a confidence of 95% is $N = 38,000,000$.

## 4   Numerical Reachability Analysis

Instead of indirectly approximating probabilities with statistical estimation procedures, we can use a numerical reachability analysis to solve Eq. 1. An efficient solution by applying standard numerical methods is not possible, since for realistic systems the state space of the system is extremely large. An efficient approximation is, however, possible as long as the total number of involved molecules is a manageable number. We describe a method that is based on a discretization of the process and numerically approximates the probabilities $p^{(t)}(x)$ at certain time instances.

**Adaptive Uniformization.** We discretize the system using *adaptive uniformization*, which has been introduced by van Moorsel [41] as a variant of *standard uniformization* [31,34,44,15,35]. Numerical methods based on uniformization have the advantage that they are numerically stable and often more efficient than other methods [37].

We inductively define a sequence $S_0, S_1, \dots$ of subsets of the state space $S$ of the CTMC $\{X(t)\}_{t \geq 0}$, as well as a sequence $p_0, p_1, \dots$ of functions such that $p_k : S \rightarrow [0, 1]$ for $k = 0, 1, \dots$. Recall that $y$ is the initial state. We define $S_0 = \{y\}, p_0(y) = 1$

---

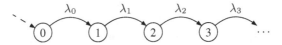

**Fig. 1.** The birth process of the adaptive uniformization procedure

and $p_0(x) = 0$ if $x \neq y$. For $k = 1, 2, \ldots$, we inductively define $S_k$ as follows. We choose a positive *uniformization rate* $\lambda_k \geq \max_{x \in S_k} \lambda_x$ and set

$$S_{k+1} = \{x' \in S \mid \exists x \in S_k : p_k(x) \cdot q_k(x, x') > 0\}, \tag{3}$$

where, for $x \in S$,

$$q_k(x, x') = \begin{cases} \sum_{m:u_m(x)=x'} \alpha_m(x)/\lambda_k & \text{if } x \neq x', \exists m : u_m(x) = x', \\ 0 & \text{if } x \neq x', \nexists m : u_m(x) = x', \\ 1 - \sum_{x' \in S: x' \neq x} q_k(x, x') & \text{if } x = x'. \end{cases} \tag{4}$$

For $x' \in S_{k+1}$ we set $p_{k+1}(x') = \sum_{x' \in S_k} p_k(x) \cdot q_k(x, x')$ and $p_{k+1}(x) = 0$ if $x \notin S_k$.

The value $p_k(x)$ is the probability of reaching state $x$ after $k$ steps in a discrete-time Markov chain $\{Y(k)\}_{k \in \mathbb{N}}$ with transition probabilities $Pr(Y(k+1) = x' \mid Y(k) = x) = q_k(x, x')$ and initial distribution $Pr(Y(0) = y) = 1$. We can reconstruct $p^{(t)}(x)$ by considering an additional process that relates steps with time. Let $\{B(t)\}_{t \geq 0}$ be a birth process with birth rates $\lambda_0, \lambda_1, \ldots$, that is, $B$ has a chain structure as illustrated in Fig. 1 and starts initially in state 0 with probability one. In [41], van Moorsel has proved that the original CTMC $\{X(t)\}_{t \geq 0}$ can be constructed from $B$ and $Y$ by setting $Y(B(t)) = X(t)$ if $B$ does not explode. Since $Y$ and $B$ are independent, the state probability $p^{(t)}(x)$ of the original CTMC can be expressed as

$$p^{(t)}(x) = \sum_{k=0}^{\infty} Pr(Y(k) = x) \cdot Pr(B(t) = k) = \sum_{k=0}^{\infty} p_k(x) \cdot Pr(B(t) = k). \tag{5}$$

Note that in Eq. 5, there are no negative summands involved. Moreover, $p_k$ can be computed inductively. Lower and upper summation bounds $L$ and $U$ can be obtained such that for each state $x$ the truncation error

$$p^{(t)}(x) - \sum_{k=L}^{U} p_k(x) \cdot Pr(B(t) = k) = \sum_{\substack{0 \leq k < L, \\ U < k < \infty}} p_k(x) \cdot Pr(B(t) = k) \leq$$

$$\sum_{\substack{0 \leq k < L, \\ U < k < \infty}} Pr(B(t) = k) = \qquad 1 - \sum_{k=L}^{U} Pr(B(t) = k) < \epsilon \tag{6}$$

can be bounded by $\epsilon > 0$. Finally, we note that from Eq. 4 it is clear that choosing the smallest possible $\lambda_k$ is advantageous since this avoids high self-loop probabilities in $q_k$.

**Standard Uniformization.** Standard uniformization is a special case of adaptive uniformization where a global uniformization rate $\lambda = \lambda_0 = \lambda_1 = \ldots$ has to be chosen. If each transition in the birth process occurs at a constant rate $\lambda$, the values $Pr(B(t) = k)$

follow a Poisson distribution with parameter $\lambda t$. They can be calculated efficiently using the iterative procedure introduced by Fox and Glynn [8]. Standard uniformization becomes inefficient whenever $\lambda$ is much larger than the exit rates $\lambda(x)$ of many states $x$ that are involved in the computation. Note that in chemically reacting systems, the dynamics of the system may change considerably. In this case the discretization using adaptive uniformization is more efficient.

**Approximate Discretization.** In its standard form, adaptive uniformization is not appropriate for Markov chains that describe biochemical reaction networks for two reasons. Firstly, the size of the sets $S_0, S_1, \ldots$ grows after each step and the computational complexity for $p_k$ becomes huge. Secondly, the birth process may explode even if the original CTMC does not. The reason is that $S_k$ approaches $S$ as $k \to \infty$. The latter problem can be circumvented by neglecting states that are very unlikely, that is, we replace Eq. 3 by

$$S_{k+1} = \{x' \in S \mid \sum_{x \in S_k} p_k(x) \cdot q_k(x, x') > \Delta\}, \tag{7}$$

where $\Delta$ is a small constant. This ensures that even in the limit $S_k$ is finite, since only a finite number of states can have a probability greater than $\delta$. Moreover, the number of states in $S_k$ is now manageable as long as the total number of molecules is manageable since only a comparatively small number of different values for each state variables have to be considered.

The error after $k$ steps introduced by the threshold $\Delta$ can be calculated as $1 - \sum_{x \in S_k} p_k(x)$. Note that the error increases monotonically in $k$ since more and more probability "gets lost". Therefore we choose $\Delta$ several orders of magnitude smaller than the desired precision. For our experimental results in Section 5 we chose different values for $\Delta$ ranging from $10^{-15}$ till $10^{-8}$ in order to obtain different precision levels.

**Approximate Solution of the Birth Process.** Finally, we discuss the computation of the values $Pr(B(t) = k)$ and how truncation bounds $L$ and $R$ are obtained. We use standard uniformization to discretize $B$, since we can afford a high global uniformization rate (and thus, high self-loop probabilities) in this case. The reason is that the simple chain structure eases the discretization and the computational effort to solve the birth process is small compared to the calculation of the $p_k$. Similar as for $Y$ we approximately solve $B$ by neglecting states that are "left behind". Informally, we use a window (a set that contains all states within a certain range) that slides from left to right to approximate $Pr(B(t) = k)$ and determines the truncation points $L$ and $R$.

**Approximation Error.** Both, the solution of $Y$ and $B$ gives an underapproximation of the values $p_k(x)$ and $Pr(B(t) = k)$. Thus, summing up their product according to Eq. 5 results in an underapproximation for $p^{(t)}(x)$. The final approximation error is obtained as $\delta = 1 - \sum_{x \in S_R} p^{(t)}(x)$ where $R$ is the right truncation bound of the birth process. The probability of states that are not in $S_R$ is approximated with zero. Note that this includes all approximation errors, i.e., the probability that is lost during the solution of the birth process, and during all steps of the discretization because of the threshold $\Delta$. The computational savings achieved by solving $Y$ as well as $B$ in the way described above are substantial. The reason is that the number of states in $B$ and $Y$ that are significant after $k$ steps is several orders of magnitudes smaller than the number of all states reachable after $k$ steps. Moreover, our experimental results show that the method yields accurate results, as the approximation error $\delta$ is small.

**Iteration Over Time.** First note that we can use the method described above for systems starting with arbitrary initial distributions as long as the number of states in the initial set $S_0$ is manageable. After computing an approximation of $p^{(t)}(x)$ for all $x \in S$ we can use it as an initial distribution for the next step to obtain an approximation for $p^{(t')}(x)$ where $t' > t$ and the step size is $t' - t$. In this way, we obtain approximations for several time instances.

**Related Work.** Other approaches for an approximate numerical solution of the underlying Markov chains can be found in [26,5]. They differ from our approach in that they compute a finite projection of the state space that is based solely on the structure of the underlying graph. In our method, we add and neglect states in an on-the-fly fashion based on the stochastic properties of the Markov chain. Therefore, we consider a significantly smaller set of states during a certain time interval, without being less accurate. The projection algorithms include all states that are reachable within a fixed path depth. In our algorithm, for each single state, we dynamically decide if it significantly contributes to the overall solution or not. We have found this dynamic adaptation of the analysis to be essential for efficiency.

## 5 Experimental Results

For our experimental results, we consider two examples from biology. One is a model for the transcription regulation of a repressor protein in bacteriophage $\lambda$ [13]. This protein is responsible for maintaining lysogeny of the $\lambda$ virus in E. coli [1]. We compute the full probability distribution for different precision levels. Our second example uses the gene expression model of Ex. 1. We calculate the distribution of the time until the number of produced proteins exceeds 500.

There is no one-to-one correspondence between the statistical accuracy of the estimates that we derive via simulation and the precision of the numerical method. However, by assuming that the smallest event probability that has to be estimated is $\gamma$ all results of the simulation have a "precision" of at least $\gamma$. Intuitively, we simulate often enough to reason about events that occur with a probability of at least $\gamma$. We therefore refer to $\gamma$ as the *single event error* (cf. Table 1 and 2). Note that the simulation results are still subject to the statistical errors since the true values may not be covered by the confidence interval (compare Section 3.2).

The approximation error $\delta$ of the numerical method is the sum of the approximation error of *all* states in the Markov chain. Note that the probabilities of states not in $S_k$ are underapproximated with zero and their true probabilities increase depending on how close they are to an attracting region. The error of a single state probability $p^{(t)}(x)$ is much smaller than $\delta$ but precise values for the single error are hard to obtain. A rough estimation of the single errors can be obtained by dividing the total error by the average size $|S_k|$ of the significant sets (cf. Table 1 and 2), even though $\delta$ may not be uniformly distributed on the significant set. On the other hand, $\delta$ also includes the

**Table 1.** Comparison of the running times for the phage $\lambda$ model

| numerical approximation | | | | Gillespie simulation | | |
|---|---|---|---|---|---|---|
| running time | total approx. error | $|S_k|$ | $\Delta$ | running time | single event error | # runs |
| 55 min 5 sec | $3 \times 10^{-6}$ | 239792 | $10^{-15}$ | > 6000 h | $10^{-8}$ | $> 3 \times 10^{10}$ |
| 39 min 16 sec | $2 \times 10^{-5}$ | 187204 | $10^{-14}$ | > 500 h | $10^{-7}$ | $> 3 \times 10^{9}$ |
| 25 min 2 sec | $2 \times 10^{-4}$ | 140969 | $10^{-13}$ | 67 h 22 min | $10^{-6}$ | $> 3 \times 10^{8}$ |
| 15 min 41 sec | $1 \times 10^{-3}$ | 101078 | $10^{-12}$ | 6 h 44 min | $10^{-5}$ | $> 3 \times 10^{7}$ |
| 6 min 33 sec | $7 \times 10^{-3}$ | 67540 | $10^{-11}$ | 40 min | $10^{-4}$ | $> 3 \times 10^{6}$ |
| 3 min 12 sec | $4 \times 10^{-2}$ | 40373 | $10^{-10}$ | 4 min | $10^{-3}$ | $> 3 \times 10^{5}$ |

**Table 2.** Comparison of the running times for the gene expression example

| numerical approximation | | | | Gillespie simulation | | |
|---|---|---|---|---|---|---|
| running time | total approx. error | $|S_k|$ | $\Delta$ | running time | single event error | # runs |
| 4.2 sec | $5 \times 10^{-6}$ | 9816 | $10^{-12}$ | > 500 h | $10^{-7}$ | $> 3 \times 10^{9}$ |
| 3.6 sec | $5 \times 10^{-5}$ | 8719 | $10^{-11}$ | > 50 h | $10^{-6}$ | $> 3 \times 10^{8}$ |
| 3.0 sec | $5 \times 10^{-4}$ | 7516 | $10^{-10}$ | 5 h 3 min | $10^{-5}$ | $> 3 \times 10^{7}$ |
| 2.4 sec | $4 \times 10^{-3}$ | 6265 | $10^{-9}$ | 30 min 18 sec | $10^{-4}$ | $> 3 \times 10^{6}$ |
| 1.9 sec | $4 \times 10^{-2}$ | 4939 | $10^{-8}$ | 3 min sec | $10^{-3}$ | $> 3 \times 10^{5}$ |

error of insignificant states and, thus, distributes among much more states than only those in $S_k$.

**Phage $\lambda$ Model.** The Phage $\lambda$ model involves 6 different species and 10 reactions. Thus, a state is a vector $x = (x_1, x_2, x_3, x_4, x_5, x_6) \in \mathbb{N}_0^6$. The transition classes $C_i = (G_i, u_i, \alpha_i)$, $1 \leq i \leq 10$ are given as follows [13].

- Production of proteins: $G_1 = \{x \in \mathbb{N}_0^6 \mid x_3 > 0\}$, $u_1(x) = (x_1 + 1, x_2, x_3, x_4, x_5, x_6)$, $\alpha_1(x) = c_1 x_3$.
- Degradation of proteins: $G_2 = \{x \in \mathbb{N}_0^6 \mid x_1 > 0\}$, $u_2(x) = (x_1 - 1, x_2, x_3, x_4, x_5, x_6)$, $\alpha_2(x) = c_2 x_1$.
- Production of mRNA: $G_3 = \{x \in \mathbb{N}_0^6 \mid x_5 > 0\}$, $u_3(x) = (x_1, x_2, x_3 + 1, x_4, x_5, x_6)$, $\alpha_3(x) = c_3 x_5$.
- Degradation of mRNA: $G_4 = \{x \in \mathbb{N}_0^6 \mid x_3 > 0\}$, $u_4(x) = (x_1, x_2, x_3 - 1, x_4, x_5, x_6)$, $\alpha_4(x) = c_4 x_3$.
- First dimer binding at operator site: $G_5 = \{x \in \mathbb{N}_0^6 \mid x_2, x_4 > 0\}$, $u_5(x) = (x_1, x_2 - 1, x_3, x_4 - 1, x_5 + 1, x_6)$, $\alpha_5(x) = c_5 x_2 x_4$.the simulation results are still subject to the statistical errors since the true values may not be covered by the confidence interval (compare Section 3.2).

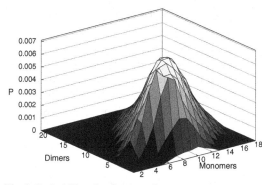

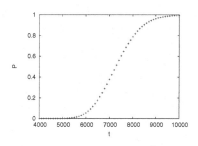

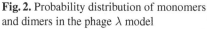

**Fig. 2.** Probability distribution of monomers and dimers in the phage $\lambda$ model

**Fig. 3.** Cumulative probability distribution of the time until the number of proteins reaches 500 for the first time in the gene expression example

- First dimer unbinding: $G_6 = \{x \in \mathbb{N}_0^6 \mid x_5 > 0\}$, $u_6(x) = (x_1, x_2 + 1, x_3, x_4 + 1, x_5 - 1, x_6)$, $\alpha_6(x) = c_6 x_5$.
- Second dimer binding at operator site: $G_7 = \{x \in \mathbb{N}_0^6 \mid x_2, x_5 > 0\}$, $u_7(x) = (x_1, x_2 - 1, x_3, x_4, x_5 - 1, x_6 + 1)$, $\alpha_7(x) = c_7 x_2 x_5$.
- Second dimer unbinding: $G_8 = \{x \in \mathbb{N}_0^6 \mid x_6 > 0\}$, $u_8(x) = (x_1, x_2 + 1, x_3, x_4, x_5 + 1, x_6 - 1)$, $\alpha_8(x) = c_8 x_6$.
- Dimerization: $G_9 = \{x \in \mathbb{N}_0^6 \mid x_1 > 1\}$, $u_9(x) = (x_1 - 2, x_2 + 1, x_3, x_4, x_5, x_6)$, $\alpha_9(x) = c_9 x_1 (x_1 - 1)/2$.
- Dissociation into monomers: $G_{10} = \{x \in \mathbb{N}_0^6 \mid x_2 > 0\}$, $u_{10}(x) = (x_1 + 2, x_2 - 1, x_3, x_4, x_5, x_6)$, $\alpha_{10}(x) = c_{10} x_2$.

For $c_1, \ldots, c_{10}$, we choose $c_1 = 0.043$, $c_2 = 0.0007$, $c_3 = 0.0715$, $c_4 = 0.0039$, $c_5 = 1.992647 \times 10^{-2}$, $c_6 = 0.4791$, $c_7 = 1.992647 \times 10^{-4}$, $c_8 = 8.765 \times 10^{-12}$, $c_9 = 8.30269 \times 10^{-2}$, and $c_{10} = 0.5$ (see [13,5]). The initial state of the system is given by $y = (2, 6, 0, 2, 0, 0)$ and the time horizon is $t = 300$. We approximate the probability distributions of the underlying CTMC at 100 equidistant time instances. Fig. 2 shows a plot of the distribution of dimers and monomers at time instant $t = 300$. In Table 1, we list the running times of our numerical method as well as the running time of the simulation. The column with header $|S_k|$ lists the average number of states in the sets $S_0, S_1, \ldots$ and $\Delta$ is the threshold in Eq. 7.

**Gene Expression.** For the transition classes of the gene expression example we refer to Ex. 1. For the rate constants, we choose $c_1 = 0.05$, $c_2 = 0.0058$, $c_3 = 0.0029$, and $c_4 = 10^{-4}$, where $c_3$ and $c_4$ correspond to a half-life of 4 minutes for mRNA and 2 hours for the protein [39]. We compute the probability that at least 500 proteins are in the system at 100 equidistant time instances. Fig 3 shows the cumulative probability distribution of the time until the number of proteins reaches 500 for the first time (note that eventually the threshold of 500 is reached with probability one). In Table 2, we list the results for the gene expression example, where, as above, $|S_k|$ denotes the average number of states in the sets $S_0, S_1, \ldots$ and $\Delta$ is the threshold in Eq. 7.

**Discussion.** Even if we consider the total approximation error $\delta$ as a rough bound for the single error of each state probability, thus favoring simulation, the speed-up factor of the numerical approximation is large, especially if the precision increases. The necessary precision level up to which probability distributions are approximated may depend on the system under study. It is, however, important to note that the occurrence of rare biochemical events can have important effects. For instance, the spontaneous, epigenetic switching rate from the lysogenic state to the lytic state in phage $\lambda$-infected E. coli is experimentally estimated to be in the order of $10^{-7}$ per cell per generation [21].

## 6  Conclusion

We have demonstrated that, for the computation of event probabilities, a numerical reachability analysis provides an efficient alternative to simulation-based methods.

Even though simulation is widely used, the advantages of numerical methods increase as more sophisticated techniques become available. They reduce the computational effort, especially if accurate results are desired. Moreover, for the calibration of parameters many instances of the model have to be solved and in this case short running times for a single solution are necessary.

Until now we have analyzed examples of intrinsically stochastic systems that have been published in the literature. As future work, we are planning to apply our numerical reachability algorithm in collaboration with experimentalists working on new stochastic models. Moreover, we are planning to combine our numerical method with parameter estimation techniques.

Standard numerical reachability analysis methods are inefficient for large state spaces (in the case of high dimension and/or many molecules) and inapplicable for unbounded state spaces, and thus one resorts to simulation. We have demonstrated that certain optimization techniques from computer science - localization, on the fly abstraction - put many examples within the reach of numerical reachability analysis. Indeed, when high accuracy is required these methods outperform simulation-based techniques.

## References

1. Arkin, A., Ross, J., McAdams, H.H.: Stochastic kinetic analysis of developmental pathway bifurcation in phage $\lambda$-infected E. coli cells. Genetics 149, 1633–1648 (1998)
2. Barkai, N., Leibler, S.: Biological rhythms: Circadian clocks limited by noise. Nature 403, 267–268 (2000)
3. Blake, W.J., Kaern, M., Cantor, C.R., Collins, J.J.: Noise in eukaryotic gene expression. Nature 422, 633–637 (2003)
4. Bremaud, P.: Markov Chains. Springer, Heidelberg (1998)
5. Burrage, K., Hegland, M., Macnamara, F., Sidje, R.: A Krylov-based finite state projection algorithm for solving the chemical master equation arising in the discrete modelling of biological systems. In: Proc. of the Markov 150th Anniversary Conference, Boson Books, pp. 21–38 (2006)
6. Elowitz, M.B., Levine, M.J., Siggia, E.D., Swain, P.S.: Stochastic gene expression in a single cell. Science 297, 1183–1186 (2002)
7. Fedoroff, N., Fontana, W.: Small numbers of big molecules. Science 297, 1129–1131 (2002)

8. Fox, B.L., Glynn, P.W.: Computing Poisson probabilities. Communications of the ACM 31(4), 440–445 (1988)
9. Gillespie, D.T.: Exact stochastic simulation of coupled chemical reactions. J. Phys. Chem. 81(25), 2340–2361 (1977)
10. Gillespie, D.T.: Markov Processes. Academic Press, New York (1992)
11. Gonze, D., Halloy, J., Goldbeter, A.: Robustness of circadian rhythms with respect to molecular noise. PNAS, USA 99(2), 673–678 (2002)
12. Gonze, D., Halloy, J., Goldbeter, A.: Stochastic models for circadian oscillations: Emergence of a biological rhythm. Quantum Chemistry 98, 228–238 (2004)
13. Goutsias, J.: Quasiequilibrium approximation of fast reaction kinetics in stochastic biochemical systems. J. Chem. Phys. 122(18), 184102 (2005)
14. Hasty, J., Pradines, J., Dolnik, M., Collins, J.J.: Noise-based switches and amplifiers for gene expression. PNAS, USA 97, 2075 (2000)
15. Hellander, A.: Efficient computation of transient solutions of the chemical master equation based on uniformization and quasi-Monte carlo. J. Chem. Phys. 128(15), 154109 (2008)
16. Henderson, D.A., Boys, R.J., Proctor, C.J., Wilkinson, D.J.: Linking systems biology models to data: a stochastic kinetic model of p53 oscillations. In: O'Hagan, A., West, M. (eds.) Handbook of Applied Bayesian Analysis. Oxford University Press, Oxford (2009)
17. Henzinger, T., Mateescu, M., Wolf, V.: Sliding window abstraction for infinite Markov chains. In: Proc. CAV. LNCS. Springer, Heidelberg (to appear, 2009)
18. van Kampen, N.G.: Stochastic Processes in Physics and Chemistry, 3rd edn. Elsevier, Amsterdam (2007)
19. Kierzek, A., Zaim, J., Zielenkiewicz, P.: The effect of transcription and translation initiation frequencies on the stochastic fluctuations in prokaryotic gene expression. Journal of Biological Chemistry 276(11), 8165–8172 (2001)
20. Law, A., Kelton, D.: Simulation Modelling and Analysis. McGraw-Hill Education, New York (2000)
21. Little, J.W., Shepley, D.P., Wert, D.W.: Robustness of a gene regulatory circuit. The EMBO Journal 18(15), 4299–4307 (1999)
22. Losick, R., Desplan, C.: Stochasticity and Cell Fate. Science 320(5872), 65–68 (2008)
23. Maamar, H., Raj, A., Dubnau, D.: Noise in gene expression determines cell fate in Bacillus subtilis. Science 317(5837), 526–529 (2007)
24. McAdams, H.H., Arkin, A.: Stochastic mechanisms in gene expression. PNAS, USA 94, 814–819 (1997)
25. McAdams, H.H., Arkin, A.: It's a noisy business! Trends in Genetics 15(2), 65–69 (1999)
26. Munsky, B., Khammash, M.: The finite state projection algorithm for the solution of the chemical master equation. J. Chem. Phys. 124, 044144 (2006)
27. Ozbudak, E.M., Thattai, M., Kurtser, I., Grossman, A.D., van Oudenaarden, A.: Regulation of noise in the expression of a single gene. Nature Genetics 31(1), 69–73 (2002)
28. Patel, P., Arcangioli, B., Baker, S., Bensimon, A., Rhind, N.: DNA replication origins fire stochastically in fission yeast. Mol. Biol. Cell 17, 308–316 (2006)
29. Paulsson, J.: Summing up the noise in gene networks. Nature 427(6973), 415–418 (2004)
30. Rao, C., Wolf, D., Arkin, A.: Control, exploitation and tolerance of intracellular noise. Nature 420(6912), 231–237 (2002)
31. Sandmann, W.: Stochastic simulation of biochemical systems via discrete-time conversion. In: Proceedings of the 2nd Conference on Foundations of Systems Biology in Engineering, pp. 267–272. Fraunhofer IRB Verlag (2007)
32. Sandmann, W., Maier, C.: On the statistical accuracy of stochastic simulation algorithms implemented in Dizzy. In: Proc. WCSB, pp. 153–156 (2008)

33. Sandmann, W., Wolf, V.: A computational stochastic modeling formalism for biological networks. Enformatika Transactions on Engineering, Computing and Technology 14, 132–137 (2006)
34. Sandmann, W., Wolf, V.: Computational probability for systems biology. In: Fisher, J. (ed.) FMSB 2008. LNCS (LNBI), vol. 5054, pp. 33–47. Springer, Heidelberg (2008)
35. Sidje, R., Burrage, K., MacNamara, S.: Inexact uniformization method for computing transient distributions of Markov chains. SIAM J. Sci. Comput. 29(6), 2562–2580 (2007)
36. Srivastava, R., You, L., Summers, J., Yin, J.: Stochastic vs. deterministic modeling of intracellular viral kinetics. Journal of Theoretical Biology 218, 309–321 (2002)
37. Stewart, W.J.: Introduction to the Numerical Solution of Markov Chains. Princeton University Press, Princeton (1995)
38. Swain, P.S., Elowitz, M.B., Siggia, E.D.: Intrinsic and extrinsic contributions to stochasticity in gene expression. PNAS, USA 99(20), 12795–12800 (2002)
39. Thattai, M., van Oudenaarden, A.: Intrinsic noise in gene regulatory networks. PNAS, USA 98(15), 8614–8619 (2001)
40. Turner, T.E., Schnell, S., Burrage, K.: Stochastic approaches for modelling in vivo reactions. Computational Biology and Chemistry 28, 165–178 (2004)
41. van Moorsel, A., Sanders, W.: Adaptive uniformization. ORSA Communications in Statistics: Stochastic Models 10(3), 619–648 (1994)
42. Warmflash, A., Dinner, A.: Signatures of combinatorial regulation in intrinsic biological noise. PNAS 105(45), 17262–17267 (2008)
43. Wilkinson, D.J.: Stochastic Modelling for Systems Biology. Chapman & Hall, Boca Raton (2006)
44. Zhang, J., Watson, L.T., Cao, Y.: A modified uniformization method for the solution of the chemical master equation. TR-07-31, Computer Science, Virginia Tech. (2007)

# Equivalence and Discretisation in Bio-PEPA

Vashti Galpin[1] and Jane Hillston[1,2]

[1] LFCS, School of Informatics, University of Edinburgh
[2] Centre for Systems Biology (CSBE), University of Edinburgh
Vashti.Galpin@ed.ac.uk, Jane.Hillston@ed.ac.uk

**Abstract.** Bio-PEPA is a process algebra for modelling biological systems. An important aspect of Bio-PEPA is the ability it provides to discretise concentrations resulting in a smaller, more manageable state space. The discretisation is based on a step size which determines the size of each discrete level and also the maximum number of levels. This paper considers the relationship between two discretisations of the same Bio-PEPA model that differ only in the step size and hence the maximum number of levels, by using the idea of equivalence from concurrency and process algebra. We present a novel behavioural semantic equivalence, compression bisimulation, that equates two discretisations of the same model and we show that this equivalence is a congruence with respect to the synchronisation operator.

## 1 Introduction

The use of process algebras for modelling biological systems has become a popular technique [1,2,3,4,5]. Some approaches use the process algebra as originally defined for description of computer systems and in others a process algebra is tailored to be specific to systems biology. One of the latter class is Bio-PEPA [6] which was developed from the stochastic process algebra PEPA [7] and has been successfully used to model Goldbeter's model of cyclin oscillation [8,9], the Repressilator [10], genetic networks [6], the MAPK model [11], the Neurospora circadian clock [12] and the gp130/JAK/STAT pathway [13]. This paper investigates a semantic equivalence for Bio-PEPA.

An important aspect of Bio-PEPA is the ability it provides for the discretisation of concentrations. Instead of working with a "process-as-molecule" approach, it uses a "process-as-species" approach whereby a process can either be parameterised by concentration or by a discrete level which is obtained from dividing the concentration into a discrete number of chunks. Typically, there is a fixed step size and each species has a maximum number of levels dependent on its maximum concentration. For a given step size, we call a system with levels a discretisation.

Bio-PEPA distinguishes itself from many other process algebras for modelling biological systems by providing multiple analyses including continuous-time Markov chains (CTMCs), ordinary differential equations (ODEs), stochastic simulation and model checking. By developing a semantic equivalence for Bio-PEPA, we have a new type of analysis based on behaviour that can be used to compare models.

Semantic equivalences are an important technique in process algebra for specifying notions of equivalent behaviour. They equate processes that have the same behaviour,

P. Degano and R. Gorrieri (Eds.): CMSB 2009, LNBI 5688, pp. 189–204, 2009.

and can be divided into qualitative equivalences which only consider the behaviour in terms of which actions can be performed and quantitative equivalences which consider the rates at which actions can happen as part of the behaviour. In this paper we consider a qualitative equivalence. Adding quantitative aspects is ongoing research.

Semantic equivalences typically have two important properties – they are equivalence relations (hence the name) and congruence relations. A congruence relation is a relation that is preserved by the operators of the process algebra. For example, if $\phi$ is a binary operator, then an equivalence $\equiv$ is preserved by this operator if $P \equiv Q$ implies that $P \phi R \equiv Q \phi R$ and $R \phi P \equiv R \phi Q$ for every process $R$.

These properties are important when modelling and evaluating the concurrent behaviour of computer systems as they let us substitute like for like thereby exploiting the compositionality provided by a process algebra. This allows for the substitution of a system with a smaller state space or other desirable properties, and make the analysis of the system easier. In applying the idea of a semantic equivalence in systems biology, similar advantages will be gained hence the importance and timeliness of this research. Moreover, since the semantic equivalence we define is based upon what reactions can be observed, it is well suited for biological modelling.

In searching for a suitable equivalence, there are at least two approaches that can be taken. One is to consider existing equivalences from the literature. The other is to consider what behaviours we want to consider as identical and to develop an equivalence from this starting point. This is the approach taken here.

We have a immediate candidate for what we want to consider the same. For a Bio-PEPA system, we can consider two different discretisations of that system. Since they both represent the same system, we want their behaviours to be identified (assuming neither have few enough levels to give pathological behaviour). This approach is suitable since semantic equivalences are used to identify the same behaviour in different abstractions of a system, and clearly two discretisations are two abstractions.

Starting from this point, we define an equivalence relation over the states of the model that relates states that have the same possible reactions. This equivalence relation defines equivalence classes of states with the same behaviour and from this we can use a classical notion of equivalence, bisimulation, to define our semantic equivalence.

This new semantic equivalence, compression bisimulation, has not been chosen randomly but through understanding the differences between discretisations and ensuring that the semantic equivalence has the desirable properties mentioned above. Ensuring a semantic equivalence is an equivalence relation is not hard. Establishing congruence is much harder because stoichiometry coefficients greater than one lead to a complex transition system. Being able to prove congruence played a major role in the selection of compression equivalence. Moreover, since two discretisations of a single species should have the same behaviour under the equivalence, it was necessary to prove this as well.

The first result of this paper shows that in the sequential case, a single species, two discretisations are related by compression bisimulation. The second shows that compression bisimulation is a congruence with respect to the cooperation operator. A corollary of this is that in the general case of a Bio-PEPA system, namely with multiple species, two discretisations are related by compression bisimulation.

The structure of the document is as follows. First, we introduce Bio-PEPA, then we present some basic ideas relating to semantic equivalences and then we present compression bisimulation and show it has the desired properties. We then give an example.

## 2    Bio-PEPA

This section presents an overview of Bio-PEPA [6]. The main components of a Bio-PEPA system are the sequential components describing the behaviour of each of the species and the model component describing the interactions between the species and initial amounts. Additionally, a context is defined, including functional rates, compartments, parameters. The syntax of the sequential (species) components is defined as

$$S ::= (\alpha, \kappa) \text{ op } S \mid S + S \mid C \qquad \qquad \text{op} ::= \downarrow \mid \uparrow \mid \oplus \mid \ominus \mid \odot.$$

In the prefix term $(\alpha, \kappa) \text{ op } S$, $\alpha$ is an action name and can be viewed as the name or label of a reaction, $\kappa$ is the stoichiometry coefficient of the species and the prefix combinator op represents the role of the element in the reaction. Specifically, $\downarrow$ is a reactant, $\uparrow$ a product, $\oplus$ an activator, $\ominus$ an inhibitor and $\odot$ a generic modifier. The operator $+$ expresses the choice between possible actions and the constant $C$ is defined by an equation $C \stackrel{def}{=} S$. The syntax of model components is defined as

$$P ::= P \bowtie_{\mathcal{L}} P \mid S(x)$$

The process $P \bowtie_{\mathcal{L}} Q$ denotes the synchronisation between components $P$ and $Q$ and the set $\mathcal{L}$ specifies those activities on which the components must synchronise. In the model component $S(x)$, the parameter $x \in \mathbb{R}$ represents the concentration or level. We work with a constrained set of Bio-PEPA model components as given by the following definition which specifies a well-formed set of components. We ensure that a species consists of a choice between reactions, and no reaction name is repeated within a species. At the model level, there can only be one species component for each species.

**Definition 1.** *A Bio-PEPA sequential component $C$ is* well-defined *if it has the form*

$$C \stackrel{def}{=} (\alpha_1, \kappa_1) \text{ op}_1 C + \ldots + (\alpha_n, \kappa_n) \text{ op}_n C \quad written\ as \quad C \stackrel{def}{=} \sum_{i=1}^{n} (\alpha_i, \kappa_i) \text{ op}_i C$$

*where $\alpha_i \neq \alpha_j$ for $i \neq j$.*

*A Bio-PEPA model component $P$ is* well-defined *if it has the form*

$$P \stackrel{def}{=} C_1(x_1) \bowtie_{\mathcal{L}_1} \ldots \bowtie_{\mathcal{L}_{p-1}} C_p(x_p),$$

*each $C_i$ is a well-defined sequential component, the elements of each $\mathcal{L}_j$ appear in $P$ and if $i \neq j$ then $C_i \neq C_j$.*

We define a Bio-PEPA system, consisting of a set of well-defined sequential components, a well-defined model component and context, as follows.

prefixReac
$$\frac{}{(\alpha,\kappa)\downarrow S(l) \xrightarrow{(\alpha,[S:\downarrow(l,\kappa)])}_c S(l-\kappa)} \quad \kappa \le l \le N_S$$

prefixProd
$$\frac{}{(\alpha,\kappa)\uparrow S(l) \xrightarrow{(\alpha,[S:\uparrow(l,\kappa)])}_c S(l+\kappa)} \quad 0 \le l \le N_S - \kappa$$

prefixMod
$$\frac{}{(\alpha,\kappa) \text{ op } S(l) \xrightarrow{(\alpha,[S:op(l,\kappa)])}_c S(l)} \quad \begin{array}{l} 0 < l \le N_S \text{ if op} = \oplus \\ 0 \le l \le N_S \text{ if op} \in \{\ominus,\odot\} \end{array}$$

choice1
$$\frac{S_1(l) \xrightarrow{(\alpha,w)}_c S_1'(l)}{(S_1 + S_2)(l) \xrightarrow{(\alpha,w)}_c S_1'(l)}$$

choice2
$$\frac{S_2(l) \xrightarrow{(\alpha,w)}_c S_2'(l)}{S_1 + S_2(l) \xrightarrow{(\alpha,w)}_c S_2'(l)}$$

constant
$$\frac{S(l) \xrightarrow{(\alpha,[S:\text{ op }(l,\kappa)])}_c S'(l)}{C(l) \xrightarrow{(\alpha,[C:\text{ op }(l,\kappa)])}_c S'(l)} \quad C \stackrel{def}{=} S$$

coop1
$$\frac{P_1 \xrightarrow{(\alpha,w)}_c P_1'}{P_1 \bowtie_M P_2 \xrightarrow{(\alpha,w)}_c P_1' \bowtie_M P_2} \quad \alpha \notin M$$

coop2
$$\frac{P_2 \xrightarrow{(\alpha,w)}_c P_2'}{P_1 \bowtie_M P_2 \xrightarrow{(\alpha,w)}_c P_1 \bowtie_M P_2'} \quad \alpha \notin M$$

coop3
$$\frac{P_1 \xrightarrow{(\alpha,w_1)}_c P_1' \quad P_2 \xrightarrow{(\alpha,w_2)}_c P_2'}{P_1 \bowtie_M P_2 \xrightarrow{(\alpha,w_1::w_2)}_c P_1' \bowtie_M P_2'} \quad \alpha \in M$$

**Fig. 1.** Operational semantics of Bio-PEPA

Final
$$\frac{P \xrightarrow{(\alpha,w)}_c P'}{\langle \mathcal{V},\mathcal{N},\mathcal{K},\mathcal{F},Comp,P \rangle \xrightarrow{(\alpha,r_\alpha[w,\mathcal{N},\mathcal{K}])}_s \langle \mathcal{V},\mathcal{N},\mathcal{K},\mathcal{F},Comp,P' \rangle}$$

Enrich
$$\frac{P \xrightarrow{(\alpha,w)}_c P'}{\langle \mathcal{V},\mathcal{N},\mathcal{K},\mathcal{F},Comp,P \rangle \xrightarrow{(\alpha,w)}_{sc} \langle \mathcal{V},\mathcal{N},\mathcal{K},\mathcal{F},Comp,P' \rangle}$$

**Fig. 2.** Operational semantics of Bio-PEPA (continued)

**Definition 2.** *A Bio-PEPA system $\mathcal{P}$ is a 6-tuple $\langle \mathcal{V},\mathcal{N},\mathcal{K},\mathcal{F},Comp,P \rangle$, where $\mathcal{V}$ is the set of compartments, $\mathcal{N}$ is the set of quantities describing each species, $\mathcal{K}$ is the set of parameters, $\mathcal{F}$ is the set of functional rates, $Comp$ is the set of well-defined sequential components and $P$ is a well-defined model component.*

*Elements of $\mathcal{N}$ have the form $C : H = h, N = n, M = m, V = v, unit = u$ where $C$ is a species name that is defined in $Comp$, $H = h$ defines the step size, $N = n$ defines the maximum number of levels for $C$, $M = m$ defines the maximum concentration for $C$, $V = v$ names the compartment in which $C$ appears and $unit = u$ defines the measurement unit of the concentration.*

The notation $\langle \mathcal{T}, P \rangle$ will be used for $\langle \mathcal{V}, \mathcal{N}, \mathcal{K}, \mathcal{F}, Comp, P \rangle$ since the details of the tuple are not relevant here. The model component is defined in terms of concentrations, but can be expressed terms of levels where these are a discretisation of the concentration. In $\mathcal{N}$, for each species, there are three elements $H$, $N$ and $M$ which represent the step size, the maximum number of levels and the maximum concentration respectively. Their relationship is defined as $N = \lceil M/H \rceil$. A species with a concentration then has an associated level in the range $0, 1, \ldots, N - 1, N$, giving a total of $N + 1$ possible levels. We call the system obtained in this way a *Bio-PEPA system with levels*. We will assume that the step size $H$ is the same for all species to ensure conservation of mass.

The operational semantics for Bio-PEPA systems with levels is given in Tables 1 and 2. In the first table, $N_S$ refers to the maximum number of levels for the species $S$. In the rule coop3, $w_1::w_2$ represents list concatenation. For the rule Final, $r_\alpha[w, \mathcal{N}, \mathcal{K}] = f_\alpha[w, \mathcal{N}, \mathcal{K}]/H \in (0, \infty)$ where $f_\alpha$ is the functional rate for the reaction $\alpha$ from $\mathcal{F}$ and $H$ is the step size. We do not discuss this or the string $w$ further as the equivalences in this paper only consider the action $\alpha$ and ignore the rest of the transition label.

The operational semantics creates three different transition systems. The rules for $\rightarrow_c$ define the capability relation. The rule Final defines the system/stochastic relation $\rightarrow_s$ which includes the context, and the rate at which the transition takes place appears together with the action. The rule Enrich defines the system-capability relation $\rightarrow_{sc}$. This relation is necessary since it contains the context information as well as the detailed information that is captured in the list of strings $w$.

The following definition describes the derivative set for the relation $\rightarrow_{sc}$. In this paper, we work almost exclusively with this relation since it provides the necessary information about the context.

**Definition 3.** *The system-capability derivative set ds($\mathcal{P}$) is the smallest set such that* $\mathcal{P} \in ds(\mathcal{P})$ *and if* $\mathcal{P}' \in ds(\mathcal{P})$ *and* $\mathcal{P}' \xrightarrow{(\alpha,r)}_{sc} \mathcal{P}''$ *then* $\mathcal{P}'' \in ds(\mathcal{P})$.

The next definition captures the reactions that are immediately possible with respect to the operational semantics. This means it takes into account the stoichiometry of a reaction as well as the current level of a species.

**Definition 4.** *The set of* current actions *enabled in* $\langle \mathcal{T}, P \rangle$ *is defined as* $\mathcal{A}(\langle \mathcal{T}, P \rangle) = \mathcal{A}(P)$ *where* $N_S$ *is the maximum number of levels for species component* $S$.

$$\mathcal{A}(((\alpha, \kappa) \downarrow S)(l)) = \{\alpha\} \ \text{if} \ \kappa \leq l \leq N_S \ \text{otherwise} \ \emptyset$$
$$\mathcal{A}(((\alpha, \kappa) \uparrow S)(l)) = \{\alpha\} \ \text{if} \ 0 \leq l \leq N_S - \kappa \ \text{otherwise} \ \emptyset$$
$$\mathcal{A}(((\alpha, \kappa) \oplus S)(l)) = \{\alpha\} \ \text{if} \ 0 < l \leq N_S \ \text{otherwise} \ \emptyset$$
$$\mathcal{A}(((\alpha, \kappa) \ominus S)(l)) = \{\alpha\} \ \text{if} \ 0 \leq l \leq N_S \ \text{otherwise} \ \emptyset$$
$$\mathcal{A}(((\alpha, \kappa) \odot S)(l)) = \{\alpha\} \ \text{if} \ 0 \leq l \leq N_S \ \text{otherwise} \ \emptyset$$
$$\mathcal{A}((S_1 + S_2)(l)) = \mathcal{A}(S_1(l)) \cup \mathcal{A}(S_2(l))$$
$$\mathcal{A}(C(l)) = \mathcal{A}(S(l)) \ \text{where} \ C \stackrel{\text{def}}{=} S$$
$$\mathcal{A}(P_1 \bowtie_L P_2) = \mathcal{A}(P_1) \setminus L \cup \mathcal{A}(P_2) \setminus L \cup (\mathcal{A}(P_1) \cap \mathcal{A}(P_2) \cap L)$$

The stoichiometry plays a role in defining the set of current actions. A species definition specifies a set of actions (reactions), but the current action set may be a subset if the current level is insufficient to satisfy the constraints imposed by the stoichiometry.

Since we are working with Bio-PEPA systems that vary only in step size and maximum numbers of levels, we require notation to capture this. Given a Bio-PEPA system $\mathcal{P}$ we can define a system where the lowest maximum number of levels for any species is $n$. As mentioned previously, we assume that $H$ is identical for all components.

**Definition 5.** *Let* $\mathcal{P} = \langle \mathcal{V}, \mathcal{N}, \mathcal{K}, \mathcal{F}, Comp, P \rangle$ *be a Bio-PEPA system with well-defined P parameterised by concentration. For* $n \in \mathbb{N}$, *the Bio-PEPA system with levels* $\mathcal{P}^n$ *is defined as* $\mathcal{P}^n = \langle \mathcal{V}, \mathcal{N}', \mathcal{K}, \mathcal{F}, Comp, P' \rangle$ *where*

1. $\gamma = (1/n) \cdot \min\{m \mid C : H = h, N = n', M = m, V = v, unit = u \in \mathcal{N}\}$
2. $C : H = h, N = n', M = m, V = v, unit = u \in \mathcal{N} \ \Rightarrow$
$$C : H = \gamma, N = \lceil m/\gamma \rceil, M = m, V = v, unit = u \in \mathcal{N}'$$
3. $P \stackrel{\text{def}}{=} C_1(x_1) \bowtie_{L_1} \ldots \bowtie_{L_{p-1}} C_p(x_p) \Rightarrow P' \stackrel{\text{def}}{=} C_1(\lceil x_1/\gamma \rceil) \bowtie_{L_1} \ldots \bowtie_{L_{p-1}} C_p(\lceil x_p/\gamma \rceil)$

$\mathcal{N}$ contains information about each species. The definition above identifies the species with the smallest concentration, determines the new step size that will ensure $n$ levels for that species and then adjusts the other species to use the same step size (to conserve mass) hence modifying $\mathcal{N}$. Since $\mathcal{P}$ is a Bio-PEPA system with species components parameterised by concentration and we wish to work with a system with levels, the initial concentrations in the third point are transformed to initial levels. We will use the notation $\mathcal{P}^n = \langle \mathcal{T}^n, P \rangle$ to indicate that the lowest number of maximum levels for any species is $n$ and refer to $\mathcal{P}^n$ as a discretisation of $\mathcal{P}$.

## 3   Semantic Equivalences

In process algebras, a semantic equivalence defines what it means for two models to have the same behaviour. A classical notion of equivalence is that of bisimulation [14].

**Definition 6.** *A binary relation* $\mathcal{R}$ *is a bisimulation if for any* $(P, Q) \in \mathcal{R}$ *and for any* $\theta$ *whenever*

1. $P \xrightarrow{\theta} P'$, *there exists* $Q'$ *such that* $Q \xrightarrow{\theta} Q'$ *and* $(P', Q') \in \mathcal{R}$, *and*
2. $Q \xrightarrow{\theta} Q'$, *there exists* $P'$ *such that* $P \xrightarrow{\theta} P'$ *and* $(P', Q') \in \mathcal{R}$

*P and Q are* bisimilar, $P \sim Q$ *if* $(P, Q) \in \mathcal{R}$ *for some bisimulation* $\mathcal{R}$.

This leads to the definition $\sim \ = \ \bigcup\{\mathcal{R} \mid \mathcal{R}$ a bisimulation$\}$ and one can show that $\sim$ is the largest bisimulation. Moreover, it can also be shown that bisimulation is an equivalence relation therefore it is reflexive, symmetric and transitive. Bisimulation is a fine-grained notion of behaviour and equates far fewer models than language/trace equivalence, for example. It requires that related models can match each other's transitions and that the resultant models also have this property. Consider the labelled transition systems in Figure 3. They generate the same strings/traces but they are not bisimilar because we cannot find anything to match with $Q_1$. $Q_2$ is not suitable since it only has a $b$ transition and $Q_2'$ is not suitable since it only has a $c$ transition.

As mentioned in the introduction, we also wish that our new semantic equivalence be a congruence with respect to the language we use.

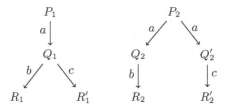

**Fig. 3.** Example of transition systems that are not bisimilar

$$\langle \mathcal{T}^2, A(2) \rangle \xrightarrow{(\alpha, v_1)} \langle \mathcal{T}^2, A(1) \rangle \xrightarrow{(\alpha, v_0)} \langle \mathcal{T}^2, A(0) \rangle$$

$$\langle \mathcal{T}^3, A(3) \rangle \xrightarrow{(\alpha, u_2)} \langle \mathcal{T}^3, A(2) \rangle \xrightarrow{(\alpha, u_1)} \langle \mathcal{T}^3, A(1) \rangle \xrightarrow{(\alpha, u_0)} \langle \mathcal{T}^3, A(0) \rangle$$

**Fig. 4.** Example of discretisations that are not bisimilar

### 3.1   Compression Bisimulation

We now define the new equivalence. As noted in the introduction, our approach here is to consider the systems we want to be equivalent and to work from there. We want our equivalence to be a congruence and we also want to equate discretisations with sufficiently large numbers of levels because this is our starting point. However, having said that, we are still interested in an equivalence that is similar to classical equivalences such as bisimulation. Note that we cannot use bisimulation directly here. This can be shown by the species component $A \stackrel{def}{=} (\alpha, 1)\uparrow A$. Figure 4 gives the transition system for two different discretisation, one where the maximum level is 2 and the other where it is 3. We can relate $\langle \mathcal{T}^2, A(i) \rangle$ and $\langle \mathcal{T}^3, A(i) \rangle$ for $0 \le i \le 2$ but we cannot relate $\langle \mathcal{T}^3, A(3) \rangle$ to any of $\langle \mathcal{T}^2, A(i) \rangle$.

However, although we cannot use bisimulation directly, we are able to use it indirectly over equivalence classes and achieve the goals of congruence and equating discretisations. We now present definitions that allow us to achieve that.

We first need to define the equivalence relation that will define the relevant equivalence classes. The current level together with the stoichiometry associated with a reaction determine which reactions can occur, therefore we are interested in grouping together those states of the Bio-PEPA system for which the same reactions can take place. The collection of enabled reactions becomes our underlying notion of behaviour. This captures the similarities that we see between different discretisations. From a biological point of view, this is sensible because it is an observational notion of equivalence.

In light of this, we can define an equivalence relation over Bio-PEPA systems that depends on $\mathcal{A}$ which defines the actions that are currently enabled. Two processes are related if their current action sets are the same.

**Definition 7.** *The current action relation $\mathcal{H}$ over well-defined Bio-PEPA systems is defined as $\mathcal{H} = \{(\mathcal{P}_1, \mathcal{P}_2) \mid \mathcal{A}(\mathcal{P}_1) = \mathcal{A}(\mathcal{P}_2)\}$.*

**Proposition 1.** *$\mathcal{H}$ is an equivalence relation.*

Because $\mathcal{H}$ is an equivalence relation, it defines equivalence classes of Bio-PEPA systems which have the same current actions. For a set of Bio-PEPA systems $\mathcal{X}$, the equivalence classes of $\mathcal{X}$ with respect to $\mathcal{H}$ is denoted $\mathcal{X}/\mathcal{H}$. $\mathcal{A}$ can be extended to the equivalence classes in the obvious manner. Hence, for $P \in H$ an equivalence class, $\mathcal{A}(H) = \mathcal{A}(\mathcal{P})$.

We are interested in considering the equivalence classes over the derivative set of a given Bio-PEPA system $\mathcal{P}$ because we want to consider the overall behaviour of individual Bio-PEPA systems and we define $\mathcal{P}_{\mathcal{H}} = ds(\mathcal{P})/\mathcal{H}$. Since we want to define a bisimulation-style equivalence we need to define transitions between equivalence classes. The basic idea is that if there is a transition between individual members of two equivalence classes then there is a transition between those equivalence classes.

**Definition 8.** *For $H, H' \in \mathcal{P}_{\mathcal{H}}$, $H \xhookrightarrow{\alpha} H'$ if there exists $\mathcal{P} \in H$ and $\mathcal{P}' \in H'$ such that $\mathcal{P} \xrightarrow{(\alpha, w)}_{sc} \mathcal{P}'$.*

We can then finalise the definition for our new equivalence as follows. We use Definition 6 for the definition of $\sim$ but substitute $\xhookrightarrow{\alpha}$ for all instances of $\xrightarrow{\theta}$ and, moreover the relation $\sim$ is defined between equivalence classes.

**Definition 9.** *$\mathcal{P}$ and $\mathcal{Q}$ are* compression bisimilar, *$\mathcal{P} \simeq \mathcal{Q}$, if $\mathcal{P}_{\mathcal{H}} \sim \mathcal{Q}_{\mathcal{H}}$.*

### 3.2   Equivalence and Congruence Results

This section details results about the new equivalence. First we need show that it is an equivalence relation.

**Proposition 2.** *$\mathcal{P} \simeq \mathcal{Q}$ is an equivalence relation.*

*Proof.* This is straightforward because $\sim$ is an equivalence relation.     □

Next, we consider the sequential case of two discretisations and show that they are equated by the new equivalence. The sequential case consists of considering a single species and two discretisations. The first theorem of the paper shows that given a single species component and two discretisations, then the two discretisations are compression bisimilar because their induced equivalence classes are bisimilar (in fact, they are isomorphic). First, some notation and various lemmas are required. The complexity of these results is due to the fact that stoichiometry can be larger than one. For a sequential Bio-PEPA component $C \stackrel{def}{=} \sum_{i=1}^{m}(\alpha_i, \kappa_i) \, \mathrm{op}_i \, C$, let

$$T_\uparrow = \{\kappa_i \mid (\alpha_i, \kappa_i) \uparrow C \text{ appears in the definition of } C\} \qquad t_\uparrow = |T_\uparrow|$$
$$T_\downarrow = \{\kappa_i \mid (\alpha_i, \kappa_i) \downarrow C \text{ appears in the definition of } C\} \cup$$
$$\qquad \{1 \mid (\alpha_i, \kappa_i) \oplus C \text{ appears in the definition of } C\} \qquad t_\downarrow = |T_\downarrow|$$
$$\mathcal{A}_C = \{\alpha_i \mid (\alpha_i, \kappa_i) \, \mathrm{op}_i \, C \text{ appears in the definition of } C\}$$
$$k_m = \max\{k_\downarrow, k_\uparrow, 1\}, \; k_\downarrow = \max(T_\downarrow), \; k_\uparrow = \max(T_\uparrow), \text{ hence } k_m \geq k_\downarrow, k_m \geq k_\uparrow.$$

The diagram in Figure 5 illustrates the equivalence classes for two discretisations of $C \stackrel{def}{=} (\alpha, 2) \uparrow C + (\beta, 3) \uparrow C + (\gamma, 4) \downarrow C + (\delta, 1) \oplus C$ with $n = 11$ and $n' = 13$. It

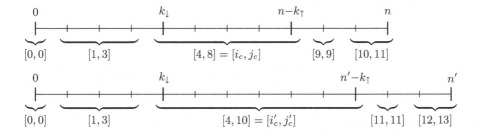

**Fig. 5.** The equivalence classes of two discretisations of a species component

also demonstrates how the various stoichiometry coefficients result in different equivalent classes. The lemmas that follow prove the properties of the equivalence classes as shown in this diagram.

The next lemma establishes that equivalence classes can be ordered which makes them easier to manipulate in later lemmas. The following lemma builds on this and shows that there are a fixed number of equivalence classes if there are sufficient levels and it contributes to the definition of the isomorphism in the first theorem.

**Lemma 1.** *For a sequential Bio-PEPA component $C \stackrel{\text{def}}{=} \sum_{i=1}^{m} (\alpha_i, \kappa_i) \, op_i \, C$ and the Bio-PEPA system $\mathcal{S}^n = \langle T^n, C \rangle$, the equivalence classes of $\mathcal{S}_{\mathcal{H}}^n$ form a strict order.*

*Proof.* The set $ds(\mathcal{S}^n)$ contains elements of the form $\langle T^n, C(l) \rangle$, where $l$ ranges over $0, \ldots, n$. Each equivalence class is a subsequence of $0, \ldots, n$ because of the side conditions of the prefix rules. Therefore a class can be described by its smallest and largest elements $[i, j]$ for $i \leq j$. These intervals do not overlap because the equivalence classes form a partition. Hence for any two equivalence classes $[i, j]$ and $[i', j']$, either $j < i'$ or $j' < i$, and this property defines a strict order over the equivalence classes.     □

**Lemma 2.** *For a sequential Bio-PEPA component $C \stackrel{\text{def}}{=} \sum_{i=1}^{m} (\alpha_i, \kappa_i) \, op_i \, C$ and the Bio-PEPA system $\mathcal{S}^n = \langle T^n, C \rangle$, if $n \geq k_\uparrow + k_\downarrow + 1$, then $\mathcal{S}_{\mathcal{H}}^n$ has $t_\uparrow + t_\downarrow + 1$ equivalence classes.*

*Proof.* By Lemma 1, a sequence of equivalence classes $[i_1, j_1], [i_2, j_2], \ldots, [i_t, j_t]$ partitioning $\mathcal{S}_{\mathcal{H}}^n$ exist. We show that there is an equivalence class $[i_c, j_c]$ with $\mathcal{A}([i_c, j_c]) = \mathcal{A}_C$, $i_c = k_\downarrow$ and $j_c = n - k_\uparrow$. This is well-defined since $n \geq k_\downarrow + k_\uparrow + 1$. Consider $l \in [i_c, j_c]$. Any production prefix $(\alpha, \kappa) \uparrow C$ is enabled since $0 \leq i_c \leq l \leq n - k_\uparrow \leq n - \kappa$. Any reactant prefix $(\alpha, \kappa) \downarrow C$ is enabled because $\kappa \leq k_\downarrow \leq l \leq n - k_\uparrow \leq n$. Prefixes containing $\ominus$ or $\odot$ can always generate transitions. Since $l \geq 1$, any prefix of the form $(\alpha, \kappa) \oplus C$ is enabled. Hence $\mathcal{A}([i_c, j_c]) = \mathcal{A}_C$ and the class cannot be larger.

Next we consider the equivalence classes that come before $[i_c, j_c]$. Order the elements of $T_\downarrow$ from smallest to largest, $\tau_1, \tau_2, \ldots, \tau_{t_\downarrow - 1}, \tau_{t_\downarrow}$ where $\tau_{t_\downarrow} = k_\downarrow$ then we have that $[i_1, j_1], [i_2, j_2], \ldots, [i_{c-1}, j_{c-1}]$ is $[0, \tau_1 - 1], [\tau_1, \tau_2 - 1], \ldots, [\tau_{t_\downarrow - 1}, \tau_{t_\downarrow} - 1]$ which gives $t_\downarrow$ equivalence classes.

Similarly, we have $t_\uparrow$ equivalence classes corresponding to the elements of $T_\uparrow$. This means that there are $t_\downarrow + t_\uparrow + 1$ equivalence classes in total.     □

**Corollary 1.** *Let* $C \stackrel{def}{=} \sum_{i=1}^{m} (\alpha_i, \kappa_i) \, op_i \, C$ *be a sequential Bio-PEPA component which has stoichiometry coefficient 1 for all reactant prefixes and product prefixes then any discretisation with $n \geq 3$ has three equivalence classes.*

The next lemma establishes a lower bound on the size of the equivalence class that is capable of performing all actions, namely the "central" class $[i_c, j_c]$ with $\mathcal{A}([i_c, j_c]) = \mathcal{A}_c$. This class is the only one that differs in cardinality for different discretisations, and it grows in size as the number of levels are increased. The other classes do not differ between different discretisations because they are defined by the same stoichiometry coefficients. The value $k_m$ is used since knowing that $k_m$ is a bound on the size of $[i_c, j_c]$ is important for a later lemma.

**Lemma 3.** *Let* $C \stackrel{def}{=} \sum_{i=1}^{m} (\alpha_i, \kappa_i) \, op_i \, C$ *be a sequential Bio-PEPA component and let* $S^n = \langle T^n, C \rangle$ *for $n \geq k_\uparrow + k_m + k_\downarrow$. Then $[i_c, j_c]$, the equivalence class of $S^n/\mathcal{H}$ with* $\mathcal{A}[i_c, j_c] = \mathcal{A}_C$, *has cardinality greater than $k_m$.*

*Proof.* It is the case that $i_c = k_\downarrow$ and $j_c = n - k_\uparrow$. The cardinality of $[i_c, j_c]$ is $j_c - i_c + 1 = n - k_\uparrow - k_\downarrow + 1 \geq k_\uparrow + k_m + k_\downarrow - k_\uparrow - k_\downarrow + 1 = k_m + 1 > k_m$. $\qquad\qquad\square$

This implies that the cardinality/width of $[i_c, j_c]$ is greater than both $k_\downarrow$ and $k_\uparrow$. The next lemma relates the equivalence classes obtained for two different values of $n$ by expressing the classes for the larger value in terms of the intervals of the smaller value.

**Lemma 4.** *Let* $C \stackrel{def}{=} \sum_{i=1}^{m} (\alpha_i, \kappa_i) \, op_i \, C$ *be a sequential Bio-PEPA component and let $S = \langle T, C \rangle$. Let $n' = n + d$. Then the equivalence classes of $S^n_\mathcal{H}$ are described by the ordered intervals $[0, j_1], \ldots, [i_c, j_c], \ldots, [i_{t-1}, j_{t-1}], [i_t, n]$ and the equivalence classes of $S^{n'}_\mathcal{H}$ are described by the ordered intervals $[0, j_1], \ldots, [i_c, j_c + d], \ldots, [i_{t-1} + d, j_{t-1} + d], [i_t + d, n + d]$ where $[i_c, j_c]$ and $[i_c, j_c + d]$ are the equivalence classes in which all actions of $C$ are possible.*

*Proof.* The elements of $S^n_\mathcal{H}$ are $[0, j_1], \ldots, [i_c, j_c], \ldots, [i_{t-1}, j_{t-1}], [i_t, n]$ and those of $S^{n'}_\mathcal{H}$ are $[0, j'_1], \ldots, [i'_c, j'_c], \ldots, [i'_{t-1}, j'_{t-1}], [i'_t, n']$. The first $t_\downarrow$ equivalence classes are the same in both cases because they are defined by the same stoichiometric coefficients so $S^n_\mathcal{H}$ is $[0, j_1], \ldots, [i_c, j'_c], \ldots, [i'_{t-1}, j'_{t-1}], [i'_t, n']$. Since $j_c = n - k_\uparrow$ and $j'_c = n' - k_\uparrow$, $j'_c = j_c + d$ and this offset is the same throughout the remaining equivalence classes which are defined by the same stoichiometric coefficients hence $S^n_\mathcal{H}$ can be written $[0, j_1], \ldots, [i_c, j_c + d], \ldots, [i_{t-1} + d, j_{t-1} + d], [i_t + d, n + d]$. $\qquad\square$

Finally the most important lemma shows that the same transitions occur between equivalence classes if the numbers of levels are large enough. This contributes to the isomorphism defined in the theorem about sequential Bio-PEPA systems.

**Lemma 5.** *Let* $C \stackrel{def}{=} \sum_{i=1}^{m} (\alpha_i, \kappa_i) \, op_i \, C$ *be a sequential Bio-PEPA component and let $S = \langle T, C \rangle$. Let $E_1, \ldots, E_t$ be the ordered equivalence classes of $S^n_\mathcal{H}$ and $E'_1, \ldots, E'_t$ be the ordered equivalence classes of $S^{n'}_\mathcal{H}$. If $n \geq k_\downarrow + k_m + k_\uparrow$ then $E_p \xrightarrow{\alpha} E_q$ if and only if $E'_p \xrightarrow{\alpha} E'_q$.*

*Proof.* Let $n' = n + d$. Lemma 4 gives notation for the two sequences of equivalence classes and let $[i_c, j_c]$ and $[i'_c, j'_c]$ be the classes whose current action set is $\mathcal{A}_C$. Consider $[i_p, j_p] \xrightarrow{\alpha} [i_q, j_q]$. Since $\mathcal{A}([i_p, j_p]) = \mathcal{A}([i'_p, j'_p])$, $[i'_p, j'_p] \xrightarrow{\alpha} [i', j']$. We need to show that $[i', j'] = [i'_q, j'_q]$. There are two transitions $\langle \mathcal{T}^n, C(l) \rangle \xrightarrow{(\alpha, w)} \langle \mathcal{T}^n, C(l_1) \rangle$ with $l \in [i_p, j_p]$ and $l_1 \in [i_q, j_p]$ and $\langle \mathcal{T}^{n'}, C(l') \rangle \xrightarrow{(\alpha, w')} \langle \mathcal{T}^{n'}, C(l'_1) \rangle$ with $l' \in [i'_p, j'_p]$ and $l'_1 \in [i', j']$. So we need to show that $l'_1 \in [i'_q, j'_q]$. We also need to consider the other direction. If $[i'_p, j'_p] \xrightarrow{\alpha} [i'_q, j'_q]$ then $[i_p, j_p] \xrightarrow{\alpha} [i, j]$ and we need to show that $l \in [i_q, j_q]$.

**Prefixes** $(\alpha, \kappa)$ op $C$ **where** op $\in \{\oplus, \ominus, \odot\}$**:** A transition does not result in a change of level therefore $p = q$ and $[i', j'] = [i'_p, j'_p]$. Similarly for the other direction.

We now consider the relative positions of $p$, $q$ and $c$. The cases ($p < c$ and $q > c$) and ($q < c$ and $p > c$) cannot occur due to Lemma 3 as no transition can change the value of $l$ or $l'$ sufficiently.

**Prefixes of the form** $(\alpha, \kappa)\uparrow C$**:** Here the level $l$ changes to $l + \kappa$ and $p \leq q$.

If $q < c$ then $l \in [i_p, j_p]$ and $l + \kappa \in [i_q, j_q]$ hence choosing $l' = l$ leads to $l' \in [i_p, j_p]$ and $l' + \kappa \in [i_q, j_q]$ as required. The other direction is similar.

For $q = c$ and $p < c$, let $l' = l$ then $l' + \kappa \in [i_c, j_c + d]$ since $l' + \kappa \in [i_c, j_c]$. For the other direction, $l' \in [i_p, j_p]$ and $l' + \kappa \in [i_c, j_c + d]$. We need to show that $l + \kappa \in [i_c, j_c]$. The largest value $l'$ can take is $i_c - 1$. The width of $[i_c, j_c]$ is $j_c - i_c + 1 > k_m$ by Lemma 3. Therefore $l + \kappa = l' + \kappa \leq i_c - 1 + \kappa \leq i_c - 1 + k_m < i_c - 1 + j_c - i_c + 1 \leq j_c$.

If $p > c$, then $l \in [i_p, j_p]$ and $l + \kappa \in [i_q, j_q]$, Let $l' = l + d$ then $l' \in [i_p + d, j_p + d]$ and $l' + \kappa \in [i_q + d, j_q + d] = [i'_q, j'_q]$ as required. For the other direction, let $l = l' - d$.

If $p = c$ and $q = c$ let $l' = l$ then $l + \kappa \in [i_c, j_c]$ implies $l' + \kappa \in [i_c, j_c + d]$. For the other direction, we need to show that $l, l + \kappa \in [i_c, j_c]$. Choose $l = i_c$ and note that $l + \kappa = i_c + \kappa \leq i_c + k_m < j_c + 1$ by Lemma 3 therefore $l + \kappa \leq j_c$.

If $p = c$ and $q > c$ then $l \in [i_c, j_c]$ and $l + \kappa \in [i_q, j_q]$. To match this, choose $l' = l + d$. For the other direction, $l' \in [i_c, j_c + d]$ and $l' + \kappa \in [i_q + d, j_q + d]$. Let $l = l' - d$, then $l \in [i_c - d, j_c]$ and $l + \kappa \in [i_q, j_q]$. We need to show $l \geq i_c$. The smallest value that $l' + \kappa$ can take is $j_c + d + 1$ which implies $l + d + \kappa = l' + \kappa \geq j_c + d + 1$. So $l \geq j_c - \kappa + 1 \geq j_c - k_m + 1 > i_c$ by Lemma 3.

**Prefixes of the form** $(\alpha, \kappa)\downarrow C$**:** The level changes from $l$ to $l - \kappa$ and $p \geq q$. These are similar to the previous case. □

The following theorem shows that for large enough value of $n$, two discretisations of a sequential Bio-PEPA system are compression bisimilar.

**Theorem 1.** *Let* $\mathcal{S} = \langle \mathcal{T}, C \rangle$ *be a well-defined Bio-PEPA system with the single species component* $C \stackrel{\text{def}}{=} \sum_{i=1}^m (\alpha_i, \kappa_i) \, op_i \, C$ *then* $\mathcal{S}^n \simeq \mathcal{S}^{n'}$ *for* $n, n' \geq k_{\downarrow} + k_m + k_{\uparrow}$.

*Proof.* Without loss of generality, assume that $n$ is the maximum number of levels for species $C$ in $\mathcal{S}^n$ and $n'$ is the maximum number of levels for species $C$ in $\mathcal{S}^{n'}$.

We will show that $\mathcal{S}^n_{\mathcal{H}}$ is isomorphic to $\mathcal{S}^{n'}_{\mathcal{H}}$ hence $\mathcal{S}^n_{\mathcal{H}} \sim \mathcal{S}^{n'}_{\mathcal{H}}$ and therefore $\mathcal{S}^n \simeq \mathcal{S}^{n'}$. Let $f : \mathcal{S}^n_{\mathcal{H}} \to \mathcal{S}^{n'}_{\mathcal{H}}$ be defined as $f(B) = D$ if $\mathcal{A}(B) = \mathcal{A}(D)$. This function is well-defined by Lemma 2 since $\mathcal{S}^n_{\mathcal{H}}$ and $\mathcal{S}^{n'}_{\mathcal{H}}$ have the same number of equivalence classes and hence for $B, B' \in \mathcal{S}^n_{\mathcal{H}}$, $f(B) = f(B')$ implies $B = B'$ and for any $D \in \mathcal{S}^{n'}_{\mathcal{H}}$, there exists $B \in \mathcal{S}^n_{\mathcal{H}}$ such that $f(B) = D$.

Additionally, define $f(B \xrightarrow{\alpha} B') = f(B) \xrightarrow{\alpha} f(B')$. This is a homomorphism because it preserves transitions. By Lemma 5, for any $D \xrightarrow{\alpha} D'$, there exist $B, B' \in \mathcal{S}_{\mathcal{H}}^n$ such that $f(B) \xrightarrow{\alpha} f(B')$. Hence $f$ is a isomorphism.    $\square$

Classically in congruence proofs, there would be a proof for each operator, hence there would be one for each of the prefix operators and then one for the choice operator. We do not need to show that the new semantic equivalence is a congruence with respect to the prefix operators and the choice operator since we work specifically with well-defined model components which give a constrained syntax that restricts how the prefix operators and the choice operator can be used.

We next consider a congruence result for the synchronisation operator. In this theorem, the notation $[\mathcal{P}]_{\mathcal{H}}$ refers to the equivalence class generated by $\mathcal{H}$, the current action relation, that contains the Bio-PEPA system $\mathcal{P}$. From this it is possible to obtain the result about compression bisimilarity between two different discretisations of a model component. First, we need a lemma and then a property describing the actions that are possible in a synchronisation. The property captures the idea that the actions in $L$ are those that are shared by both components in the synchronisation.

**Lemma 6.** *Equality with respect to $\mathcal{A}$ is preserved by cooperation. In other words,*

$$\mathcal{A}(\langle \mathcal{T}, P_1 \rangle) = \mathcal{A}(\langle \mathcal{T}, P_2 \rangle) \quad \Rightarrow \quad \begin{cases} \mathcal{A}(\langle \mathcal{T}, P_1 \bowtie_L Q \rangle) = \mathcal{A}(\langle \mathcal{T}, P_2 \bowtie_L Q \rangle) & and \\ \mathcal{A}(\langle \mathcal{T}, Q \bowtie_L P_1 \rangle) = \mathcal{A}(\langle \mathcal{T}, Q \bowtie_L P_2 \rangle) \end{cases}$$

**Definition 10.** *Given a well-defined Bio-PEPA system of the form $\langle \mathcal{T}, P \bowtie_L Q \rangle$, it has the* current action decomposition property *if $\mathcal{A}(\langle \mathcal{T}, P_1 \bowtie_L Q_1 \rangle) = \mathcal{A}(\langle \mathcal{T}, P_2 \bowtie_L Q_2 \rangle)$ implies $\mathcal{A}(\langle \mathcal{T}, P_1 \rangle) = \mathcal{A}(\langle \mathcal{T}, P_2 \rangle)$ and $\mathcal{A}(\langle \mathcal{T}, Q_1 \rangle) = \mathcal{A}(\langle \mathcal{T}, Q_2 \rangle)$ for all systems $\langle \mathcal{T}, P_1 \bowtie_L Q_1 \rangle, \langle \mathcal{T}, P_2 \bowtie_L Q_2 \rangle \in ds(\langle \mathcal{T}, P \bowtie_L Q \rangle)$.*

**Theorem 2.** *Let $\langle \mathcal{T}_1, P_1 \rangle$, $\langle \mathcal{T}_2, P_2 \rangle$, $\langle \mathcal{T}_1, Q_1 \rangle$ and $\langle \mathcal{T}_2, Q_2 \rangle$ be well-defined Bio-PEPA systems such that $\langle \mathcal{T}_1, P_1 \bowtie_L Q_1 \rangle$ and $\langle \mathcal{T}_2, P_2 \bowtie_L Q_2 \rangle$ both have the current action decomposition property. If $\langle \mathcal{T}_1, P_1 \rangle \simeq \langle \mathcal{T}_2, P_2 \rangle$ and $\langle \mathcal{T}_1, Q_1 \rangle \simeq \langle \mathcal{T}_2, Q_2 \rangle$ then it is the case that $\langle \mathcal{T}_1, P_1 \bowtie_L Q_1 \rangle \simeq \langle \mathcal{T}_2, P_2 \bowtie_L Q_2 \rangle$.*

*Proof.* Let $\mathcal{R} = \big\{ \big( [\langle \mathcal{T}_1, P_1' \bowtie_L Q_1' \rangle]_{\mathcal{H}}, [\langle \mathcal{T}_2, P_2' \bowtie_L Q_2' \rangle]_{\mathcal{H}} \big) \mid [\langle \mathcal{T}_1, P_1' \rangle]_{\mathcal{H}} \sim [\langle \mathcal{T}_2, P_2' \rangle]_{\mathcal{H}}$ and $[\langle \mathcal{T}_1, Q_1' \rangle]_{\mathcal{H}} \sim [\langle \mathcal{T}_2, Q_2' \rangle]_{\mathcal{H}} \big\}$. We want to show that $\mathcal{R}$ is a bisimulation. This proof follows the standard technique for this case but requires a few additional steps. We only consider the case of $\alpha \in L$. The other two cases are simpler.

Consider a transition from $[\langle \mathcal{T}_1, P_1' \bowtie_L Q_1' \rangle]_{\mathcal{H}}$. This may be generated by a transition from another system in this class, say $\langle \mathcal{T}_1, P_3' \bowtie_L Q_3' \rangle \xrightarrow{(\alpha, w)}_{sc} \langle \mathcal{T}_1, P_3'' \bowtie_L Q_3'' \rangle$. By shorter inferences and by the definition of transitions between equivalence classes we can infer $[\langle \mathcal{T}_1, P_3' \rangle]_{\mathcal{H}} \xrightarrow{\alpha} [\langle \mathcal{T}_1, P_3'' \rangle]_{\mathcal{H}}$ and $[\langle \mathcal{T}_1, Q_3' \rangle]_{\mathcal{H}} \xrightarrow{\alpha} [\langle \mathcal{T}_1, Q_3'' \rangle]_{\mathcal{H}}$.

From the current action decomposition property, we know that these transitions are $[\langle \mathcal{T}_1, P_1' \rangle]_{\mathcal{H}} \xrightarrow{\alpha} [\langle \mathcal{T}_1, P_3'' \rangle]_{\mathcal{H}}$ and $[\langle \mathcal{T}_1, Q_1' \rangle]_{\mathcal{H}} \xrightarrow{\alpha} [\langle \mathcal{T}_1, Q_3'' \rangle]_{\mathcal{H}}$.

By the definition of $\mathcal{R}$, there exist $P_2''$ and $Q_2''$ such that $[\langle \mathcal{T}_2, P_2' \rangle]_{\mathcal{H}} \xrightarrow{\alpha} [\langle \mathcal{T}_2, P_2'' \rangle]_{\mathcal{H}}$ and $[\langle \mathcal{T}_2, Q_2' \rangle]_{\mathcal{H}} \xrightarrow{\alpha} [\langle \mathcal{T}_2, Q_2'' \rangle]_{\mathcal{H}}$ such that $[\langle \mathcal{T}_1, P_3'' \rangle]_{\mathcal{H}} \sim [\langle \mathcal{T}_2, P_2'' \rangle]_{\mathcal{H}}$ and $[\langle \mathcal{T}_1, Q_3'' \rangle]_{\mathcal{H}} \sim [\langle \mathcal{T}_2, Q_2'' \rangle]_{\mathcal{H}}$.

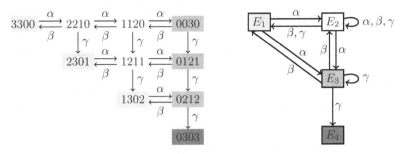

**Fig. 6.** Transition system for $MC$ when $n = 3$ (left) and associated equivalence classes (right)

We can then infer $\langle \mathcal{T}_2, P_4' \rangle \xrightarrow{(\alpha, w_2)}_{sc} \langle \mathcal{T}_2, P_4'' \rangle$ with $\mathcal{A}(\langle \mathcal{T}_2, P_4' \rangle) = \mathcal{A}(\langle \mathcal{T}_2, P_2' \rangle)$ and $\mathcal{A}(\langle \mathcal{T}_2, P_4'' \rangle) = \mathcal{A}(\langle \mathcal{T}_2, P_2'' \rangle)$ and $\langle \mathcal{T}_2, Q_2' \rangle \xrightarrow{(\alpha, v_2)}_{sc} \langle \mathcal{T}_2, Q_2'' \rangle$ with $\mathcal{A}(\langle \mathcal{T}_2, Q_4' \rangle) = \mathcal{A}(\langle \mathcal{T}_2, Q_2' \rangle)$ and $\mathcal{A}(\langle \mathcal{T}_2, Q_4'' \rangle) = \mathcal{A}(\langle \mathcal{T}_2, Q_2'' \rangle)$.

By Lemma 6 and the above equalities, $\mathcal{A}(\langle \mathcal{T}_2, P_4' \underset{L}{\bowtie} Q_4' \rangle) = \mathcal{A}(\langle \mathcal{T}_2, P_2' \underset{L}{\bowtie} Q_2' \rangle)$ and $\mathcal{A}(\langle \mathcal{T}_2, P_4'' \underset{L}{\bowtie} Q_4'' \rangle) = \mathcal{A}(\langle \mathcal{T}_2, P_2'' \underset{L}{\bowtie} Q_2'' \rangle)$.

We can infer $\langle \mathcal{T}_2, P_4' \underset{L}{\bowtie} Q_4' \rangle \xrightarrow{(\alpha, w)}_{sc} \langle \mathcal{T}_2, P_4'' \underset{L}{\bowtie} Q_4'' \rangle$ which leads to the matching transition $[\langle \mathcal{T}_2, P_2' \underset{L}{\bowtie} Q_2' \rangle]_{\mathcal{H}} \xrightarrow{\alpha} [\langle \mathcal{T}_2, P_2'' \underset{L}{\bowtie} Q_2'' \rangle]_{\mathcal{H}}$ and by definition we know that $\left( [\langle \mathcal{T}_1, P_3'' \underset{L}{\bowtie} Q_3'' \rangle]_{\mathcal{H}}, [\langle \mathcal{T}_2, P_2'' \underset{L}{\bowtie} Q_2'' \rangle]_{\mathcal{H}} \right) \in \mathcal{R}$. $\qquad \square$

**Corollary 2.** *Let* $\mathcal{P} = \langle \mathcal{T}, P \rangle$ *be a well-defined Bio-PEPA system then* $\mathcal{P}^n \simeq \mathcal{P}^{n'}$ *for* $n, n' \geq k_\downarrow + k_m + k_\uparrow$ *if the current action decomposition property applies to pairs of subcomponents of* $\mathcal{P}^n$ *and* $\mathcal{P}^{n'}$.

*Proof.* Since $\mathcal{P}$ is well-defined, $P \overset{def}{=} C_1(l_1) \underset{\mathcal{L}_1}{\bowtie} \ldots \underset{\mathcal{L}_{m-1}}{\bowtie} C_m(l_m)$ with each $C_i$ a sequential component. By Theorem 1 and $n, n' \geq k_\downarrow + k_m + k_\uparrow$, $\langle \mathcal{T}^n, C_i \rangle \simeq \langle \mathcal{T}^{n'}, C_i \rangle$ for all $i$. By repeated applications of Theorem 2, $\langle \mathcal{T}^n, P \rangle \simeq \langle \mathcal{T}^{n'}, P' \rangle$. $\qquad \square$

## 4   Example

We give an example of discretisations and the associated equivalence classes. Consider the substrate-enzyme-product reactions $S + E \leftrightarrows SE \rightarrow P + E$ which can be expressed in Bio-PEPA as

$$S \overset{def}{=} (\alpha, 1) \downarrow S + (\beta, 1) \uparrow S \qquad E \overset{def}{=} (\alpha, 1) \downarrow E + (\beta, 1) \uparrow E + (\gamma, 1) \uparrow E$$

$$SE \overset{def}{=} (\alpha, 1) \uparrow SE + (\beta, 1) \downarrow SE + (\gamma, 1) \downarrow SE \qquad P \overset{def}{=} (\gamma, 1) \uparrow P$$

$$MC \overset{def}{=} S(x) \underset{\{\alpha, \beta\}}{\bowtie} E(x) \underset{\{\alpha, \beta, \gamma\}}{\bowtie} SE(0) \underset{\{\gamma\}}{\bowtie} P(0)$$

Figure 6 gives the transition system for $MC$ when the maximum number of levels for all species is three as well as that for the equivalence classes. In the transition system for $MC$ each state is a Bio-PEPA system and is indicated by its vector representation which describes the level of each species in that system using the vector $(S, E, SE, P)$ but we omit the brackets and commas in the diagram. Additionally, we have omitted the strings

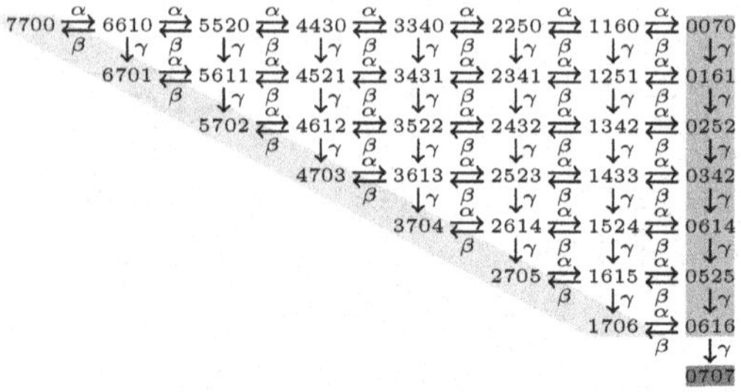

$$
\begin{array}{l}
7700 \underset{\beta}{\overset{\alpha}{\rightleftharpoons}} 6610 \underset{\beta}{\overset{\alpha}{\rightleftharpoons}} 5520 \underset{\beta}{\overset{\alpha}{\rightleftharpoons}} 4430 \underset{\beta}{\overset{\alpha}{\rightleftharpoons}} 3340 \underset{\beta}{\overset{\alpha}{\rightleftharpoons}} 2250 \underset{\beta}{\overset{\alpha}{\rightleftharpoons}} 1160 \underset{\beta}{\overset{\alpha}{\rightleftharpoons}} 0070 \\
\quad\ \downarrow\gamma \quad\ \downarrow\gamma \quad\ \downarrow\gamma \quad\ \downarrow\gamma \quad\ \downarrow\gamma \quad\ \downarrow\gamma \quad\ \downarrow\gamma \\
\quad\ 6701 \underset{\beta}{\overset{\alpha}{\rightleftharpoons}} 5611 \underset{\beta}{\overset{\alpha}{\rightleftharpoons}} 4521 \underset{\beta}{\overset{\alpha}{\rightleftharpoons}} 3431 \underset{\beta}{\overset{\alpha}{\rightleftharpoons}} 2341 \underset{\beta}{\overset{\alpha}{\rightleftharpoons}} 1251 \underset{\beta}{\overset{\alpha}{\rightleftharpoons}} 0161 \\
\qquad\quad \downarrow\gamma \quad\ \downarrow\gamma \quad\ \downarrow\gamma \quad\ \downarrow\gamma \quad\ \downarrow\gamma \quad\ \downarrow\gamma \\
\qquad\quad 5702 \underset{\beta}{\overset{\alpha}{\rightleftharpoons}} 4612 \underset{\beta}{\overset{\alpha}{\rightleftharpoons}} 3522 \underset{\beta}{\overset{\alpha}{\rightleftharpoons}} 2432 \underset{\beta}{\overset{\alpha}{\rightleftharpoons}} 1342 \underset{\beta}{\overset{\alpha}{\rightleftharpoons}} 0252 \\
\qquad\qquad\quad \downarrow\gamma \quad\ \downarrow\gamma \quad\ \downarrow\gamma \quad\ \downarrow\gamma \quad\ \downarrow\gamma \\
\qquad\qquad\quad 4703 \underset{\beta}{\overset{\alpha}{\rightleftharpoons}} 3613 \underset{\beta}{\overset{\alpha}{\rightleftharpoons}} 2523 \underset{\beta}{\overset{\alpha}{\rightleftharpoons}} 1433 \underset{\beta}{\overset{\alpha}{\rightleftharpoons}} 0342 \\
\qquad\qquad\qquad\quad \downarrow\gamma \quad\ \downarrow\gamma \quad\ \downarrow\gamma \quad\ \downarrow\gamma \\
\qquad\qquad\qquad\quad 3704 \underset{\beta}{\overset{\alpha}{\rightleftharpoons}} 2614 \underset{\beta}{\overset{\alpha}{\rightleftharpoons}} 1524 \underset{\beta}{\overset{\alpha}{\rightleftharpoons}} 0614 \\
\qquad\qquad\qquad\qquad\quad \downarrow\gamma \quad\ \downarrow\gamma \quad\ \downarrow\gamma \\
\qquad\qquad\qquad\qquad\quad 2705 \underset{\beta}{\overset{\alpha}{\rightleftharpoons}} 1615 \underset{\beta}{\overset{\alpha}{\rightleftharpoons}} 0525 \\
\qquad\qquad\qquad\qquad\qquad\quad \downarrow\gamma \quad\ \downarrow\gamma \\
\qquad\qquad\qquad\qquad\qquad\quad 1706 \underset{\beta}{\overset{\alpha}{\rightleftharpoons}} 0616 \\
\qquad\qquad\qquad\qquad\qquad\qquad\quad \downarrow\gamma \\
\qquad\qquad\qquad\qquad\qquad\qquad\quad 0707
\end{array}
$$

**Fig. 7.** The transition system for $MC$ when $n = 7$

$w$ from the diagram for reasons of space. The shading shows the different equivalence classes in both diagrams. Figure 7 gives the system when the maximum number of levels is seven and the equivalence classes are also shown. This demonstrates how the two discretisations are related by the equivalence classes given in Figure 6.

## 5   Related Work

The use of process algebras for modelling systems biology has multiplied rapidly since the first paper advocated the use of the $\pi$-calculus [15]. Approaches include the $\kappa$-calculus [2], stochastic $\pi$-calculus [3,1], Beta-binders [4] and Bio-Ambients [5]. Most of these approaches use stochastic simulation as their analysis tool, and few approaches have considered the use of semantic equivalences. Weak bisimulation is shown to be a congruence for the bio-$\kappa$-calculus as is a context bisimulation which allows for the modelling of cell interaction [16]. Observational equivalence has been used to show that CCS specifications of elements of lactose operon regulation have the same behaviour as more detailed models [17]. In an example of biological modelling using hybrid systems, bisimulation is used to quotient the state space with respect to a subset of variables as a technique for state space reduction [18]. Bisimulation has also been used in the comparison of ambient-style models and membrane-style models [19] and the comparison of a term-rewriting calculus and a simple brane calculus [20].

## 6   Conclusions and Further Research

This paper has presented a new semantic equivalence for Bio-PEPA called compression bisimulation and shown that it is a congruence and it identifies different discretisations of the same system.

A first step for further work is to find a syntactic characterisation of the Bio-PEPA systems that exhibit the current action decomposition property so that those systems with this property can be easily identified. Since reactions have unique names in Bio-PEPA, an extension that would be useful is to allow for a relation over names to relate different reactions in different models. We would also like to extend the equivalence to be quantitative and take into account reaction rates but it is not immediately obvious how to do this. Finally, we wish to apply the equivalence to various biological models and to other formalisms using discretisation.

**Acknowledgements.** Vashti Galpin is supported by EPSRC Grant EP/E031439/1. Jane Hillston is supported by EPSRC ARF EP/C543696/01. The Centre for Systems Biology at Edinburgh is a Centre for Integrative Systems Biology (CISB) funded by the BBSRC and EPSRC in 2006. We thank Maria Luisa Guerriero for her comments.

# References

1. Blossey, R., Cardelli, L., Phillips, A.: A compositional approach to the stochastic dynamics of gene networks. In: Priami, C., Cardelli, L., Emmott, S. (eds.) Transactions on Computational Systems Biology IV. LNCS (LNBI), vol. 3939, pp. 99–122. Springer, Heidelberg (2006)
2. Danos, V., Laneve, C.: Formal molecular biology. Theoretical Computer Science 325, 69–110 (2004)
3. Priami, C., Regev, A., Shapiro, E., Silverman, W.: Application of a stochastic name-passing calculus to representation and simulation of molecular processes. Information Processing Letters 80, 25–31 (2001)
4. Priami, C., Quaglia, P.: Beta binders for biological interactions. In: Danos, V., Schächter, V. (eds.) CMSB 2004. LNCS (LNBI), vol. 3082, pp. 20–33. Springer, Heidelberg (2005)
5. Regev, A., Panina, E., Silverman, W., Cardelli, L., Shapiro, E.: BioAmbients: an abstraction for biological compartments. Theoretical Computer Science 325, 141–167 (2004)
6. Ciocchetta, F., Hillston, J.: Bio-PEPA: a framework for the modelling and analysis of biological systems. Theoretical Computer Science (in press)
7. Hillston, J.: A compositional approach to performance modelling. CUP (1996)
8. Ciocchetta, F., Hillston, J.: Calculi for biological systems. In: Formal Methods for Computational Systems Biology. In: SFM 2008. LNCS, vol. 5016, pp. 265–312. Springer, Heidelberg (2008)
9. Ciocchetta, F.: Bio-PEPA with SBML-like events. In: Back, R.J., Petre, I. (eds.) Proceedings of the Workshop on Computational Models for Cell Processes, pp. 11–22 (2008)
10. Ciocchetta, F., Hillston, J.: Bio-PEPA: a framework for the modelling and analysis of biological systems. Technical Report EDI-INF-RR-1231, School of Informatics, University of Edinburgh (2008)
11. Ellavarason, K.: An automatic mapping from the Systems Biology Markup Language to the Bio-PEPA process algebra. Master's thesis, University of Trento (2008)
12. Akman, O.E., Ciocchetta, F., Degasperi, A., Guerriero, M.L.: Modelling biological clocks with Bio-PEPA: stochasticity and robustness for the Neurospora Crassa circadian network. To appear in Proceedings of CMSB 2009 (2009)
13. Guerriero, M.L.: Qualitative and quantitative analysis of a Bio-PEPA model of the gp130/JAK/STAT signalling pathway. To appear in TCSB
14. Milner, R.: Communication and concurrency. Prentice Hall, Englewood Cliffs (1989)
15. Regev, A., Shapiro, E.: Cellular abstractions: Cells as computation. Nature 419, 343 (2002)
16. Laneve, C., Tarissan, F.: A simple calculus for proteins and cells. Theoretical Computer Science 404, 127–141 (2008)
17. Pinto, M.C., Foss, L., Mombach, J.C.M., Ribeiro, L.: Modelling, property verification and behavioural equivalence of lactose operon regulation. Computers in Biology and Medicine 37, 134–148 (2007)
18. Antoniotti, M., Piazza, C., Policriti, A., Simeoni, M., Mishra, B.: Taming the complexity of biochemical models through bisimulation and collapsing: theory and practice. Theoretical Computer Science 325, 45–67 (2004)

19. Ciobanu, G., Aman, B.: On the relationship between membranes and ambients. BioSystems (91), 515–530 (2008)
20. Barbuti, R., Maggiolo-Schettini, A., Milazzo, P., Troina, A.: Bisimulation in calculi modelling membranes. Formal Aspects of Computing 20, 351–377 (2008)

# Improved Parameter Estimation for Completely Observed Ordinary Differential Equations with Application to Biological Systems

Peter Gennemark[1,2] and Dag Wedelin[3]

[1] University of Göteborg, SE-412 96 Göteborg, Sweden
[2] Uppsala University, SE-751 06 Uppsala, Sweden
peterg@chalmers.se
[3] Chalmers University of Technology, SE-412 96 Göteborg, Sweden
dag@chalmers.se

**Abstract.** We consider parameter estimation in ordinary differential equations (ODEs) from completely observed systems, and describe an improved version of our previously reported heuristic algorithm (*IET Syst. Biol.*, 2007). Basically, in that method, estimation based on decomposing the problem to simulation of one ODE, is followed by estimation based on simulation of all ODEs of the system.

The main algorithmic improvement compared to the original version, is that we decompose not only to single ODEs, but also to arbitrary subsets of ODEs, as a complementary intermediate step. The subsets are selected based on an analysis of the interaction between the variables and possible common parameters.

We evaluate our algorithm on a number of well-known hard test problems from the biological literature. The results show that our approach is more accurate and considerably faster compared to other reported methods on these problems. Additionally, we find that the algorithm scales favourably with problem size.

**Keywords:** ordinary differential equations, parameter estimation, decomposition.

**Supplementary material:** All problems, solutions, on-line software and supplementary information are available at *www.odeidentification.org*.

## 1 Introduction

This paper considers parameter estimation in ordinary differential equation (ODE) models, applied to biological systems. Usually, such models are based on reaction kinetics and are non-linear in both variables and parameters, see Fig. 1 for an example.

A general way to estimate the parameters is to define an optimisation problem in which the objective is to find parameter values minimising an error function based on the discrepancy between the observed data and the simulated model.

P. Degano and R. Gorrieri (Eds.): CMSB 2009, LNBI 5688, pp. 205–217, 2009.
© Springer-Verlag Berlin Heidelberg 2009

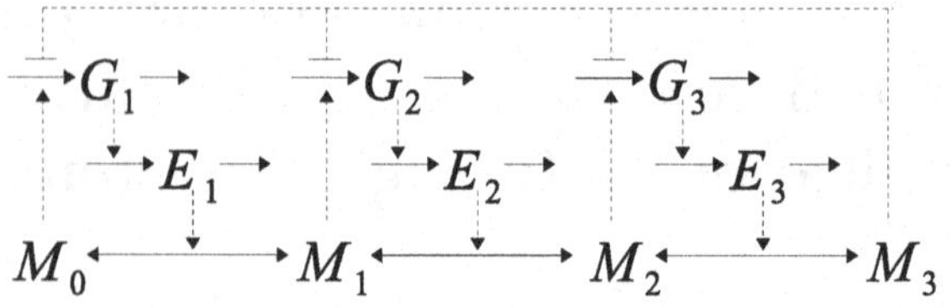

$$M_1'(t) = \frac{k_{cat1}\,E_1(t)\dfrac{1}{K_{m1}}\bigl(M_0(t)-M_1(t)\bigr)}{1+\dfrac{M_0(t)}{K_{m1}}+\dfrac{M_1(t)}{K_{m2}}} - \frac{k_{cat2}\,E_2(t)\dfrac{1}{K_{m3}}\bigl(M_1(t)-M_2(t)\bigr)}{1+\dfrac{M_1(t)}{K_{m3}}+\dfrac{M_2(t)}{K_{m4}}}$$

**Fig. 1.** Model of a gene network considered in Moles et al. [1]. $M_{0-3}$ are metabolites, $E_{1-3}$ are enzymes, and $G_{1-3}$ are mRNAs. $M_0$ are $M_3$ are input variables, while all others are dependent variables. To each of the dependent variables, there is a corresponding ODE. The ODE for $M_1$, containing the parameters $k_{cat1}$, $K_{m1}$, $K_{m2}$, $k_{cat2}$, $K_{m3}$, and $K_{m4}$, is shown for illustration. The model contains 36 parameters in total.

The evaluation of the error function is usually slow, since it requires the entire model to be simulated for each experiment. Since this has to be repeated many times, the overall method is computationally intensive for realistic problems.

We consider the case where some time course information is available for every state variable, and present an improved version of the algorithm of Gennemark and Wedelin [2], together with empirical evaluation on a number of test problems.

Motivation for this work comes from several sources. While many parameter estimation algorithms often allow data to be specified for only a subset of the variables, further work on a base case is generally of interest if this leads to significantly faster algorithms for this case.

Another specific motivation is that we are interested in the use of parameter estimation as a subroutine in a more complex algorithm for identifying also the structure of ODE systems, when the mathematical expressions of the ODEs are unknown. Such algorithms have recently become of increased interest because of the combination of large-scale experimental techniques, and increased computational power (see [3] and references therein). To be feasible, this inference problem typically relies on *complete* data sets in which all state variables are observed. For example, one may identify a gene regulatory network from a set of mRNA time profiles collected from a microarray experiment. In the majority of methods proposed in this area, a structure search heuristic iteratively proposes different ODE structures to try, and for each such model the parameters are estimated. Usually, most of the computation time is spent on parameter estimation, which hence constitutes a highly time critical subroutine. For this reason, research on parameter estimation on completely observed systems has an direct impact on the more general problem of finding both structure and parameters.

## 1.1 Related Work

For the feasibility of the parameter estimation problem see e.g. Ljung [4], who discusses how problems may not be unambiguously solved without additional data or some additional model constraints. Local optimisation methods like Gauss-Newton can be used to optimise the error function, and, generally, such methods are computationally fast but can fail because of local optima of the objective function [5, 6]. Standard softwares for local methods are publicly available, and benchmark problems to evaluate these are also available, e.g. in the data base EASY-FIT [5]. Global methods, on the other hand, may identify the global optimum but to the cost of an increased computational time [1, 7, 8, 9]. For example, one of the simplest possible global methods is to restart local estimation from several randomly chosen starting points.

For problems with complete data sets, and where the derivatives can be estimated with relatively high precision, the ODEs can be transformed to a set of algebraic equations by replacing the left-hand sides by estimated slopes from data. Then, the derivative method [10], or rigorous deterministic methods based on interval analysis [11] can be applied. Compared to these two methods, our own algorithm [2] is also dependent on complete data, but not so much on high precision estimates of derivatives.

## 2 Parameter Estimation

Since the algorithm is based on Gennemark and Wedelin [2], we refer to that reference for details. Here we briefly review the basic principle of the algorithm and then describe our improvements.

### 2.1 The Original Algorithm

Input to the algorithm consists of the structure of the model (the ODEs), lower and upper bounds for all parameters, and sets of time series experiments with standard deviations, where every variable should be measured in at least one experiment. The solution is a parameter vector $\mathbf{k}$ that minimises an error function, which in principle can take any form.

In our case we work with the log likelihood of the observed data. By assuming independent and normally distributed measurement errors and disregarding constant terms we can express the log-likelihood for one time series as

$$L(\hat{X}_j \mid \mathbf{k}) = -\frac{1}{2} \sum_i \left( \frac{X_j(t_i) - \hat{X}_j(t_i)}{\sigma_j(t_i)} \right)^2 \tag{1}$$

where $i$ indexes the measurement points, and where $X_j$, $\hat{X}_j$ and $\sigma_j$ denotes simulated data, experimental data and standard deviation for variable $j$, respectively. The total log-likelihood $L(\hat{X} \mid \mathbf{k})$ is defined by summing over all variables and all experiments. The error function to minimise is $-L(\hat{X} \mid \mathbf{k})$.

One main idea in Gennemark and Wedelin [2] is to use the complete data set to decompose the parameter estimation to single ODEs. In this way we obtain one small optimisation problem for each ODE. The decomposition is achieved by using interpolated experimental data for the other variables occurring on the right-hand side of the ODE, resulting in a decomposed problem for a single variable. Standard parameter estimation routines (local or global) can then be applied on the decomposed problem, and it is solved by applying simulation of the single ODE to evaluate the error function. As an example of the decomposition, to estimate the six parameters of $M_1'(t)$ in Fig. 1, interpolated data are used for $M_0$, $M_2$ $E_1$ and $E_2$ when simulating the ODE of $M_1$.

The advantage of this approach is twofold. First, the small problems can be solved much faster, and second, the overall stability of the estimation increases, since the decomposed problems are constrained by the empirical data and not by results from the algorithm itself. The disadvantage however, is that the precision may be lower than for estimation of the entire model.

To successfully take advantage of this potential speed gain, and at the same time achieve high precision, another idea is to use a mix of different estimation methods with different properties with respect to speed and precision. Thus, the originally proposed algorithm also occasionally applies estimation to the entire problem. Initially, we also use the very rough but fast derivative method [10], to create a starting point. Finally, we can choose between experimental and simulated data given the best parameters hitherto as input for the sub-problems as appropriate (see [2] for details). An iterative procedure is obtained as follows:

REPEAT
  1. Select input for the other variables, to be used in step 2: experimental data or simulated data.
  2. FOR each equation DO
     (a) Make a rough estimate of the parameters in the equation with the derivative method.
     (b) Improve by estimation for single equation sub-problems.
  3. Improve by estimation for the entire problem.
UNTIL convergence

In step 1, it is natural to consider experimental data in the first iteration, and simulated data in subsequent iterations. Note that simulated data is only used to estimate variables and derivatives in the decomposition of step 2, not to replace $\hat{X}_j$ in the error function. In order to avoid local minima, both step 2a and 2b can be repeated with several randomly chosen initial parameters. Typically, we use 60 random starting points in step 2a, and then feed the best parameters into step 2b which is run only once.

## 2.2   The Improved Algorithm

While the original algorithm is very fast and works well for many problems, we have observed two problems that we address in the improved version:

- For some systems, if estimation for single equation sub-problems have low precision and estimation for the entire problem is too slow, it can be difficult to find a good mix between the two.
- For systems based on traditional kinetic reactions, it is common that the same reaction term occurs in two of the ODEs, one for production and one for consumption. Using estimation for single equations, the parameters of such reactions will hence be estimated twice, leading to non-optimal use of data that may result in ambiguity and potentially divergent solutions.

To address these problems, we propose a generalisation of our algorithm. Instead of restricting the estimation to one ODE (step 2) or to all ODEs (step 3), we allow parameter estimation of the parameters *in any sub-model*, i.e. by considering any subset of the equations.

It is intuitively reasonable that one should select sub-models with equations that are strongly linked. This can be defined in different ways, for example by a known direct interaction, and/or by common parameters. Such links can be specified by the user, or be inferred automatically. Here we propose two different automatic approaches.

The first approach is intended for metabolic systems. We have found it useful to define one sub-model for every variable by starting from every single ODE, and find those other equations that have reactions (and thus also parameters) in common with this ODE. A shared parameter is usually involved in a reaction term that is consumed in one ODE and produced in another.

Sub-models that are empty or contained in a sub-model of another equation are ignored. For example, equation 1 may share a reaction with equation 2 and another with equation 3. Hence, these 3 equations form the sub-model for variable 1. Now, if equation 2 only shares parameters with equation 1, the sub-model of equation 2 is contained in the sub-model of equation 1 and can be ignored.

The second approach is intended for other types of systems that typically lack shared parameters, e.g. genetic networks. Here we let the sub-model of each ODE include the variables which occur on the right-hand side of the equation.

This new step of parameter estimation based on sub-models complements the previous parameter estimation steps. In the original algorithm, it is inserted between step 2 and 3, resulting in our modified algorithm:

REPEAT
    1. Select input for the other variables, to be used in step 2 and 3: experimental data or simulated data.
    2. FOR each equation DO
        (a) Make a rough estimate of the parameters in the equation with the derivative method.
        (b) Improve by estimation for single equation sub-problems.
    3. Improve by estimation of sub-models defined around each single ODE.
    4. Improve by estimation for the entire problem.
UNTIL convergence

In addition to these changes to the structure of the algorithm, the implementation has been modified so that common parameters are also recognised in the final step of estimation of the entire problem. The actual implementation has also been improved in numerous ways, compared to the original one.

# 3   Numerical Evaluation on Benchmark Problems

We evaluate the algorithm by comparing it with previously presented results in the literature. The problems we consider are collected from the literature based on the following requirements: (1) all state variables of the system should be observed, and (2) information on accuracy and/or efficiency for other methods solving the problem should be reported. All found problems satisfying these criteria have been included in our test set, see Table 1. For problems with noise, many publications only specify the character of the noise, without explicitly supplying data. In these cases we have reconstructed the problems according to the specifications. All test problems and our solutions are fully specified on our web site *www.odeidentification.org*. On the web site, it is also possible to run these problems or own similar problems on-line on our server.

**Table 1.** Overview of the problems considered in our study . Problems in the upper part of the table are collected from the literature, while the lower part includes problems that are designed for this study. #var and #param correspond to the number of variables and parameters, respectively. For some of the problems with noisy data, the number of model parameters is complemented with (after the plus sign) the number of data parameters (one initial value for each time series). #exp and #tp correspond to the number of experiments and the number of time-points for each variable in each experiment, respectively. Measurement noise is added from a normal distribution with mean zero and standard deviation proportional to respective data point.

| Problem | #var | #param | #exp | #tp | Noise | Reference |
|---|---|---|---|---|---|---|
| pe_3genes1 | 8 | 36 | 16 | 21 | 0% | [1] |
| pe_3genes2f | 8 | 36 | 16 | 21 | 3% | [12] |
| pe_3genes3f | 8 | 36 | 16 | 21 | 5% | [12] |
| pe_pinene | 5 | 5 | 1 | 9 | ≈5% | [13] |
| pe_ss_cascade1 | 3 | 14 | 8 | 41 | 0% | [14] |
| pe_ss_branch4 | 4 | 18 | 4 | 20 | 0% | [15] |
| pe_ss_30genes2f | 30 | 128 | 20 | 11 | 10% | [15] |
| pe_3genes2 | 8 | 36+128 | 16 | 21 | 3% | This work |
| pe_3genes3 | 8 | 36+128 | 16 | 21 | 5% | This work |
| pe_ss_30genes2 | 30 | 128+600 | 20 | 11 | 10% | This work |
| pe_4genes1 | 11 | 48 | 16 | 21 | 0% | This work |
| pe_5genes1 | 14 | 60 | 16 | 21 | 0% | This work |
| pe_6genes1 | 17 | 72 | 16 | 21 | 0% | This work |

For problems with names beginning with 'pe_ss' we have used the second approach to select sub-models. For all other problems we have used the first approach.

Since most methods in this field are dependent on random numbers, the results may differ between runs. Therefore, we run all our tests several times with different random seeds. Computation times are reported as the average computation time for runs with different random seeds, scaled to a 1GHz processor for easy comparison to other results (our actual runs were performed on a Pentium IV, 2.13GHz).

## 3.1   Problems from the Literature

An overview of the results is given in Table 2. We first consider a known hard test problem based on the model in Fig. 1, for which common estimation algorithms often fail, see [1]. The main challenge of this problem is the wide search ranges for the parameters and the non-linearities in the ODEs.

For this problem, referred to as **pe_3genes1**, our proposed algorithm returns a perfect solution ($L = 0.00$), and the efficiency is improved compared to all previously reported results on this problem, see Table 2. The parameter accuracy is in the same order as the chosen precision of the numerical integration routine. Notably, the methods we compare to are considered state of the art in the field of global estimation for ODEs, as indicated by the extensive comparison in [1], and by further comparisons in [12, 13].

Furthermore, we consider two variations of this problem with Gaussian noise added to data, **pe_3genes2f** and **pe_3genes3f**. Also on these problems our algorithm performs better than previously reported methods with respect to efficiency, see Table 2. In this case, no comparison on the solution can be made, since we have not used exactly the same data (same noise level but not the same actual data).

The next problem, **pe_pinene**, was originally studied by Box et al. [16], and represents a biochemical system with five variables. Like the method of Rodriguez-Fernandez et al. [13], our method finds the optimal solution, but with higher efficiency, as indicated in Table 2.

We also consider three problems with models defined as S-systems [17, 18], which are based on approximating kinetic laws with multivariate power-law functions. The first S-system problem **pe_ss_cascade1** was introduced by Voit [18] and applied in Tsai and Wang [14]. The model represents a cascaded pathway with three dependent variables. The second S-system problem **pe_ss_branch4** was introduced by Voit [18] and applied in Kutalik et al. [15]. It represents a branched pathway with four dependent variables. Finally, the third S-system problem **pe_ss_30genes2f** was introduced by Maki et al. [19] and applied in Kutalik et al. [15]. It represents a genetic network with 30 variables. The main challenge of this problem is the large number of parameters.

The results presented in Table 2 indicate that our approach is faster and has higher accuracy compared to the evolutionary optimisation approach used by Tsai and Wang [14]. Moreover, the results indicates that our approach is faster compared to the decomposition approach used in Kutalik et al. [15]. Again, since data

**Table 2.** Running times and accuracies for our parameter estimation method. For problems in the upper part of the table results for other methods are available and also reported. Computation time is given as the average of several runs, and scaled to a 1GHz processor. Stability measures the frequency of runs for which the best $-L$ was obtained.

| Problem | Reference | Best solution | | Performance | |
|---------|-----------|---------------|---|-------------|---|
| | | $-L$ | Accuracy | Time (s) | Stability |
| pe_3genes1 | External [1] | | <16% | 120000 | |
| | External [12] | | <0.02% | 13000 | |
| | Best external [13] | | <6x10$^{-3}$% | 540 | |
| | This work | 0.00 | ≈10$^{-9}$% | 57 | 6/10 |
| pe_3genes2f | Best external [12] | | | 13000 | |
| | This work | 1286 | | 90 | 5/10 |
| pe_3genes3f | Best external [12] | | | 13000 | |
| | This work | 3128 | | 100 | 1/10 |
| pe_pinene | Best external [13] | 9.936 | | 30 | |
| | This work | 9.936 | | 0.1 | 10/10 |
| pe_ss_cascade1 | Best external [14] | | <112% | 660 | |
| | This work | 0.00 | ≈10$^{-9}$% | 13 | 10/10 |
| pe_ss_branch4 | Best external [15] | 0.00 | | 250 | |
| | This work | 0.00 | ≈10$^{-9}$% | 0.5 | 10/10 |
| pe_ss_30genes2f | Best external [15] | | | 1700 | |
| | This work | 3240 | | 210 | 10/10 |
| pe_3genes2 | This work | 1214 | | 140 | 5/10 |
| pe_3genes3 | This work | 2309 | | 170 | 1/10 |
| pe_ss_30genes2 | This work | 3004 | | 1000 | 10/10 |
| pe_4genes1 | This work | 0.00 | ≈10$^{-9}$% | 610 | 7/10 |
| pe_5genes1 | This work | 0.00 | ≈10$^{-9}$% | 840 | 10/10 |
| pe_6genes1 | This work | 0.00 | ≈10$^{-9}$% | 1100 | 8/10 |

is noisy in **pe_ss_30genes2f**, we have not used exactly the same data and no comparison on the solution can be made. As Tsai and Wang, and Kutalik et al. test their methods on few problems with relatively rich data sets, it is difficult to make a more thorough comparison with respect to these methods than done here.

For completeness, we also report our results for the noisy test problems, when the initial value for the variable in each time series is considered a parameter that is estimated along with the kinetic rate constants. We refer to those problems as **pe_3genes2**, **pe_3genes2** and **pe_ss_30genes2**, respectively. As expected, the additional degrees of freedom result in improved solutions.

The speed improvement compared to our original algorithm [2] varies between problems, but is roughly a factor of four, split on a factor of two for the new subproblem estimation, and a factor of two for the improved implementation, see the Appendix for further details.

## 3.2 Problems of Increasing Size

The computational complexity is typically difficult to explicitly express for heuristic algorithms like the one presented. From the outline of the algorithm in Section 2 we see that step 1 and 2 are linear with respect to the number of variables/parameters, while step 3 and 4 involve local optimisation with polynomial or in worst case exponential time complexity with respect to the number of variables/parameters.

As a general test to evaluate the computational complexity of this and other algorithms, we have therefore defined a series of benchmark problems with the same general properties but with varying size. We prepare the problems from a generalised version of the gene regulatory network, see Fig. 2.

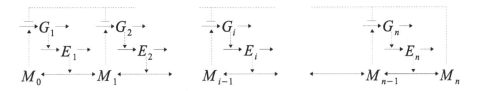

**Fig. 2.** The generalised gene network based on the n=3 case considered in [1]. $M_i$ are metabolites, $E_i$ are enzymes, and $G_i$ are mRNAs. $M_0$ are $M_n$ are input variables, while all others are dependent variables.

The size of the system is determined by a constant $n$ corresponding to the number of different mRNA species and enzymes in the system. The ODEs are given as

$$G'_i(t) = \frac{V_i^G}{1+\left(M_n/K_{I_i}\right)^{z_i}+\left(K_{A_i}/M_{i-1}\right)^{w_i}} - k_i^G G_i \qquad i = 1 \ldots n \qquad (2)$$

$$E'_i(t) = \frac{V_i^E \, G_i}{K_i^E + G_i} - k_i^E E_i \qquad i = 1 \ldots n \qquad (3)$$

$$v_i^+ = \frac{k_i^{M1} \, E_i \left(k_i^{M2}\right)^{-1} (M_{i-1}-M_i)}{1+M_{i-1}/k_i^{M2}+M_i/k_i^{M3}} \qquad i = 1 \ldots n-1 \qquad (4)$$

$$v_i^- = \frac{k_{i+1}^{M1} \, E_{i+1} \left(k_{i+1}^{M2}\right)^{-1} (M_i-M_{i+1})}{1+M_i/k_{i+1}^{M2}+M_{i+1}/k_{i+1}^{M3}} \qquad i = 1 \ldots n-1 \qquad (5)$$

$$M'_i(t) = v_i^+ - v_i^- \qquad i = 1 \ldots n-1 \qquad (6)$$

where $M_0$ and $M_n$ corresponds to the input variables, and where $V_i^G$, $K_{I_i}$, $z_i$, $K_{A_i}$, $w_i$, $k_i^G$, $V_i^E$, $K_i^E$ and $k_i^E$ are parameters defined for $i = 1 \ldots n$, and $k_i^{M1}$, $k_i^{M2}$, $k_i^{M3}$ are parameters defined for $i = 1 \ldots n-1$.

Using this system with $n = 3, 4, 5$ and $6$, we have created parameter estimation problems with 36, 48, 60 and 72 parameters, respectively. The original problem

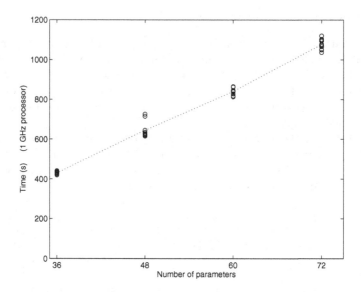

**Fig. 3.** Computational time (scaled to a 1GHz processor) for noise-free data on different instances of the generalised benchmark problem originally suggested in [1]. For each considered size of the system, 10 runs with varying random seeds were executed. The means are connected by a dotted line.

presented in [1] corresponds to $n = 3$. For each model, we have generated data with the same sampling schedules as the benchmark problem given for the $n = 3$ system, and we have also used the same parameter bounds. In this way, the four problems have basically the same properties but for the size. We refer to the problems as pe_3genes1, pe_4genes1, pe_5genes1, and pe_6genes1, respectively.

For each problem, we ran our algorithm 10 times with varying random seeds, see Table 2. The algorithm settings were chosen so that the largest system was successfully identified in the majority of runs. Hence, for the smaller problems, we used more random starting points than required for successful identification. Data in Fig. 3 indicates that the time complexity is dominated by a linear trend for these problems. The result is conservative in the sense that data is slightly less informative for the larger systems, since variables are influenced by the variations in input variable $M_0$ to a smaller extent.

Experiments of this kind have not, to our knowledge, been done for other optimisation methods in this field, so no direct comparison can be made. As already mentioned, however, many global methods fail already on the $n = 3$ case [1].

## 4   Discussion

We have presented an improved decomposition algorithm for parameter estimation of ODEs. The improved decomposition is based on considering estimation of arbitrary subsets of ODEs, suitably mixed with considering estimation of single ODEs and estimation of the entire model.

The presented results show that for the tested problems, our improved algorithm is significantly faster than other reported methods, with excellent accuracy. The total effect of the decomposition and the mix of different stages is to dramatically reduce the running time, compared to other approaches without decomposition. We also show that the improved algorithm can scale favourably with the size of the problem. While it is difficult to draw general quantitative conclusions from a limited test set, we can qualitatively say that our approach is clearly competitive.

As mentioned, parameter estimation is a critical subroutine when identifying both the structure and the parameters of an ODE model. Preliminary evaluation on a subset of the benchmark problems of Gennemark and Wedelin [3], indicate that the running time is improved in the order of 40% compared to the results presented in [3]. Some of these problems take a long time to run so any improvement is useful in practice.

All problems and their solutions are available on-line in an easy to read format, facilitating for others to test and compare using exactly the same problems. This is important, since many problems and solutions discussed in other publications are not well documented, which creates difficulties for further research.

Along the lines of the ideas presented here, several directions of future development can be considered, for example:

- Sub-problems can be defined in different ways combining the interaction structure, common parameters, and the strength of the interactions. By estimating subsets of tightly coupled variables together, one can potentially further improve performance since the relative impact of interpolated data is reduced in the simulation.
- For problems with incomplete data, in which some state variables are not observed, most or all current identification methods rely on simulation of the entire model, and not decomposition. However, by simultaneously analysing the dependency graph of the system and the observation functions, parameter estimation based on simulation of subsets of variables may become feasible, and hence reduce work.

## Acknowledgements

PG was supported by the Göteborg University quantitative biology platform and the Swedish Strategic Research Foundation through the Göteborg Mathematical Modeling Center. Thanks to the reviewers for valuable comments.

## References

[1] Moles, C.G., Mendes, P., Banga, J.R.: Parameter estimation in biochemical pathways: a comparison of global optimization methods. Genome Res. 13(11), 2467–2474 (2003)

[2] Gennemark, P., Wedelin, D.: Efficient algorithms for ordinary differential equation model identification of biological systems. IET Syst. Biol. 1(2), 120–129 (2007)

[3] Gennemark, P., Wedelin, D.: Benchmarks for identification of ordinary differential equations from time series data. Bioinformatics 25(6), 780–786 (2009)

[4] Ljung, L., Glad, T.: On global identifiability for arbitrary model parametrizations. Automatica 30(2), 265–276 (1994)

[5] Schittkowski, K.: Numerical data fitting in dynamical systems: a practical introduction with applications and software. Kluwer Academic, Dordrecht (2002)

[6] Polisetty, P.K., Voit, E.O., Gatzke, E.P.: Identification of metabolic system parameters using global optimization methods. Theor. Biol. Med. Model. 3(4) (2006)

[7] Mendes, P., Kell, D.: Non-linear optimization of biochemical pathways: applications to metabolic engineering and parameter estimation. Bioinformatics 14(10), 869–883 (1998)

[8] Peifer, M., Timmer, J.: Parameter estimation in ordinary differential equations for biochemical processes using the method of multiple shooting. IET Syst. Biol. 1, 78–88 (2007)

[9] Balsa-Canto, E., Peifer, M., Banga, J.R., Timmer, J., Fleck, C.: Hybrid optimization method with general switching strategy for parameter estimation. BMC Syst. Biol. 2(26) (2008)

[10] Englezoz, P., Kalogerakis, N.: Applied parameter estimation for chemical engineers. Marcel Dekker, Inc., New York (2001)

[11] Tucker, W., Moulton, V.: Parameter reconstruction for biochemical networks using interval analysis. Reliable computing 12(5), 389–402 (2006)

[12] Rodriguez-Fernandez, M., Mendes, P., Banga, J.R.: A hybrid approach for efficient and robust parameter estimation in biochemical pathways. Biosystems 83(2-3), 248–265 (2006a)

[13] Rodriguez-Fernandez, M., Egea, J.A., Banga, J.R.: Novel metaheuristic for parameter estimation in nonlinear dynamic biological systems. BMC Bioinformatics 7, 483 (2006b)

[14] Tsai, K.Y., Wang, F.S.: Evolutionary optimization with data collocation for reverse engineering of biological networks. Bioinformatics 21(7), 1180–1188 (2005)

[15] Kutalik, Z., Tucker, W., Moulton, V.: S-system parameter estimation for noisy metabolic profiles using newton-flow analysis. IET Syst. Biol. 1, 174–180 (2007)

[16] Box, G.E.P., Hunter, W.G., MacGregor, J.F., Erjavec, J.: Some problems associated with the analysis of multiresponse data. Technometrics 15, 33–51 (1973)

[17] Savageau, M.A.: Biochemical systems analysis: a study of function and design in molecular biology. Addison-Wesley, Reading (1976)

[18] Voit, E.O.: Computational analysis of biochemical systems. A practical guide for biochemists and molecular biologists, pp. 176–184. Cambridge University Press, Cambridge (2000)

[19] Maki, Y., Tominaga, D., Okamoto, M., Watanabe, S., Eguchi, Y.: Development of a system for the inference of large scale genetic networks. In: Pac. Symp. Biocomput. 2001, pp. 446–458 (2001)

# Appendix

Error function and running times for our parameter estimation method, with and without the estimation of submodels. Each problem was run 10 times, and the best, median and worst errors and computation times are reported. Computation times are scaled to a 1GHz processor. Stability measures the frequency of runs for which the best $-L$ was obtained.

| Problem | Sub-models | Solution $-L$ | | | Time (s) | | | Stab. |
|---|---|---|---|---|---|---|---|---|
| | | Best | Median | Worst | Best | Median | Worst | |
| pe_3genes1 | No | 0.00 | 0.00 | 136.3 | 45 | 62 | 140 | 8/10 |
| | Yes | 0.00 | 0.00 | 486.2 | 24 | 43 | 140 | 6/10 |
| pe_3genes2f | No | 1335 | $> 10^4$ | $> 10^4$ | 120 | 170 | 220 | 1/10 |
| | Yes | 1286 | 1827 | $> 10^4$ | 56 | 76 | 160 | 5/10 |
| pe_3genes3f | No | 3225 | $> 10^4$ | $> 10^4$ | 150 | 180 | 260 | 1/10 |
| | Yes | 3128 | 3748 | $> 10^4$ | 83 | 97 | 120 | 1/10 |
| pe_pinene | No | 9.936 | 9.936 | 9.936 | 0.2 | 0.2 | 0.2 | 9/10 |
| | Yes | 9.936 | 9.936 | 9.936 | 0.1 | 0.1 | 0.1 | 9/10 |
| pe_ss_cascade1 | No | 0.00 | 0.00 | 0.00 | 9 | 11 | 14 | 10/10 |
| | Yes | 0.00 | 0.00 | 0.00 | 11 | 13 | 15 | 10/10 |
| pe_ss_branch4 | No | 0.00 | 0.00 | 0.00 | 2.2 | 2.2 | 2.2 | 10/10 |
| | Yes | 0.00 | 0.00 | 0.00 | 0.5 | 0.5 | 0.5 | 10/10 |
| pe_ss_30genes2f | No | 3240 | 3240 | 3240 | 310 | 330 | 360 | 10/10 |
| | Yes | 3240 | 3240 | 3240 | 190 | 210 | 240 | 10/10 |
| pe_3genes2 | No | 2719 | $> 10^4$ | $> 10^4$ | 240 | 380 | 520 | 1/10 |
| | Yes | 1214 | 3815 | $> 10^4$ | 88 | 110 | 220 | 5/10 |
| pe_3genes3 | No | 3217 | $> 10^4$ | $> 10^4$ | 210 | 350 | 530 | 1/10 |
| | Yes | 2309 | 3188 | $> 10^4$ | 70 | 160 | 270 | 1/10 |
| pe_ss_30genes2 | No | 3004 | 3004 | 3004 | 1900 | 2000 | 2300 | 10/10 |
| | Yes | 3004 | 3004 | 3004 | 860 | 910 | 1200 | 10/10 |
| pe_4genes1 | No | 0.00 | 0.00 | $> 10^4$ | 1300 | 1300 | 1600 | 8/10 |
| | Yes | 0.00 | 0.00 | $> 10^4$ | 610 | 630 | 720 | 7/10 |
| pe_5genes1 | No | 0.00 | 0.00 | 0.00 | 1700 | 1700 | 1800 | 10/10 |
| | Yes | 0.00 | 0.00 | 0.00 | 810 | 840 | 860 | 10/10 |
| pe_6genes1 | No | 0.00 | 0.00 | $> 10^4$ | 2100 | 2200 | 2800 | 8/10 |
| | Yes | 0.00 | 0.00 | $> 10^4$ | 1100 | 1100 | 1200 | 8/10 |

# A Bayesian Approach to
# Model Checking Biological Systems[*]

Sumit K. Jha[1], Edmund M. Clarke[1], Christopher J. Langmead[1,2],
Axel Legay[3], André Platzer[1], and Paolo Zuliani[1]

[1] Computer Science Department, Carnegie Mellon University, USA
[2] Lane Center for Computational Biology, Carnegie Mellon University, USA
[3] Institut d'Informatique INRIA, Rennes, France

**Abstract.** Recently, there has been considerable interest in the use of
Model Checking for Systems Biology. Unfortunately, the state space of
stochastic biological models is often too large for classical Model Checking techniques. For these models, a statistical approach to Model Checking has been shown to be an effective alternative. Extending our earlier
work, we present the first algorithm for performing statistical Model
Checking using Bayesian Sequential Hypothesis Testing. We show that
our Bayesian approach outperforms current statistical Model Checking
techniques, which rely on tests from Classical (aka Frequentist) statistics, by requiring fewer system simulations. Another advantage of our
approach is the ability to incorporate prior Biological knowledge about
the model being verified. We demonstrate our algorithm on a variety
of models from the Systems Biology literature and show that it enables
faster verification than state-of-the-art techniques, even when no prior
knowledge is available.

## 1 Introduction

Computational models are increasingly used in the field of Systems Biology to
examine the dynamics of biological processes (e.g., [8,10,20,30,34,37]). By 'computational', we mean discrete-variable and continuous or discrete-time models
[4], where the components of the system interact and evolve by obeying a set
of instructions or rules. In contrast to differential equation-based models, which
are also widely used in Systems Biology, computational models can provide insights into the role of stochastic effects over discrete-populations of molecules or
cells. Recently, there has been considerable interest in the application of Model

[*] This research was sponsored by the GSRC (University of California) under contract
no. SA423679952, National Science Foundation under contracts no. CCF0429120,
no. CNS0411152, and no. CCF0541245, Semiconductor Research Corporation under contract no. 2005TJ1366, Air Force (University of Vanderbilt) under contract
no. 18727S3, International Collaboration for Advanced Security Technology of the
National Science Council, Taiwan, under contract no. 1010717, the U.S. Department
of Energy Career Award (DE-FG02-05ER25696), and a Pittsburgh Life-Sciences
Greenhouse Young Pioneer Award.

P. Degano and R. Gorrieri (Eds.): CMSB 2009, LNBI 5688, pp. 218–234, 2009.
© Springer-Verlag Berlin Heidelberg 2009

Checking [15] as a powerful tool for *formally* reasoning about the dynamic properties of such models (e.g., [1,6,9,11,14,18,24,38]). This paper presents a new Model Checking algorithm that is well-suited for verifying properties of very large stochastic models, such as those created and used in Systems Biology.

The stochastic nature of most computational models from Systems Biology gives rise to an instance of the *Probabilistic Model Checking* (PMC) problem [13,15,31]. Suppose $\mathcal{M}$ is a stochastic model over a set of states $S$, $s_0$ is a starting state, $\phi$ is a dynamic property expressed as a formula in temporal logic, and $\theta \in [0,1]$ is a probability threshold. The PMC problem is: given the 4-tuple $(\mathcal{M}, s_0, \phi, \theta)$, to decide algorithmically whether $\mathcal{M}, s_0 \models P_{\geq \theta}(\phi)$. In this paper, property $\phi$ is expressed in BLTL - Bounded Linear Temporal Logic [36,35,19]. Given these, PMC algorithms decide whether the model satisfies the property with at least probability $\theta$.

Existing algorithms for solving the PMC problem fall into one of two categories. The first category comprises numerical methods (e.g. [2,3,12,16,31]) which can compute the probability with which the property holds with high precision. Numerical methods are generally only suitable for small systems ($\approx 10^6$ to $10^7$ states). In a Biological System, the number of states can easily exceed this limit, which motivates the need for algorithms for solving the PMC problem in an approximate fashion. Approximate methods (e.g., [23,26,39,46]) work by sampling a set of *traces* from the model. Each trace is then evaluated to determine whether it satisfies the property. The number of satisfying traces is used to (approximately) decide whether $\mathcal{M}, s_0 \models P_{\geq \theta}(\phi)$.

Approximate PMC methods can be further divided into two sub-categories: (i) those that seek to *estimate* the probability that the property holds and then compare that estimate to $\theta$ (e.g., [26,39]), and (ii) those that reduce the PMC problem to a *hypothesis testing* problem (e.g., [46,47]). That is, deciding between two hypotheses — $H_0 : P_{\geq \theta}(\phi)$ versus $H_1 : P_{<\theta}(\phi)$. Hypothesis-testing based methods are more efficient than those based on estimation when $\theta$ (which is specified by the user) is significantly different than the true probability that the property holds (which is determined by $\mathcal{M}$ and $s_0$) [45].

Existing PMC methods based on hypothesis testing rely on *Classical* (aka *Frequentist*) statistical procedures, like Wald's Sequential Probability Ratio Test (SPRT) [42], to answer the decision problem. Our algorithm performs hypothesis testing, but uses *Bayesian* statistical procedures. This distinction is not trivial, as Bayesian and Classical statistics are two very different fields. We will show that in practice, our Bayesian approach requires fewer samples than Wald's SPRT. Finally, we note that because we adopt a Bayesian approach, our algorithm can incorporate prior knowledge, in the form of a probability distribution, $P(\theta)$, when available. This is relevant because in a Biological setting, it is often the case that prior knowledge is available.

The contributions of this paper are as follows:

- The first application of Bayesian Sequential Hypothesis Testing to statistical Model Checking,

- The first hypothesis-testing based statistical Model Checking algorithm designed for composite hypotheses, which can in particular include prior knowledge via a mixture of prior distributions,
- A theorem proving that our algorithm terminates with probability 1,
- Error bounds for our algorithm, and
- A series of case studies using Systems Biology models demonstrating that our method is empirically more efficient than existing algorithms for statistical Model Checking.

## 2   Background and Related Work

Our algorithm can be applied to any stochastic model $\mathcal{M}$ with a well-defined probability space over traces. Several well-studied stochastic models like (discrete and continuous) Markov Chains satisfy this property [47]. We assume that each execution of the system can be represented by a sequence of states and the time spent in these states. The sequence $\sigma = (s_0, t_0), (s_1, t_1), \ldots$ denotes an execution of the system along states $s_0, s_1, \ldots$ with durations $t_0, t_1, \ldots \in \mathbb{R}$. The system stays in state $s_i$ for duration $t_i$ and makes a transition to $s_{i+1}$. We require that the sum $\sum_i^\infty t_i$ must diverge, that is, the system can not make infinitely many state switches in finite time.

### 2.1   Specifying Properties in Temporal Logic

Our algorithm verifies properties of $\mathcal{M}$ expressed as formulas in *Probabilistic Bounded Linear Temporal Logic* (PBLTL). We first define the syntax and semantics of *Bounded Linear Temporal Logic* (BLTL) [36,35,19] and then extend that logic to PBLTL.

For a stochastic model $\mathcal{M}$, let the set of state variables $SV$ be a finite set of real-valued variables. A Boolean predicate over $SV$ is a constraint of the form $x \sim v$, where $x \in SV$, $\sim \in \{\geq, \leq, =\}$, and $v \in \mathbb{R}$. A BLTL property is built on a finite set of Boolean predicates over $SV$ using Boolean connectives and temporal operators. The syntax of the logic is given by the following grammar:

$$\phi ::= x \sim v \mid (\phi_1 \vee \phi_2) \mid (\phi_1 \wedge \phi_2) \mid \neg \phi_1 \mid (\phi_1 \mathbf{U^t} \phi_2),$$

where $\sim \in \{\geq, \leq, =\}$, $x \in SV$, $v \in \mathbb{Q}$, and $t \in \mathbb{Q}_{\geq 0}$. We can define additional temporal operators such as $\mathbf{F^t}\psi = \mathbf{True}\, \mathbf{U^t}\, \psi$, or $\mathbf{G^t}\psi = \neg \mathbf{F^t} \neg \psi$ in terms of the bounded until $\mathbf{U^t}$.

We define the semantics of BLTL with respect to executions of $\mathcal{M}$. The fact that an execution $\sigma$ satisfies property $\phi$ is denoted by $\sigma \models \phi$. Let $\sigma = (s_0, t_0), (s_1, t_1), \ldots$ be an execution of the model along states $s_0, s_1, \ldots$ with durations $t_0, t_1, \ldots \in \mathbb{R}$. We denote the execution trace starting at state $i$ by $\sigma^i$ (in particular, $\sigma^0$ denotes the original execution $\sigma$). The value of the state variable $x$ in $\sigma$ at the state $i$ is denoted by $V(\sigma, i, x)$. The semantics of BLTL for a trace $\sigma^k$ starting at the $k^{th}$ state ($k \in \mathbb{N}$) is defined as follows:

- $\sigma^k \models x \sim v$ if and only if $V(\sigma, k, x) \sim v$;
- $\sigma^k \models \phi_1 \vee \phi_2$ if and only if $\sigma^k \models \phi_1$ or $\sigma^k \models \phi_2$;

- $\sigma^k \models \phi_1 \wedge \phi_2$ if and only if $\sigma^k \models \phi_1$ and $\sigma^k \models \phi_2$;
- $\sigma^k \models \neg\phi_1$ if and only if $\sigma^k \models \phi_1$ does not hold (written $\sigma^k \not\models \phi_1$);
- $\sigma^k \models \phi_1 \mathbf{U}^t \phi_2$ if and only if there exists $i \in \mathbb{N}$ such that $(a)$ $\sum_{0 \leq l < i} t_{k+l} \leq t$, $(b)$ $\sigma^{k+i} \models \phi_2$ and $(c)$ for each $0 \leq j < i$, $\sigma^{k+j} \models \phi_1$.

Statistical Model Checking is based on evaluating whether $\sigma \models \phi$ holds on sample simulations $\sigma$ of the system. In practice, sample simulations only have a finite duration. The question is how long these simulations have to be for the formula $\phi$ to have a well-defined semantics such that $\sigma \models \phi$ can be checked. If $\sigma$ is too short, say of duration 2, the semantics of $\phi_1 \mathbf{U}^5 \phi_2$ may be unclear. But at what duration of the simulation can we stop because we know that the truth-value for $\sigma \models \phi$ will never change by continuing the simulation? In [29], we prove that finite simulations of bounded duration are always sufficient for Model Checking BLTL on traces.

We can now define Probabilistic Bounded Linear Temporal Logic.

**Definition 1.** *A Probabilistic Bounded LTL (PBLTL) formula is a formula of the form* $P_{\geq\theta}(\phi)$*, where* $\phi$ *is a BLTL formula and* $\theta \in (0,1)$*.*

We say that $\mathcal{M}$ satisfies PBLTL property $P_{\geq\theta}(\phi)$, denoted by $\mathcal{M} \models P_{\geq\theta}(\phi)$, if and only if the probability that an execution of $\mathcal{M}$ satisfies BLTL property $\phi$ is greater than or equal to $\theta$. The problem is well-defined [47] since one can always assign a unique probability measure to the set of executions of $\mathcal{M}$ that satisfy a formula in BLTL. Note that counterexamples to the BLTL property $\phi$ are *not* counterexamples to the PBLTL property $P_{\geq\theta}(\phi)$, because the truth of $P_{\geq\theta}(\phi)$ depends on the likelihood of all counterexamples to $\phi$. This makes PMC more difficult than standard Model Checking, because one counterexample to $\phi$ is not enough to answer $P_{\geq\theta}(\phi)$.

## 2.2 Existing Statistical Probabilistic Model Checking Algorithms

As outlined in the introduction, Probabilistic Model Checking algorithms can either be exact (e.g. [2,3,12,16,31]), or statistical in nature. In practice, statistical methods (e.g., [23,26,32,39,46]), which iteratively draw sample traces from the model, are generally better suited to Model Checking Biological systems because they scale better. Our method is statistical, and so we will compare and contrast our method to existing statistical methods in this section.

Existing PMC methods based on hypothesis testing rely on *Classical* (aka *Frequentist*) statistical procedures, like Wald's Sequential Probability Ratio Test (SPRT) [42], to answer the decision problem. Younes and Simmons introduced the first algorithm for statistical Model Checking [45,46,47] for verifying probabilistic temporal properties of stochastic systems. Their work uses the SPRT, which is designed for *simple* hypothesis testing[1]. Specifically, the SPRT decides between

---

[1] A simple hypothesis completely specifies a distribution. For example, a Bernoulli distribution of parameter $p$ is fully specified by the hypothesis $p = 0.5$ (or some other fixed value). A composite hypothesis has instead free parameters, *e.g.* the hypothesis $p < 0.3$, for a Bernoulli distribution.

the simple null hypothesis $H'_0 : \mathcal{M}, s_0 \models P_{=\theta_0}(\phi)$ against the simple alternate hypothesis $H'_1 : \mathcal{M}, s_0 \models P_{=\theta_1}(\phi)$, where $\theta_0 < \theta_1$. It can be shown that the SPRT is optimal for simple hypothesis testing, in the sense that it minimizes the expected number of samples among all the tests satisfying the same Type I and II errors [43], when either $H'_0$ or $H'_1$ is true. The PMC problem is instead a choice between two *composite* hypotheses $H_0 : \mathcal{M}, s_0 \models P_{\geq\theta}[\phi]$ versus $H_1 : \mathcal{M}, s_0 \models P_{<\theta}[\phi]$. The SPRT is not defined unless $\theta_0 \neq \theta_1$, so Younes and Simmons overcome this problem by separating the two hypotheses by an *indifference region* $(\theta - \delta, \theta + \delta)$, where $0 < \delta < 1$ is a user-specified parameter. It can be shown that the SPRT with indifference region can be used for testing composite hypotheses, while respecting the same Type I and II errors of a standard SPRT [21, Section 3.4]. However, in this case the test is no longer optimal, and the maximum expected sample size may be much bigger than the optimal fixed sample size sampling test - see [7] and [21, Section 3.6]. We note that our algorithm solves the composite hypothesis testing problem, but does so using Bayesian statistics, and thus requires no indifference region.

The method of [26] uses a fixed number of samples and estimates the probability the property holds as the number of satisfying traces divided by the number of sampled traces. Their algorithm guarantees the accuracy of the results using Chernoff-Hoeffding bounds. In particular, their algorithm can guarantee that the difference in the estimated and the true probability is less than $\epsilon$, with probability $\rho$, where $\rho < 1$ and $\epsilon > 0$ are user-specified parameters. Grosu and Smolka use a similar technique for verifying formulas in LTL [23]. Their algorithm randomly samples lassos from a Büchi automaton in an on-the-fly fashion. The method of [32] is also Bayesian, like the algorithm in this paper, but estimates the probability the property holds and does not invoke hypothesis testing. Unlike the algorithm in this paper, [32] is fully Bayesian in the sense that it explicitly considers the prior distributions over the initial state and parameters of the model, in addition to the prior over the property.

Finally, Sen *et al.* [39,40] used the *p-value* for the null hypothesis as a statistic for hypothesis testing. The *p*-value is defined as the probability of obtaining observations at least as extreme as the one that was actually seen, given that the null hypothesis is true. It is important to realize that a *p*-value is *not* the probability that the null hypothesis is true. Sen *et al.*'s method does not have a way to control the Type I and II errors.

## 3    Bayesian Statistical Model Checking

In this section, we first review some important concepts from statistical Model Checking, and then introduce theory and terminology from Bayesian statistics. We then present our algorithm in Sec. 3.2.

Recall that the PMC problem is to decide whether $\mathcal{M} \models P_{\geq\theta}(\phi)$, where $\theta \in (0,1)$ and $\phi$ is a BLTL formula. Let $p$ be the (unknown but fixed) probability of the model satisfying $\phi$: thus, the PMC problem can now be stated as deciding between two hypotheses:

$$H_0 : p \geqslant \theta \qquad H_1 : p < \theta.$$

For any trace $\sigma_i$ of the system, we can deterministically decide whether $\sigma_i$ satisfies $\phi$. Therefore, we can define a Bernoulli random variable $X_i$ denoting the outcome of $\sigma_i \models \phi$. The probability mass function associated with $X_i$ is thus:

$$f(x_i|u) = p^{x_i}(1-p)^{1-x_i}$$

where $x_i = 1$ iff $\sigma_i \models \phi$, otherwise $x_i = 0$. Note that the $X_i$ are independent and identically distributed, as each trace is given by an independent execution of the model. Since $p$ is unknown, we assume that it is given by a random variable, whose density $g(\cdot)$ is called the *prior* density. The prior is usually based on our previous experiences and beliefs about the system. A complete lack of information about the probability of the system satisfying the formula is usually summarized by a *non-informative* or *objective* prior probability.

## 3.1   Bayesian Statistics

Suppose we have a sequence of random variables $X_1, \ldots, X_n$ defined as above, and let $d = (x_1, \ldots, x_n)$ denote a sample of those variables. Then Bayes' theorem states that the *posterior odds* are

$$P(H_0|d) = \frac{P(d|H_0)P(H_0)}{P(d)} \qquad P(H_1|d) = \frac{P(d|H_1)P(H_1)}{P(d)}$$

where $P(d) = P(d|H_0)P(H_0) + P(d|H_1)P(H_1)$, which in our case is always non-zero. The ratio of the posterior odds for hypotheses $H_0$ and $H_1$ given data $d$ is

$$\frac{P(H_0|d)}{P(H_1|d)} = \frac{P(d|H_0)}{P(d|H_1)} \frac{P(H_0)}{P(H_1)} . \tag{1}$$

**Definition 2.** *The Bayes factor $\mathcal{B}$ of sample $d$ and hypotheses $H_0$ and $H_1$ is*

$$\mathcal{B} = \frac{P(d|H_0)}{P(d|H_1)} .$$

For fixed priors in a given example, the Bayes factor is directly proportional to the posterior odds ratio by Equation (1). Thus, it may be used as a measure of relative confidence in $H_0$ vs. $H_1$, as proposed by Jeffreys [28]. In particular, he suggested that a value of the Bayes factor greater than 100 provides decisive evidence in favor of $H_0$. To test $H_0$ vs. $H_1$ we compute the Bayes factor $\mathcal{B}$ of the available data and then compare it against a fixed threshold $T > 1$: we shall accept $H_0$ iff $\mathcal{B} > T$. Jeffreys interprets the value of the Bayes factor as a measure of the evidence in favor of $H_0$ (dually, $\frac{1}{\mathcal{B}}$ is the evidence in favor of $H_1$).

We now show how to compute the Bayes factor. According to Definition 2, we have to calculate the probability of the observed sample $d = (x_1, \ldots, x_n)$ given $H_0$ and $H_1$. They are given by integrating the joint density $h(d|\cdot)$ with respect to the prior $g(\cdot)$, and since we assume that the sample is drawn from iid variables, we have that $h(d|\cdot) = f(x_1|\cdot) \cdots f(x_n|\cdot)$. Therefore, the Bayes factor is the ratio:

$$\mathcal{B} = \frac{P(x_1, \ldots, x_n | H_0)}{P(x_1, \ldots, x_n | H_1)} = \frac{\int_\theta^1 f(x_1|u) \cdots f(x_n|u) \cdot g(u) \ du}{\int_0^\theta f(x_1|u) \cdots f(x_n|u) \cdot g(u) \ du} . \qquad (2)$$

We observe that the Bayes factor depends on the data $d$ and on the prior $g$, so it may be considered a measure of confidence in $H_0$ vs. $H_1$ provided by the data $x_1, \ldots, x_n$, and "weighted" by the prior $g$. Hence, the choice of the threshold Bayes Factor $(T)$ in Sec. 3.2 also indicates an objective degree of confidence in the accepted hypothesis when the Bayesian Statistical Model Checking algorithm stops.

## 3.2   Algorithm

Our algorithm is essentially a sequential version of Jeffreys' test. Remember we want to establish whether $\mathcal{M} \models P_{\geqslant\theta}(\phi)$, where $\theta \in (0, 1)$ and $\phi$ is a BLTL formula. Like all statistical Model Checking algorithms, we assume that it is possible to generate unbiased samples from the model. The algorithm iteratively draws independent and identically distributed sample traces $\sigma_1, \sigma_2, \ldots$, and checks whether they satisfy $\phi$. As explained above, we can model this procedure as independent sampling from a Bernoulli distribution $X$ of unknown parameter $p$ - the actual probability of the model satisfying $\phi$. At stage $n$ the algorithm has drawn samples $x_1, \ldots, x_n$ iid like $X$. It then computes the Bayes factor $\mathcal{B}_n$ according to (2), and it stops iff $(\mathcal{B}_n > T \ \lor \ \mathcal{B}_n < \frac{1}{T})$. When this occurs, it will accept $H_0$ iff $\mathcal{B}_n > T$, and will accept $H_1$ iff $\mathcal{B}_n < \frac{1}{T}$. The algorithm is shown below.

From (2) we see that the algorithm can incorporate prior knowledge through $g$, when computing the Bayes factor. Our examples focus on Beta priors which are defined over the $(0, 1)$ interval by the following probability density (for real parameters $\alpha, \beta > 0$):

$$\forall u \in (0, 1) \quad g(u, \alpha, \beta) \cong \frac{1}{B(\alpha, \beta)} u^{\alpha-1}(1 - u)^{\beta-1} \qquad (3)$$

where the Beta function $B(\alpha, \beta)$ is defined as:

$$B(\alpha, \beta) \cong \int_0^1 t^{\alpha-1}(1 - t)^{\beta-1} dt . \qquad (4)$$

By varying the parameters $\alpha$ and $\beta$, one can approximate other smooth unimodal densities on $(0, 1)$ by a Beta density (*e.g.*, the uniform density over $(0, 1)$ is a Beta with $\alpha = \beta = 1$). We also define the Beta distribution function $F_{(\alpha,\beta)}(u)$:

$$\forall u \in (0, 1) \quad F_{(\alpha,\beta)}(u) \cong \int_0^u g(t, \alpha, \beta) \ dt = \frac{1}{B(\alpha, \beta)} \int_0^u t^{\alpha-1}(1 - t)^{\beta-1} \ dt \qquad (5)$$

which is just the usual distribution function for a Beta random variable of parameters $\alpha, \beta$ (*i.e.*, the probability that it takes values less than or equal to $u$).

---

**Algorithm 1.** Bayesian Statistical Model Checking

---

**Require:** PBLTL Property $P_{\geqslant \theta}(\phi)$, Threshold $T > 1$, Prior density $g$ for unknown parameter $p$

$n := 0$        {*number of traces drawn so far*}
$x := 0$        {*number of traces satisfying $\phi$ so far*}
**repeat**
    $\sigma :=$ draw a sample trace of the system (iid)
    $n := n + 1$
    **if** $\sigma \models \phi$ **then**
        $x := x + 1$
    **end if**
    $\mathcal{B}_n :=$ BayesFactor$(n, x)$        {*compute according to Equation (2)*}
**until** $(\mathcal{B}_n > T \vee \mathcal{B}_n < \frac{1}{T})$
**if** $(\mathcal{B}_n > T)$ **then**
    **return** $H_0$ accepted
**else**
    **return** $H_1$ accepted
**end if**

---

The choice of the Beta density is not arbitrary. It is well-known that the Beta distribution is the *conjugate prior* to the Bernoulli distribution[2]. This relationship gives rise to closed-form solutions to the *posterior* density over $\theta$ (i.e., $P(\theta|d)$), thus avoiding numerical integration when calculating the Bayes factor. Our data $(x_1, \ldots, x_n)$ are assumed to be iid samples drawn from a Bernoulli distribution of unknown parameter $p$. We write $x = \sum_{i=1}^{n} x_i$ for the number of successes in $(x_1, \ldots, x_n)$. The prior density $g(\cdot)$ is assumed to be a Beta density with fixed parameters $\alpha, \beta > 0$. In [29], we show that the Bayes factor $\mathcal{B}_n$ at stage $n$ can be computed in terms of the Beta distribution function:

$$\mathcal{B}_n = \frac{1}{F_{(x+\alpha, n-x+\beta)}(\theta)} - 1 .$$

The Beta distribution function can be computed with high accuracy by standard mathematical libraries (*e.g.* the GNU Scientific Library) or software (*e.g.* Matlab). Hence, the Beta distribution is the appropriate choice for summarizing the prior probability distribution in Statistical Model Checking.

We present the following two Theorems:

**Theorem 1 (Termination).** *The Bayesian Statistical Model Checking algorithm terminates with probability one, for Beta priors and Bernoulli samples. (See [29] for a proof.)*

---

[2] A distribution $P(\theta)$ is said to be a conjugate prior for a likelihood function, $P(d|\theta)$, if the posterior, $P(\theta|d)$ is in the same family of distributions.

**Theorem 2.** *If the Bayesian Model Checking algorithm terminates after observing $n$ sample traces, an upper bound on the probability of the Type I error is*

$$\sum_{x=0}^{n} I_{\{\mathcal{B}(n,x) < 1/T\}}(x) \binom{n}{x} t_{max}^{x}(1 - t_{max})^{n-x}$$

*where $t_{max}$ is the value of $t$ that maximizes the expression $t^i(1-t)^{n-i}$ defined on $[0,1]$, $T$ is the Bayes Factor threshold used in the Bayesian Model Checking algorithm, and $I$ is the indicator function. (See [29] for a proof.)*

### 3.3   Verification over General Priors

The use of conjugate priors does not pose restrictions, in practice. It is known that any prior distribution (with or without a density) can be well approximated by a *finite* mixture of conjugate priors [17]. Thus, we can approximate an arbitrary prior over $(0,1)$ by constructing a density $G(\cdot)$ of the form:

$$G(u) \triangleq \sum_{i=1}^{N} r_i \cdot g_i(u, \alpha_i, \beta_i)$$

where $N$ is a positive integer which depends on the level of accuracy required, the $g_i$'s are Beta densities (of possibly different parameters $\alpha_i, \beta_i$), and the $r_i$'s are positive reals summing up to 1 - this ensures that $G$ is a proper density.

For such priors, the computation of the Bayes factor is slightly more complicated. In [29], we show that the Bayes factor at stage $n$ is given by:

$$\mathcal{B}_n = \frac{\sum_{i=1}^{N} r_i' \cdot B(x + \alpha_i, n - x + \beta_i)}{\sum_{i=1}^{N} r_i' \cdot B(x + \alpha_i, n - x + \beta_i) \cdot F_{(x+\alpha_i, n-x+\beta_i)}(\theta)} - 1$$

where $r_i' = \frac{r_i}{B(\alpha_i, \beta_i)}$. Again, we see that the Bayes factor can be computed by means of standard, well-known numerical methods, thereby simplifying the implementation of the algorithm. Theorem 1 can be extended to handle this case, too [29].

## 4   Benchmarks

In this section, we analyze the performance of our algorithm on five benchmark models from the Systems Biology literature. Three of the models are written in the PRISM Model Checking tool's specification language [27,31], and the remaining two are written in SBML and were obtained from the Matlab Systems Biology Toolbox. The PRISM Model Checker tool is capable of both symbolic (i.e., exact) Probabilistic Model Checking, and statistical Probabilistic Model Checking. PRISM's statistical Probabilistic Model Checking Algorithm implements the algorithm of [26] which uses a fixed sized sampling approach and estimates the true probability as the number of satisfying traces over the number of sampled traces. We note that for each of the benchmark sets, we consider models that are too large for symbolic model checking.

Our experiments demonstrate two important properties of our algorithm: (i) we show that our algorithm requires fewer traces than either the algorithm of [26] implemented in PRISM or Wald's SPRT algorithm - while retaining the same bounds on the frequentist Type-I and Type-II error probabilities. (ii) The performance of both the Wald's algorithm [42] and our Bayesian Model Checking algorithm degrades as the threshold probability (i.e., $\theta$) in the PBLTL temporal logic formula gets close to the actual probability of the model satisfying the BLTL formula. However, the Bayesian algorithm shows a more graceful degradation compared to Wald's SPRT approach.

## 4.1 PRISM Benchmarks

We studied three large PRISM benchmarks which are not well suited for numerical approaches to Probabilistic Model Checking. In our experiments, the Bayesian Model Checking algorithm used uniform priors, and accepted a hypothesis when it was 10000 times more likely than the other hypothesis (Bayes Factor threshold $T = 10000$). Our experiments with Wald's SPRT used Type I and II error bounds of 0.01. We chose an indifference region $\delta$ so as to make the Type I and Type II errors for both the Wald's Test and the Bayes Factor test equal. The statistical estimation engine of the PRISM model checker always needed 92042 samples to estimate the probability of the BLTL formulae being true.

The results of experiments with the Fibroblast Growth Factor Signaling Model (see [24], [25] for details) are presented. We checked the property whether the probability that Grb2 binds to FRS2 within 20 time units exceeds $\theta$ (for several values of $\theta$):

$$H_0 : \mathcal{M} \models P_{\geq \theta}[\ \mathbf{F^{20}}\ (FRS2\_GRB > 0\ )]$$

The power curves and the number of samples for this benchmark are plotted in Fig. 2(a) and Fig. 2(b) respectively. A power curve indicates the probability of accepting the null hypothesis for various values of the threshold probability $\theta$ in the PBLTL formula. We chose the Wald's Test so that its power curve matched that of the Bayesian Test at the 0.01 and 0.99 acceptance probability. The goal is to make sure that the two tests have equal statistical power. From Figure 2(b), it is clear that both the power curves are almost on top of each other and hence, both the tests have indeed been calibrated to be equally powerful. The Bayesian algorithm needs fewer samples than Wald's SPRT test for this benchmark. This shows that the Bayesian Statistical Model Checking performs better than an approach based on Wald's SPRT.

We also studied the continuous time Markov Chain model [5,41] for circadian rhythm. We checked the property that the probability of the number of activated messenger RNAs exceeding 5 units within 0.25 time units is more than $\theta$ (for various values of $\theta$):

$$H_0 : \mathcal{M} \models P_{\geq \theta}[\ \mathbf{F^{0.25}}\ (ma > 5)\ ]$$

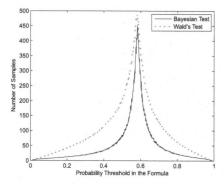
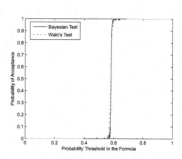

(a) Number of Samples for various probability (b) Power Curve of the Bayesian and
thresholds in the formula                          Wald's approach

**Fig. 1. Fibroblast Growth Factor Signaling Model:** The system satisfies the
formula with probability 0.58. (Bayes Factor=10000)

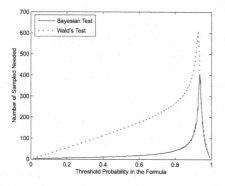
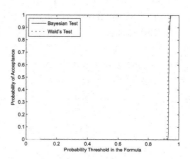

(a) Number of Samples for various probability (b) Power Curve of the Bayesian and
thresholds in the formula                          Wald's approach

**Fig. 2. Circadian Rhythm:** The system satisfies the formula with probability 0.93.
(Bayes Factor=10000)

The power curves and the number of samples for this benchmark are plotted
in Fig. 2(b) and Fig. 2(a) respectively. We calibrated Wald's test so that its
power curve closely matched that of the Bayesian Test so as to make a fair
comparison. From the figure, we observe that the Bayesian algorithm always
needs fewer samples than the Wald's SPRT test for this benchmark.

We also analyzed the model on Cell cycle control [33] and studied the probability that Cyclin gets bound within the first 0.5 time units. We check the
property that the probability of the number of bound Cyclin molecules exceeds
3 units within 0.5 time units exceeds $\theta$ (for various values of $\theta$):

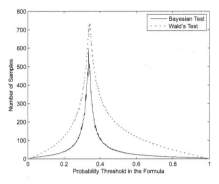 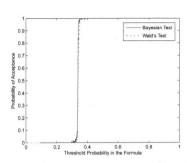

(a) Number of Samples for various probability (b) Power Curve of the Bayesian and thresholds in the formula                              Wald's approach

**Fig. 3. Cell Cycle Control:** The system satisfies the formula with probability 0.34. (Bayes Factor=10000)

$$H_0 : \mathcal{M} \models P_{\geq \theta}[\ \mathbf{F}^{0.5}\ (cyclin\_bound > 3)\ ]$$

The results of our experiment are presented in Fig. 3(a). The Bayesian Statistical Model Checking algorithm usually required fewer samples than the approach based on Wald's SPRT.

## 4.2   SBML Experiments

We also studied SBML models using the implementation of Gillespie's Stochastic Simulation Algorithm in Matlab's Systems Biology Toolbox. We analyzed two large models with over $10^8$ and $10^{17}$ species. We used monitors written in Matlab to verify the BLTL properties on traces. Our analysis of the experiments in this section is purely Bayesian, i.e., we have studied the performance of the algorithm over only one run (using uniform priors). In the previous sections, we had compared the performance of our algorithm with Wald's SPRT by running the algorithm several times on the same model - a frequentist approach.

We analyzed the Yeast Heterotrimeric G Protein Cycle benchmark [44]. We analyzed the property that the G protein stays above the threshold of 6000 units for 2 time units and falls below 6000 before 20 time units.

$$H_0 : \mathcal{M} \models P_{\geq \theta}[\ \mathbf{G}^2(GProtein > 6000) \text{ and } \mathbf{F}^{20}(GProtein < 6000)]\ .$$

We also ran experiments using the Lotka model [22] and verified the property that the number of copies of the $x$ species rises to a threshold level within 0.01 time units.

$$H_0 : \mathcal{M} \models P_{\geq \theta}[\ \mathbf{F}^{0.01}(x > 1.4 * 10^7)]$$

**Table 1.** Performance on the G Protein (left) and Lotka Benchmark (right)

| Probability | # Samples Needed |
| --- | --- |
| 0.2 | 3 |
| 0.6 | 8 |
| 0.8 | 14 |
| 0.9 | 23 |
| 0.9999 | 99 |

| Probability | # Samples Needed |
| --- | --- |
| 0.1 | 2 |
| 0.5 | 6 |
| 0.7 | 10 |
| 0.9 | 23 |
| 0.99 | 69 |

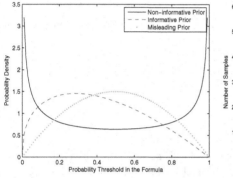
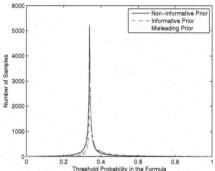

(a) Shape of the Priors used in our Exper-  (b) Number of Samples with Different
iments                                       Classes of Priors

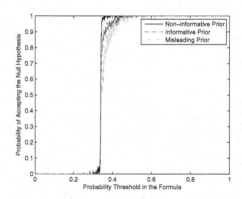

(c) Power curve of the tests of the Algorithm

**Fig. 4.** Different Classes of Priors

The results of our experiments are shown in Table 1: both hypotheses are always accepted, although the number of samples increases with the probability threshold of the temporal formula.

### 4.3   Experiment with Different Classes of Priors

We investigated the effect of priors on the performance of the Bayesian Model Checking algorithm. We used three different priors - non-informative prior, an informative prior and a misleading prior. The priors, the number of samples needed by the Bayesian algorithm for these priors, and the power curve for each of these priors is also plotted in Fig. 4(a), Fig. 4(b) and Fig. 4(c) respectively. The priors used are Beta distributions with different shape parameters: (i) $\alpha = 1/2, \beta = 1/2$: non-informative prior, (ii) $\alpha = 1.4, \beta = 2$ : informative prior with a peak around 0.34 (iii) $\alpha = 2, \beta = 2$: a misleading prior with peak around 0.5.

Fig. 4(b) shows that the number of samples needed by the Bayesian algorithm becomes smaller when the prior probability distribution is informative and supports the true hypothesis. Also, the power curve (see Fig. 4(c)) becomes sharper when the Bayesian algorithm is given a correct and informative prior probability distribution. A completely non-informative prior also performs well both in the number of samples and the power of the test. Strongly misleading priors make the power curve less steep. However, the algorithm still performs quite well when the actual probability of the system is away from the threshold probability in the formula.

## 5   Conclusions and Future Work

We have introduced the first algorithm for Probabilistic Model Checking based on Bayesian Sequential Hypothesis Testing. Our algorithm terminates with probability 1, and provides bounds on the probability of returning an incorrect answer. Empirically, we have shown that our algorithm requires fewer traces to terminate than techniques based on Classical Statistics. This is not surprising as the Bayesian method comparing composite hypotheses whereas techniques like Wald's SPRT are comparing simple hypotheses. This advantage in efficiency is important in the context of Systems Biology as the cost of generating traces is not necessarily negligible. Bayesian methods also afford a convenient means for incorporating domain knowledge through the prior distributions.

Our algorithm is presently limited to incorporating prior information on the probability that the property is true. A more fully Bayesian approach would incorporate prior information on not just the property, but also the starting state and parameters of the model. We are presently extending our method to address this limitation.

## Acknowledgments

The authors would like to thank H.L.S. Younes for comments on a draft of this paper.

# References

1. Antoniotti, M., Policriti, A., Ugel, N., Mishra, B.: Model building and model checking for biochemical processes. Cell Biochem. Biophys. 38(3), 271–286 (2003)
2. Baier, C., Clarke, E.M., Hartonas-Garmhausen, V., Kwiatkowska, M.Z., Ryan, M.: Symbolic model checking for probabilistic processes. In: Degano, P., Gorrieri, R., Marchetti-Spaccamela, A. (eds.) ICALP 1997. LNCS, vol. 1256, pp. 430–440. Springer, Heidelberg (1997)
3. Baier, C., Haverkort, B.R., Hermanns, H., Katoen, J.-P.: Model-checking algorithms for continuous-time markov chains. IEEE Trans. Software Eng. 29(6), 524–541 (2003)
4. Bailey, N.: The Elements of Stochastic Processes with Applications to the Natural Sciences. Wiley-IEEE (1990)
5. Barkai, N., Leibler, S.: Biological rhythms: Circadian clocks limited by noise. Nature 403, 267–268 (2000)
6. Batt, G., Ropers, D., de Jong, H., Geiselmann, J., Mateescu, R., Page, M., Schneider, D.: Validation of qualitative models of genetic regulatory networks by model checking: analysis of the nutritional stress response in Escherichia coli. Bioinformatics 25(1), i19–i28 (2005)
7. Bechhofer, R.: A note on the limiting relative efficiency of the Wald sequential probability ratio test. J. Amer. Statist. Assoc. 55, 660–663 (1960)
8. Calder, M., Gilmore, S., Hillston, J.: Modelling the influence of RKIP on the ERK signalling pathway using the stochastic process algebra PEPA. Transactions on Computational Systems Biology (2006) (in press)
9. Calder, M., Vyshemirsky, V., Gilbert, D., Orton, R.: Analysis of signalling pathways using the PRISM model checker. In: Proc. Computational Methods in Systems Biology (CMSB 2005), pp. 179–190 (2005)
10. Cardelli, L.: Abstract machines of systems biology. In: Priami, C., Merelli, E., Gonzalez, P., Omicini, A. (eds.) Transactions on Computational Systems Biology III. LNCS (LNBI), vol. 3737, pp. 145–168. Springer, Heidelberg (2005)
11. Chabrier, N., Fages, F.: Symbolic Model Checking of Biochemical Networks. In: Priami, C. (ed.) CMSB 2003. LNCS, vol. 2602, pp. 149–162. Springer, Heidelberg (2003)
12. Ciesinski, F., Größer, M.: On probabilistic computation tree logic. In: Baier, C., Haverkort, B.R., Hermanns, H., Katoen, J.-P., Siegle, M. (eds.) Validation of Stochastic Systems. LNCS, vol. 2925, pp. 147–188. Springer, Heidelberg (2004)
13. Clarke, E.M., Emerson, E.A.: Design and synthesis of synchronization skeletons using branching-time temporal logic. In: Logic of Programs, Workshop, London, UK, pp. 52–71. Springer, Heidelberg (1982)
14. Clarke, E.M., Faeder, J.R., Langmead, C.J., Harris, L.A., Jha, S.K., Legay, A.: Statistical model checking in biolab: Applications to the automated analysis of t-cell receptor signaling pathway. In: Heiner, M., Uhrmacher, A.M. (eds.) CMSB 2008. LNCS (LNBI), vol. 5307, pp. 231–250. Springer, Heidelberg (2008)
15. Clarke, E.M., Grumberg, O., Peled, D.A.: Model Checking. MIT Press, Cambridge (1999)
16. Courcoubetis, C., Yannakakis, M.: The complexity of probabilistic verification. Journal of the ACM 42(4), 857–907 (1995)
17. Diaconis, P., Ylvisaker, D.: Quantifying prior opinion. In: Bernardo, J.M., De Groot, M.H., Lindley, D.B., Smith, A.F.M. (eds.) Bayesian Statistics 2: Proceedings of the 2nd Valencia International Meeting. Elsevier Science Publisher, Amsterdam (1985)

18. Fages, F.: Temporal logic constraints in the biochemical abstract machine biocham. In: Hill, P.M. (ed.) LOPSTR 2005. LNCS, vol. 3901, pp. 1–5. Springer, Heidelberg (2006)

19. Finkbeiner, B., Sipma, H.: Checking finite traces using alternating automata. In: Proceedings of Runtime Verification (RV 2001), pp. 44–60 (2001)

20. Fisher, J., Piterman, N., Hubbard, E.J., Stern, M.J., Harel, D.: Computational insights into caenorhabditis elegans vulval development. Proc. Natl. Acad. Sci. U S A 102(6), 1951–1956 (2005)

21. Ghosh, B., Sen, P. (eds.): Handbook of sequential analysis. Dekker, New York (1991)

22. Gillespie, D.T.: Exact stochastic simulation of coupled chemical reactions. The Journal of Physical Chemistry 81(25), 2340–2361 (1977)

23. Grosu, R., Smolka, S.: Monte Carlo Model Checking. In: CAV, pp. 271–286 (2005)

24. Heath, J., Kwiatkowska, M., Norman, G., Parker, D., Tymchyshyn, O.: Probabilistic model checking of complex biological pathways. In: Priami, C. (ed.) CMSB 2006. LNCS (LNBI), vol. 4210, pp. 32–47. Springer, Heidelberg (2006)

25. Heath, J., Kwiatkowska, M., Norman, G., Parker, D., Tymchyshyn, O.: Probabilistic model checking of complex biological pathways. Theoretical Computer Science 319(3), 239–257 (2008)

26. Hérault, T., Lassaigne, R., Magniette, F., Peyronnet, S.: Approximate probabilistic model checking. In: Steffen, B., Levi, G. (eds.) VMCAI 2004. LNCS, vol. 2937, pp. 73–84. Springer, Heidelberg (2004)

27. Hinton, A., Kwiatkowska, M., Norman, G., Parker, D.: PRISM: A tool for automatic verification of probabilistic systems. In: Hermanns, H., Palsberg, J. (eds.) TACAS 2006. LNCS, vol. 3920, pp. 441–444. Springer, Heidelberg (2006)

28. Jeffreys, H.: Theory of Probability. Clarendon Press, Oxford (1961)

29. Jha, S.K., Clarke, E.M., Langmead, C.J., Legay, A., Platzer, A., Zuliani, P.: A bayesian approach to model checking biological systems. Technical Report CMU-CS-09-110, Computer Science Department, Carnegie Mellon University (2009)

30. Kam, N., Harel, D., Cohen, I.R.: Modeling biological reactivity: Statecharts vs. boolean logic. In: Proceedings of the Second International Conference on Systems Biology (2001)

31. Kwiatkowska, M.Z., Norman, G., Parker, D.: Prism 2.0: A tool for probabilistic model checking. In: QEST, pp. 322–323. IEEE, Los Alamitos (2004)

32. Langmead, C.J.: Generalized Queries and Bayesian Statistical Model Checking in Dynamic Bayesian Networks: Application to Personalized Medicine. In: Proc. 8th Ann. Intnl Conf. on Comput. Sys. Bioinf. (CSB), pp. 201–212 (2009)

33. Lecca, P., Priami, C.: Cell cycle control in eukaryotes: A BioSpi model. In: Proc. Workshop on Concurrent Models in Molecular Biology (BioConcur 2003). ENTCS (2003)

34. McAdams, H., Shapiro, L.: Circuit simulation of genetic networks. Science 269, 650–656 (1995)

35. Owicki, S.S., Lamport, L.: Proving liveness properties of concurrent programs. ACM Trans. Program. Lang. Syst. 4(3), 455–495 (1982)

36. Pnueli, A.: The temporal logic of programs. In: FOCS, pp. 46–57. IEEE, Los Alamitos (1977)

37. Priami, C., Regev, A., Shapiro, E., Silverman, W.: Application of a stochastic name-passing calculus to representation and simulation of molecular processes. Inf. Process. Lett. 80(1), 25–31 (2001)

234    S.K. Jha et al.

38. Sadot, A., Fisher, J., Barak, D., Admanit, Y., Stern, M.J., Hubbard, E.J.A., Harel, D.: Toward verified biological models. IEEE/ACM Transactions on Computational Biology and Bioinformatics 5(2), 223–234 (2008)
39. Sen, K., Viswanathan, M., Agha, G.: Statistical model checking of black-box probabilistic systems. In: Alur, R., Peled, D.A. (eds.) CAV 2004. LNCS, vol. 3114, pp. 202–215. Springer, Heidelberg (2004)
40. Sen, K., Viswanathan, M., Agha, G.: On statistical model checking of stochastic systems. In: Etessami, K., Rajamani, S.K. (eds.) CAV 2005. LNCS, vol. 3576, pp. 266–280. Springer, Heidelberg (2005)
41. Vilar, J., Kueh, H.-Y., Barkai, N., Leibler, S.: Mechanisms of noise-resistance in genetic oscillators. Proc. Nat. Acad. Sci. USA 99(9), 5988–5992 (2002)
42. Wald, A.: Sequential tests of statistical hypotheses. Annals of Mathematical Statistics 16(2), 117–186 (1945)
43. Wald, A.: Sequential Analysis. Dover Publications (June 2004)
44. Yi, T.M., Kitano, H., Simon, M.I.: A quantitative characterization of the yeast heterotrimeric g protein cycle. Proc. Natl. Acad. Sci. USA 100(19), 10764–10769 (2003)
45. Younes, H.L.S., Kwiatkowska, M.Z., Norman, G., Parker, D.: Numerical vs. statistical probabilistic model checking. STTT 8(3), 216–228 (2006)
46. Younes, H.L.S., Simmons, R.G.: Probabilistic verification of discrete event systems using acceptance sampling. In: Brinksma, E., Larsen, K.G. (eds.) CAV 2002. LNCS, vol. 2404, pp. 223–235. Springer, Heidelberg (2002)
47. Younes, H.L.S., Simmons, R.G.: Statistical probabilistic model checking with a focus on time-bounded properties. Information and Computation 204(9), 1368–1409 (2006)

# Dynamic Compartments in the Imperative π-Calculus

Mathias John[1], Cédric Lhoussaine[2,4], and Joachim Niehren[3,4]

[1] University of Rostock, Computer Science, Modeling and Simulation Group
[2] University of Lille 1
[3] INRIA, Lille, Mostrare
[4] BioComputing, LIFL (CNRS UMR8022) and IRI (CNRS USR3078)

**Abstract.** Dynamic compartments with mutable configurations and variable volumes are of basic interest for the stochastic modeling of biochemistry in cells. We propose a new language to express dynamic compartments that we call the *imperative π-calculus*. It is obtained from the *attributed π-calculus* by adding imperative assignment operations to a global store. Previous approaches to dynamic compartments are improved in flexibility or efficiency. This is illustrated by an appropriate model of osmosis and a correct encoding of BioAmbients.

## 1 Introduction

Concurrent control is crucial for the stochastic modeling of biochemical processes in living cells [19,2,13]. The regulation of such systems depends on all kinds of physical or chemical aspects, such as volume, surface, temperature, pressure, pH value, spatial coordinates and structures. Most of these aspects are of global nature, so they require modeling languages in which global concurrent control can be expressed [20]. In this paper, we present a new modeling language, that permits to express many aspects with global control in a uniform manner, and illustrate its usefulness by modeling dynamic compartments with mutable volumes and surfaces.

Dynamic compartments may change their nesting structure dynamically, by applying operations for compartment creation, removal and merging. These operations may influence the speed of diverse reactions within compartments, in particular when compartment volumes change (global to local interactions). Vice versa, local reactions within a single compartment may effect global numeric attributes such as volume and surface (local to global interaction). Various languages for modeling systems with dynamic compartments were proposed for systems biology [18,14,21], but none of them can express physical, chemical, and compartimental aspects in a uniform manner, while providing efficient stochastic simulation. Spatial languages such as the Brane Calculi [2] or BioAmbients [18] fix a particular set of operators on compartments, and provide a special purpose solution for these operations. The π-calculus with polyadic synchronization and global priorities π@ is more flexible, in that it permits to

P. Degano and R. Gorrieri (Eds.): CMSB 2009, LNBI 5688, pp. 235–250, 2009.

encode all kinds of compartment structures, including those of Brane Calculi and BioAmbients [20]. Unfortunately, such priority-based encodings are complex, low level, and inefficient. Consider e.g. the dissolving of a compartment with $n$ equal molecules. Informing all of them requires $O(n)$ interactions rather than $O(1)$ by updating all at once. Furthermore, $\pi@$ lacks general support for stochastic rates and numeric attributes such as volumes and pH-values. The only solution to compartments with variable volumes so far [21] was expressed in the special purpose dialect called $S\pi@$. Numerical attributes of compartments are equally lacking in *Bigraphs* [14,12], a modeling language for spatial dynamics based on a particular form of hypergraph rewriting. Thus, the question is whether there exists a better general purpose language for expressing dynamic compartments.

In this paper, we start from the *attributed $\pi$-calculus* [11], and enrich it by an imperative store for global control. The attributed $\pi$-calculus is parametrized by a sequential higher-order language $\mathcal{L}$ for describing all kinds of values (symbolic and numeric) and constraints. It features "attributed" processes $A(e_1,\ldots,e_n)$ with values defined by expressions $e_1,\ldots,e_n$ of $\mathcal{L}$. For instance, cells with variable volumes $vol$ can be modeled by using a single attribute:

$$Cell(vol) \triangleq enter[\lambda r. \text{ if } r{<}0.1 \text{ then } (\text{val } enter)]?(v).Cell(vol + v)$$

The input prefix contains a function in square brackets, that tests for every matching output prefix, whether the reaction is permitted and returns its stochastic rate in this case. Cells as above can be entered by elements $Ele(r,v)$ of radius $r$ and volume $v$, if $r$ is smaller than 0.1:

$$Ele(r,v) \triangleq enter[r]!(v).0$$

Under this condition, the stochastic rate of the *enter* reaction is obtained by evaluating the expression $(\text{val } enter)$, i.e., by accessing the value of channel *enter* from the environment. As a result of the reaction, the cell volume is increased by $v$. The entered elements disappear, since we chose to not represent elements in cells explicitly here.

We obtain the *imperative $\pi$-calculus* $\pi^{imp}(\mathcal{L})$, by allowing imperative programming languages $\mathcal{L}$ as attribute language. Thereby, we enrich the $\pi$-calculus by a global imperative store. More precisely, we add assignment expressions to $\mathcal{L}$ by which to change the values of channels dynamically, such as for instance $enter := \text{val } enter + 1.5$, whose evaluation increases the value of channel *enter* by 1.5. The expressions of $\mathcal{L}$ are evaluated as transactions, so that the evaluator cannot be interrupted by any other process. We present a stochastic semantics for $\pi^{imp}(\mathcal{L})$ that properly accounts for transactions with imperative assignments. We show how to compile processes of $\pi^{imp}(\mathcal{L})$ to stochastic simulators, independently of the choice of parameter $\mathcal{L}$. We have implemented the compiler and can report on first experimental results. To this purpose, we model a simple example of osmosis in $\pi^{imp}(\mathcal{L})$ where variable volumes and surfaces matter. Practical simulation experiments confirm higher accuracy compared to [21] due to variable surfaces (not only volumes) and good efficiency.

In order to provide a more systematic treatment of dynamic compartments, we present a compositional encoding of BioAmbients in $\pi^{imp}(\mathcal{L})$ and prove its correctness. The constraints of $\pi^{imp}(\mathcal{L})$ permit us to express the application conditions of BioAmbients operators on compartment level. This way, we obtain a stochastic simulator for BioAmbients, without special purpose implementation as in [15]. We finally discuss how to extend our encoding to a stochastic version of BioAmbients that accounts for variable volumes.

*Omitted details and proofs can be found in the extended version.*

*Related work.* Existing stochastic semantics of BioAmbients as in [1,15] consider only local stochastic aspects ignoring variable volumes or surfaces. The rates of compartment operations simply are assigned to the interaction channel, rather than depending on the compartements volume as one might expect.

Bigraphs [14] are able to express compartment merging as in BioAmbients [18] but no variable volumes. Kappa [6] is a graph rewrite language (without hypergraphs), which seems to be too limited for expressing compartment merging. Modeling languages with model checking facilities, such as BIOCHAM [3] and BioPEPA [4] are less expressive by design. BioPEPA allows for the representation of variable compartment volumes but not dynamic structures, see [5]. BlenX (or Beta binders) [7] supports compartments with some global dynamics but no variable volumes or surfaces. Stochastic simulators are available for all these languages.

## 2   Imperative π-Calculus

We introduce the imperative π-calculus $\pi^{imp}(\mathcal{L})$ by extending the attributed π-calculus with imperative assignments. As vocabulary, we fix an infinite set *Chans* whose elements $x, y, z$ are called channels. They will name communication channels in the π-calculus (and thus chemical reactions) and serve as variables in $\mathcal{L}$.

**Values and Expressions.** An attribute language over *Chans* is a triple $\mathcal{L} = (Consts, Succ, \Downarrow)$. It defines a call-by-value lambda calculus, whose values $v \in Vals$ and expressions $e \in Exprs$ are given in Fig. 1. Besides the usual concept of variables $x \in Chans$, abstractions $\lambda x.e$, and applications, there are expressions $e_1 := e_2$ for imperative assignments. Additionally, we assume function constants val, ref* $\in Consts$ in order to access values of variables in the environment. Furthermore, we include pairs $\langle e_1, e_2 \rangle$ with selectors fst, snd and conditionals if $e$ then $e_1$ else $e_2$ with Boolean constants true, false $\in Consts$. Equality tests on constants are provided by a constant $=$ of type $Consts \times Consts \to \mathbb{B}$. There may be many further constants in *Consts* such as for arithmetics. As usual, we write $fn(e)$ and $bn(e)$ for the sets of free and bound variables in $e$. We use infix syntax without extra notice, for instance, writing $e_1 = e_2$ instead of $= \langle e_1, e_n \rangle$. The shortcuts in Fig. 2 provide let expressions, sequential composition, conditionals without else, and simple pattern matching functions.

| Channels in *Chans* | $x ::= \ldots$ |
|---|---|
| Constants in *Consts* | $c ::= \mathtt{val} \mid \mathtt{ref}^* \mid \mathtt{fst} \mid \mathtt{snd} \mid \mathtt{true} \mid \mathtt{false} \mid \mathtt{=} \mid \mathtt{unit} \mid \ldots$ |
| Values | $v ::= x \mid c \mid \langle v_1, v_2 \rangle \mid \lambda x.e$ |
| Expressions | $e ::= x \mid c \mid \langle e_1, e_2 \rangle \mid \lambda x.e \mid ee' \mid e_1 := e_2 \mid \text{if } e \text{ then } e_1 \text{ else } e_2$ |

**Fig. 1.** Values and expressions of the imperative call-by-value lambda calculus

$$\text{LET } x = e_1 \text{ IN } e_2 =_{df} (\lambda x.e_2)e_1 \qquad \text{IF } e \text{ THEN } e_1 =_{df} \text{ if } e \text{ then } e_1 \text{ else } \mathtt{false}$$

$$e_1; e_2 =_{df} \text{LET } \_ = e_1 \text{ IN } e_2 \qquad \text{IF NOT } e \text{ THEN } e_1 =_{df} \text{ if } e \text{ then } \mathtt{false} \text{ else } e_1$$

$$\lambda\langle c, x \rangle.e =_{df} \lambda p. \text{ IF } (\mathtt{fst}\ p) \mathtt{=} c \text{ THEN } (\lambda x.e)(\mathtt{snd}\ p)$$

**Fig. 2.** Shortcuts for expressions

$$\frac{(e_1, \rho) \Downarrow (v_1, \rho_1) \quad (e_2, \rho_1) \Downarrow (v_2, \rho_2) \quad (v_1 v_2, \rho_2) \Downarrow (v, \rho')}{(e_1 e_2, \rho) \Downarrow (v, \rho')} \qquad \frac{c \in \mathit{Consts}}{(c, \rho) \Downarrow (c, \rho)}$$

$$\frac{(e_1, \rho) \Downarrow (x, \rho_1) \quad (e_2, \rho_1) \Downarrow (v, \rho')}{(e_1 := e_2, \rho) \Downarrow (v, \rho'[x \mapsto v])} \qquad \frac{x \in \mathit{Chans} \quad \rho(x) = v}{(\mathtt{val}\ x, \rho) \Downarrow (v, \rho)}$$

$$\frac{\rho(x) \notin \mathit{Chans}}{(\mathtt{ref}^* x, \rho) \Downarrow (x, \rho)} \qquad \frac{(\mathtt{ref}^* \rho(x), \rho) \Downarrow (y, \rho)}{(\mathtt{ref}^* x, \rho) \Downarrow (y, \rho)}$$

$$\frac{(e_1, \rho) \Downarrow (v_1, \rho_1) \quad (e_2, \rho_1) \Downarrow (v_2, \rho')}{(\langle e_1, e_2 \rangle, \rho) \Downarrow \langle v_1, v_2 \rangle, \rho')} \qquad \frac{true}{(\mathtt{fst}\ \langle v_1, v_2 \rangle, \rho) \Downarrow (v_1, \rho)}$$

$$\frac{(e, \rho) \Downarrow (\mathtt{true}, \rho_1) \quad (e_1, \rho_1) \Downarrow (v_1, \rho')}{(\text{if } e \text{ then } e_1 \text{ else } e_2, \rho) \Downarrow (v_1, \rho')} \qquad \frac{(e, \rho) \Downarrow (\mathtt{false}, \rho_2) \quad (e_2, \rho_2) \Downarrow (v_2, \rho')}{(\text{if } e \text{ then } e_1 \text{ else } e_2, \rho) \Downarrow (v_2, \rho')}$$

$$\frac{(e_1, \rho) \Downarrow (v, \rho_1) \quad (e_2, \rho_1) \Downarrow (v, \rho') \quad v \in \mathit{Chans} \cup \mathit{Consts}}{(e_1 \mathtt{=} e_2, \rho) \Downarrow (\mathtt{true}, \rho')}$$

(and $(\mathtt{snd}\ \langle v_1, v_2 \rangle, \rho) \Downarrow (v_2, \rho)$)

**Fig. 3.** Big-step evaluator for call-by-value lambda calculus

An environment for an expression $e \in \mathit{Exprs}$ is a total function $\rho : \mathit{fn}(e) \to \mathit{Vals}$ that maps free variables of $e$ to values. We write $\mathit{dom}(\rho) = \mathit{fn}(e)$ for the domain of $\rho$ and let $\mathit{Env}$ be the set of all environments for arbitrary expressions. We write $\rho[x_1 \mapsto v_1, \ldots, x_n \to v_n]$ for the environment that maps distinct variable $x_i$ to $v_i$ for all $1 \leq i \leq n$ and all other variables $y$ in the domain of $\rho$ to $\rho(y)$. Environments such as $[x \mapsto \langle x, y \rangle, y \mapsto \langle x, x \rangle]$ can store any type of data structure, including graphs and hypergraphs. In a stochastic setting, they are useful to assign rates to reactions.

The third component of $\mathcal{L}$, the big-step evaluator $\Downarrow$, is a binary relation of type $(\mathit{Exprs} \times \mathit{Env}) \times (\mathit{Vals} \times \mathit{Env})$. It fixes the semantics of all expressions. A relationship $(e, \rho) \Downarrow (v, \rho')$ states that expression $e$ in environment $\rho$ evaluates to value $v$ with new environment $\rho'$. The big-step evaluator must satisfy the rules in Fig. 3. Assignments $x := v$ change the value of $x$ in the current environment to $v$. Function $\mathtt{val}$ returns the value of a channel in the current environment. Function $\mathtt{ref}^*$ serves for dereferentiation, i.e. it returns the last channel of acyclic reference chains. In the environment $[x_1 \mapsto x_2, \ldots, x_{n-1} \mapsto x_n, x_n \mapsto v]$, $(\mathtt{ref}^\star x_i)$ e.g. evaluates to $x_n$ for all $1 \leq i \leq n$ if $v \notin Chans$, while evaluation does not terminate if $v = x_n$.

The second component of $\mathcal{L}$ is a subset $Succ \subseteq Vals$. We call the elements of $Succ$ successful values. Their role in $\pi^{imp}(\mathcal{L})$ is to describe the rate constants of communication actions. Considering a stochastic semantics, $Succ$ equals $\mathbb{R}^+$. Otherwise, it typically contains $\mathtt{true}$ but not $\mathtt{false}$.

**Processes.** The syntax of $\pi^{imp}(\mathcal{L})$, as given in Fig. 4, is equal to that of the attributed $\pi$-calculus [11], except that we now permit imperative assignments in $\mathcal{L}$. It extends on the usual syntax of the stochastic $\pi$-calculus [17,16,13], by permitting expressions to describe channel values, adding conditions to receivers and senders, and generalizing stochastic rate constants of channels to arbitrary values.

We assume a set of process names ranged over by $A$, each with a fixed arity $ar(A) \geq 0$. Furthermore, we freely use sequence notion, writing $\tilde{e}$ for a sequence of expressions, $\tilde{x}$ and $\tilde{y}$ for a sequence of channels, and $\tilde{v}$ for a sequence of values. Their lengths are denoted by $|\tilde{e}|$, $|\tilde{v}|$, and $|\tilde{x}|$, respectively.

A program consists of an initial process $P_0$ and a set of process definitions $\{D_1, \ldots, D_n\}$, exactly one per process name in $P_0$. A definition $D$ of $A$ has the form $A(\tilde{x}) \triangleq P$, where $P$ is a process and $|\tilde{x}| = ar(A)$. Process $P$ is a parallel composition of sums, channel creators, and defined processes. A channel creator $(\nu x{:}v)\, P$ asks for the creation of a new channel $x$ with scope $P$ that is mapped to $v$ by the global environment. A sender $v[e]!\tilde{v}$, which conveys a sequence of values $\tilde{v}$ on channel $v$, is constrained by expression $e$. A receiver $v[e]?\tilde{y}$ of a sequence of values for parameters $\tilde{y}$ on channel $v$ is conditioned by expression $e$. A call of a defined process $A(\tilde{e})$ consists of a process name $A$ and a sequence $\tilde{e} \in Exprs$ where $|\tilde{e}| = ar(A)$. A sum $\Sigma$ offers a choice $\pi_1.P_1 + \ldots + \pi_n.P_n$ between senders or receivers $\pi_i.P_i$, i.e., where $\pi_i$ it either a sender or receiver prefix.

| Processes | $P, Q ::= A(\tilde{e})$ | defined process |
|---|---|---|
| | $\mid\ P_1 \mid P_2$ | parallel composition |
| | $\mid\ (\nu x{:}v)\, P$ | channel creation |
| | $\mid\ \Sigma$ | sums |
| | $\mid\ \mathbf{0}$ | empty solution |
| Sums | $\Sigma ::= \pi.P$ | prefixed process |
| | $\mid\ \Sigma + \Sigma'$ | summation |
| Prefixes | $\pi ::= v[e]?\tilde{y}$ | receiver |
| | $\mid\ v[e]!\tilde{v}$ | sender |
| Definitions | $D ::= A(\tilde{x}) \triangleq P$ | parametric process definition |

**Fig. 4.** Syntax of $\pi^{imp}(\mathcal{L})$: $e, \tilde{e}$ are expressions and $v, \tilde{v}$ values of $\mathcal{L}$, and $x, \tilde{x} \in Chans$

**Nondeterministic Operational Semantics.** We start with a nondeterministic operational semantics for $\pi^{imp}(\mathcal{L})$ with an arbitrary attribute language $\mathcal{L}$. The sets of free and bound names of processes $fn(P)$ and $bn(P)$ are defined as usual, except that free and bound names in expressions are to be considered too.

The usual structural congruence on $\pi$-calculus processes $P \equiv P'$ is the least congruence containing alpha conversion $P =_\alpha P'$, where summation $+$ and parallel composition $|$ are associative and commutative, the latter with neutral element $\mathbf{0}$, and satisfy the usual scoping rules of $\nu$-binders:

$$(\nu x{:}v)\ (P_1 \mid P_2) \equiv (\nu x{:}v)\ P_1 \mid P_2 \quad \text{if } x \notin fn(P_2)$$
$$(\nu x{:}v)\ (\nu y{:}v')\ P \equiv (\nu y{:}v')\ (\nu x{:}v)\ P \ \text{if } x \notin fn(v') \text{ and } y \notin fn(v)$$

An environment for a process $P$ is a function $\rho : fn(P) \to Vals$.

The nondeterministic operational semantics in Fig. 5 defines judgements $(P_1, \rho_1) \to (P_2, \rho_2)$ meaning that a process $P_1$ in environment $\rho_1$ reduces in one step to process $P_2$ while changing the environment to $\rho_2$. The structural congruence may silently be applied at any point (CONTEXT). A step may either be a communication or an application of a defined process. A communication step (COM) applies to a sender and a receiver on the same channel $x$. Let $e_1$ and $e_2$ be the conditions of sender and receiver, respectively, and $\rho$ the current environment. The communication step is enabled if $(e_1 e_2, \rho)$ reduces to $(v, \rho')$ for some successful value $v \in Succ$. In this case, the resulting process continues in environment $\rho'$, which may have been altered by assignment operations in $e_1 e_2$. In practice, a big step evaluator for $e_1 e_2$ may first have to change the environment and then run into an irreducible expression (a program error) or an unsuccessful value (where the communication constraint fails). In these cases, all changes done to the environment are to be backtracked. Furthermore, it may happen that the big step evaluator does not terminate (another kind of program error). An application step (REC) of a defined process $A(\tilde{e})$ evaluates all expression in $\tilde{e}$ from the left to the right while threading the environment changes, and if successful, applies the definition of $A$ to the resulting values $\tilde{v}$. Parallel compositions (PAR) may be evaluated in arbitrary order even though the changes of the environment may depend on it. Rule (RES) for channel creation $(\nu x{:}v)\ P$ in environment $\rho$ first adds $[x \mapsto v]$ to the environment, then reduces $(P, \rho[x \mapsto v])$ to some $(P', \rho'[x \mapsto v'])$, and continues with $((\nu x{:}v')\ P', \rho')$ where the new value $v'$ of $x$ is put back into a $\nu$ binder.

$$(\text{COM})\ \frac{(e_1 e_2, \rho) \Downarrow (v, \rho') \qquad v \in Succ}{(x[e_1]?\tilde{y}.P + \Sigma_1 \mid x[e_2]!\tilde{v}.Q + \Sigma_2, \rho) \to (P[\tilde{v}/\tilde{y}] \mid Q, \rho')}$$

$$(\text{REC})\ \frac{(\tilde{e}, \rho) \Downarrow (\tilde{v}, \rho') \qquad A(\tilde{x}) \triangleq P}{(A(\tilde{e}), \rho) \to (P[\tilde{v}/\tilde{x}], \rho')} \qquad (\text{PAR})\ \frac{(P, \rho) \to (P', \rho')}{(P \mid Q, \rho) \to (P' \mid Q, \rho')}$$

$$(\text{RES})\ \frac{(P, \rho[x \mapsto v]) \to (P', \rho'[x \mapsto v']) \qquad x \notin dom(\rho) \cup dom(\rho')}{((\nu x{:}v)\ P, \rho) \to ((\nu x{:}v')\ P', \rho')}$$

$$(\text{CONTEXT})\ \frac{P \equiv P' \qquad (P', \rho) \to (Q', \rho') \qquad Q' \equiv Q}{(P, \rho) \to (Q, \rho')}$$

**Fig. 5.** Nondeterministic operational semantics

**Stochastic Operational Semantics.** In the stochastic operational semantics all redexes must be computed before reducing one of them. The computation of redexes requires to evaluate $\mathcal{L}$ expressions, which may fail with program errors or nontermination. If the computation of a single redex fails, the whole process is considered erroneous. In any case, all state changes during redex computation need to be backtracked before verifying the next redex candidate. Only the finally selected redex is permitted to definitely commit its changes to the environment.

The stochastic semantics in Fig. 6 applies to programs in *biochemical form* and preserves these forms by reduction. A solution $S$ is a process in biochemical form $N\Pi_{i=1}^{m}A_i(\tilde{e}_i)$, where $N$ is a quantifier prefix $(\nu x_1{:}v_1)\ldots(\nu x_n{:}v_n)$ and $\Pi_{i=1}^{m}A_i(\tilde{e}_i) = A_1(\tilde{e}_1) \mid \ldots \mid A_n(\tilde{e}_m)$ a parallel composition of so called *molecules*. Molecules $A_i(\tilde{e}_i)$ must have definitions in biochemical form $A_i(\tilde{x}_i) \triangleq N_i\Sigma_i$ where $\Sigma_i$ is a sum of prefixed processes in biochemical form. As usual, all process can be brought into *biochemical form* by flattening out nested sums into intermediate definitions.

The stochastic semantics of a program in biochemical normal form is a Markov chain, whose states are pairs $([S]_\equiv, \rho)$, where $[S]_\equiv$ is a class of a solution $S$ wrt. structural congruence $\equiv$, and $\rho$ is an environment for $S$. In order to compute a

---

**Redexes** $(1 \leq j \leq m,\ i_1, j_1, i_2, j_2 \in \mathbb{N})$

$$\text{(CHOOSE)} \quad \frac{(\tilde{e}, \rho) \Downarrow (\tilde{v}, \rho') \qquad A(\tilde{x}) \triangleq N(\pi_1.S_1 + \ldots + \pi_m.S_m)}{\text{choose}_j(A(\tilde{e}), \rho) = (N(\pi_j.S_j)[\tilde{v}/\tilde{x}], \rho')}$$

$$\text{(REDEX)} \quad \frac{\text{choose}_{j_1}(A_{i_1}(\tilde{e}_{i_1}), \rho) =_\alpha (S'_1, \rho_1) \quad S'_1 = (\nu\widetilde{y_1{:}v_1})\ (x[e'_1]?\tilde{y}.S_1)}{\text{choose}_{j_2}(A_{i_2}(\tilde{e}_{i_2}), \rho_1) =_\alpha (S'_2, \rho') \quad S'_2 = (\nu\widetilde{y_2{:}v_2})\ (x[e'_2]!\tilde{v}.S_2) \quad i_1 \neq i_2}{(S'_1, S'_2, \rho') \in \text{redex}_{(i_1, j_1, i_2, j_2)}(\prod_{i=1}^{n} A_i(\tilde{e}_i), \rho)}$$

where $x \in Chans$, $x \notin \{\tilde{y_1}\} \cup \{\tilde{y_2}\}$ and $\{\tilde{y_1}\} \cap \{\tilde{y_2}\} = \emptyset$.

**Labeled reduction** $(r \in \mathbb{R}^+ \text{ and } \ell = (i_1, j_1, i_2, j_2) \in \mathbb{N}^4)$

$$\text{(COM)} \quad \frac{(N_1(x[e'_1]?\tilde{y}.S_1), N_2(x[e'_2]!\tilde{v}.S_2), \rho_1) \in \text{redex}_\ell(\prod_{i=1}^{n} A_i(\tilde{e}_i), \rho) \quad (e'_1 e'_2, \rho_1) \Downarrow (r, \rho') \quad r \in Succ}{(\prod_{i=1}^{n} A_i(\tilde{e}_i), \rho) \xrightarrow[\ell]{r} (\prod_{i=1, i \neq i_1, i_2}^{n} A_i(\tilde{e}_i) \mid N_1 N_2(S_1[\tilde{v}/\tilde{y}] \mid S_2), \rho')}$$

$$\text{(NEW)} \quad \frac{(S, \rho[v/x]) \xrightarrow[\ell]{r} (S', \rho'[v'/x]) \qquad x \notin dom(\rho) \cup dom(\rho')}{((\nu x{:}v)\ S, \rho) \xrightarrow[\ell]{r} ((\nu x{:}v')\ S', \rho')}$$

**Markov chain** $(r, r' \in \mathbb{R}^+)$

$$\text{(CONV)} \quad \frac{\forall \ell \in \mathbb{N}^4 \forall (N_1(x[e'_1]?\tilde{y}.S_1), N_2(x[e'_2]!\tilde{v}.S_2)) \in \text{redex}_\ell(S, \rho) \quad \exists v \in Vals \exists \rho' : (e'_1 e'_2, \rho) \Downarrow (v, \rho')}{(S, \rho) \Downarrow}$$

$$\text{(SUM)} \quad \frac{(S, \rho) \Downarrow \quad S \equiv S_1 \quad r = \sum_{\{\ell | (S_1, \rho) \xrightarrow[\ell]{r'} (S_2, \rho') \text{ and } S_2 \equiv S'\}} r' \quad r \neq 0}{(S, \rho) \xrightarrow{r} (S', \rho')}$$

---

**Fig. 6.** Stochastic operational semantics

transition for such pairs, we need to compute all potential reductions of $(S, \rho)$ and sum up their stochastic rates (SUM). The computation of all redexes must converge (CONV) before applying any reduction step. A label $\ell = (i_1, j_1, i_2, j_2) \in \mathbb{N}^4$ fixes the $j_1$'th alternative of molecules $A_{i_1}(\tilde{e}_{i_1})$ of $S$ and the $j_2$'th alternative of molecule $A_{i_2}(\tilde{e}_{i_2})$. Label $\ell$ distinguishes a *redex candidate* if these molecules have distinct indexes $i_1 \neq i_2$, and if the selected alternatives consist of a sender and a receiver on the same channel (REDEX). Label $\ell$ defines a *redex*, if the sequences of expressions $\tilde{e}_1$ and $\tilde{e}_2$ can be evaluated successfully from left to right (CHOOSE), while starting with environment $\rho$, threading changes, and ending in some environment $\rho'$. In this case, we can apply the definitions of $A_{i_1}$ and $A_{i_2}$ to the resulting values, and instantiate the alternatives with index $j_1$ and $j_2$ to $S_1'$ and $S_2'$. Note that the triple $(S_1', S_2', \rho') \in \mathrm{redex}_\ell(S, \rho)$ is unique up to alpha renaming. Rule (COM) performs the actual communication step for a redex with label $\ell$ under the condition that the constraint of the redex is successful. Rule (NEW) is as for the nondeterministic case.

Consider e.g. a solution $A(x:=1) \mid B(x:=2)$. The evaluation order for the two assignments may vary with the redex candidate. For candidates where $A(x:=1)$ provides the receiver and $B(x:=2)$ the sender, we have to evaluate $x:=1$ before $x:=2$, so that we have to test the communication constraint with store $[x \mapsto 2]$. In the symmetric case, we will have to evaluate in the opposite order and to test the constraint with store $[x \mapsto 1]$.

**Stochastic Simulation.** A stochastic simulator for $\pi^{imp}(\mathcal{L})$ can be derived from the stochastic operational semantics independently of $\mathcal{L}$. The main difference to the attributed $\pi$-calculus [11] is the treatment of imperative expressions, which can either occur in constraints or in applications. Assignments in constraints increase computational complexity, since they force us to not only compare values but also environments for grouping senders and receivers. Furthermore, senders and receivers can not be evaluated separately anymore, but only in combination. However, computational complexity can be reduced by storing differences between environments before and after evaluation, i.e. the set of executed assignments, since then only the latter need to be compared. Assignments in applications make the extraction of multisets from solutions less effective and therefore negatively affect simulation efficiency.

## 3     A Model of Osmosis: Variable Volumes and Surfaces

Osmosis is a simple example for concurrent systems with compartments of variable volumes. It was modeled already in [21] based on a special purpose dialect $S\pi@$ of $\pi@$ with variable volumes. Here we show how to simulate osmosis in the imperative $\pi$-calculus with an attribute language that provides arithmetics. Our solution is more flexible and accurate, in that it accounts for dynamic changes of compartment surfaces, which cannot be expressed in $S\pi@$.

We consider a very simple system which consists of a sphere filled with water ($H_2O$), sodium ($Na^+$), and chlorine ($Cl^-$). The system contains a membrane through which water may diffuse. This membrane separates an inner compartment Inn of spherical shape, from an outer compartment Out, which has the form of a sphere shell (a ring in 2D). The center point equals for both compartments. The precise values of these parameters are listed in the extended version.

For simplicity, we adopt the assumption of [21], that the volume of a compartment is determined by summing up the volumes of the contained molecules. However, in general, $\mathcal{L}$ allows for the definition of complex functions to obtain compartment volumes that e.g. consider atomic forces between particles. The volumes of Inn and Out change with water moving through the membrane. The radius of Inn may thus vary with diffusion, while the outer radius R of Out always remains fixed. Fig. 7 shows our model of the system in $\pi^{imp}(\mathcal{L}(\mathbb{R}, \mathsf{V}, \mathsf{C}))$. Its attribute language provides real number arithmetics with function constants for division /, multiplication *, and subtraction -, and numeric constants such

---

**Parameters**
N: $\{H_2O, Na^+, Cl^-\} \times \{Inn, Out\} \rightarrow \mathbb{N}$ // copy numbers of molecules
**Constants**
V: $\{H_2O, Na^+, Cl^-\} \rightarrow \mathbb{R}^+$    // molecule volumes
$C \in \mathbb{R}$                        // diffusion coefficient of water
**Expressions**
RAD $=_{df} \lambda v.((3*v)/(4*\pi))^{\frac{1}{3}}$    // volume to radius
SURF $=_{df} \lambda r.4*\pi*r^2$            // radius to surface
DIST $=_{df} \lambda r_1 \lambda r_2. r_1 + ((r_2-r_1)/2)$ // diffusion distance
R $=_{df}$ RAD$(\sum_{c\in\{Inn,Out\}} \sum_{m\in\{H_2O,Na^+,CL^-\}} V(m)*N(m,c))$ // outer radius of
                                            // sphere shell
**Public channels** // initialize volumes of compartments
inn: $\sum_{m\in\{H_2O,Na^+,CL^-\}}$ $V(m)*N(m, Inn)$ // inner sphere
out: $\sum_{m\in\{H_2O,Na^+,CL^-\}}$ $V(m)*N(m, Out)$ // outer sphere shell
diffuse: **unit**    // diffusion channel
**Process definitions**
$H_2O(ori, des) \triangleq$
    diffuse$[\lambda\_.$ LET // diffusion from origin to destination
      r = RAD (**val** inn)        // radius of inner sphere
      a = (SURF r)/10            // diffusion area
      s = DIST r R            // diffusion distance
      diff = a*C/(s*(**val** ori)) // diffusion rate
    IN
      ori := **val** ori $-$ $V(H_2O)$; // update volume of origin
      des := **val** des $+$ $V(H_2O)$; // update volume of destination
      diff        // return diffusion rate
    $]?(). H_2O(des, ori)$
Membrane$() \triangleq$ diffuse$[$unit$]!().$Membrane$()$
**Solution**
$\prod_{i=1}^{N(H_2O,Inn)} H_2O(inn, out) \mid \prod_{i=1}^{N(H_2O,Out)} H_2O(out, inn) \mid$ Membrane$()$

---

**Fig. 7.** Modeling osmosis

as 2, 10, or $\pi$. Furthermore, there are three problem specific constants, the diffusion coefficient C of $H_2O$, the constant V for the function that maps molecules to their volumes, and the constant N for the function assigning copy numbers to molecules in compartments. The big-step evaluator for $\mathcal{L}(\mathbb{R}, \text{V}, \text{C})$ is defined as usual. Nonzero positive real numbers are successful, $Succ = \mathbb{R}^+$.

The diffusion rate of $H_2O$ is determined by $\frac{a*C}{d*v}$, where $a$ is the diffusion area, $d$ the diffusion distance, and $v$ the volume of the compartment that the molecule leaves, see [8]. We assume that 1/10 of Inn's surface serves as diffusion area. The radius and surface of Inn are computed from its volume by functions RAD and SURF, see Fig. 7. The diffusion distance represents the average way a molecule travels from one compartment to the other. Following the approach in [8], we assume the diffusion distance to be the distance between the two compartment centers. In the model, it is determined by function DIST applied to the constant outer radius of Out and the variable radius of Inn.

In our model, we represent the compartments Inn and Out as public channels inn and out, respectively, each referring to the variable volume of the corresponding compartment. The public channel diffuse with the dummy value unit represents diffusion reactions. Three processes are defined: $H_2O$ (inn,out), which describes a water molecule in Inn that may diffuse to Out, $H_2O$ (out,inn), its symmetric variant, and Membrane(), which enables diffusion on channel diffuse at all times.

The parametric processes $H_2O$ (ori,des) may perform diffusion by communication on channel diffuse and then continue with $H_2O$ (des,ori). The speed of this reaction is given by the diffusion rate, which varies with volumes and surfaces and is therefore consecutively recomputed. This is done by applying the function in the brackets diffuse[...]?. Every application of this function performs volume changes by assignments ori := val ori - V($H_2O$) and des := val ori + V($H_2O$). Since the simulator needs to compute the diffusion rates for all possible interactions in the system (there are at most two, water moving in or out), it has to reset the environment every time. Only once some interaction is chosen by the Stochastic Simulation Algorithm [9], it can commit to the changes required by this interaction.

By adapting the diffusion area and distance at each diffusion event, we extend the model presented in [21], where only volume changes are considered. In order

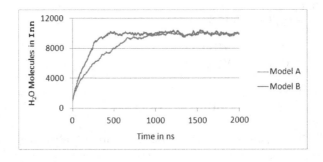

**Fig. 8.** Experiment results without (Model A) and with (Model B) variable surfaces

to compare both versions of the model, we implemented and simulated them in our tool, which is part of the modeling and simulation framework JamesII [10]. The results can be seen in Fig. 3. Model B, being the one that considers updates of the diffusion area and distance, features a steeper slope. This is due to the fact that with the increasing volume of Inn, the diffusion area grows faster than the distance, which raises the resulting diffusion rates.

## 4   Programming BioAmbients

We encode BioAmbients [18] in the imperative π-calculus, in order to show how to express concurrent systems with compartments and dynamic rearrangement systematically. In a first step, we ignore local stochastic aspects as in [1,15] which would not impose any particular problem, since these do not account for volume changes. See below for a discussion of extensions.

The syntax of BioAmbients is recalled in Fig. 9. It has the same syntactic categories as the π-calculus. Processes $P$ can be enclosed by ambients $[P]$ whose nesting structure restricts interaction capacities similarly to compartments. There are prefixes for two kinds of interactions: communication and rearrangement. *Communication prefixes* "$d\ x?(\tilde{y})$" and "$d\ x!(\tilde{y})$" are prefixes of senders or receivers annotated by a communication direction $d$, which is either local, s2s, c2p, or p2c. They enable message sending either locally in an ambient, between sibling ambients, from a parent to a child, or vice versa. Similarly, there are *rearrangement prefixes*, prefixes of senders "$c\ x!$" and receivers "$c\ x?$" without arguments and annotated by rearrangement capacity $c$, either merge, in, or out. Rearrangement operations with these prefixes serve for ambient merging, entering into siblings, or exiting the current ambient. The reduction rules of the (nondeterministic) operational semantics of BioAmbients are given in Fig. 10. We refer the reader to [18] for the full operational semantics.

In order to encode BioAmbients, we identify every ambient using a channel $r$ that gives reference to the characteristic values (CV) of the ambient. The CV $\langle n, r' \rangle$ consists of a unique name $n$ naming the ambient and the reference $r'$ of its parent (which is unit at the top-level). The ambient is encoded by a store binding $r$ to the CV possibly via a reference chain: $[r \mapsto r_1, \ldots, r_{n-1} \mapsto r_n, r_n \mapsto \langle n, r' \rangle]$. The elements in ambient $r$ will be encoded by defined processes $A(r)$.

Characteristic values can be changed by assignments $r := v$. When assignments are executed, the simulation algorithm automatically updates the communication

| | |
|---|---|
| Processes | $P, Q ::= [P] \mid A(\tilde{e}) \mid P \mid Q \mid (\nu x{:}v)\ P \mid \Sigma \mid \mathbf{0}$ |
| Sums | $\Sigma, \Sigma' ::= \pi.P \mid \Sigma + \Sigma'$ |
| Prefixes | $\pi ::= d\ x!\tilde{z} \mid d\ x?\tilde{z} \mid c\ x! \mid c\ x?$ |
| Communication directions | $d ::= \text{local} \mid \text{s2s} \mid \text{c2p} \mid \text{p2c}$ |
| Rearrangment capacities | $c ::= \text{merge} \mid \text{in} \mid \text{out}$ |
| Definitions | $D ::= A(\tilde{x}) \triangleq P$ |

**Fig. 9.** Syntax of BioAmbients

$Communication :$ local $x!\tilde{z}.P + \Sigma \mid$ local $x?\tilde{y}.Q + \Sigma' \to P \mid Q\{\tilde{z}/\tilde{y}\}$

$[Q \mid$ c2p $x!\tilde{z}.P + \Sigma] \mid$ c2p $x?\tilde{y}.P' + \Sigma' \to [Q \mid P] \mid P'\{\tilde{z}/\tilde{y}\}$

$[Q \mid$ p2c $x?\tilde{y}.P + \Sigma] \mid$ p2c $x!\tilde{z}.P' + \Sigma' \to [Q \mid P\{\tilde{z}/\tilde{y}\}] \mid P'$

$[Q \mid$ s2s $x!\tilde{z}.P + \Sigma] \mid [Q' \mid$ s2s $x?\tilde{y}.P' + \Sigma'] \to [P \mid Q] \mid [Q' \mid P'\{\tilde{z}/\tilde{y}\}]$

$Rearrangement :$ $[Q \mid$ merge $x!P + \Sigma] \mid [Q' \mid$ merge $x?P' + \Sigma'] \to [Q \mid P \mid Q' \mid P']$

$[Q \mid$ in $x!P + \Sigma] \mid [Q' \mid$ in $x?P' + \Sigma'] \to [[Q \mid P] \mid Q' \mid P']$

$[[Q \mid$ out $x!P + \Sigma] \mid Q' \mid$ out $x?P' + \Sigma'] \to [Q \mid P] \mid [Q' \mid P']$

**Fig. 10.** Reduction rules of BioAmbients

(a) enter/accept rearrangement leading to the update of entering CV.

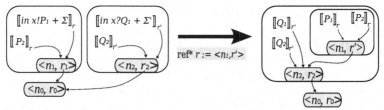

(b) exit/expel rearrangement leading to the update of exiting CV.

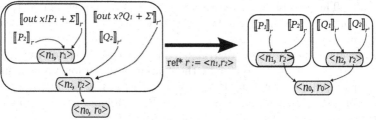

(c) merging rearrangement leading to the update of merged CV.

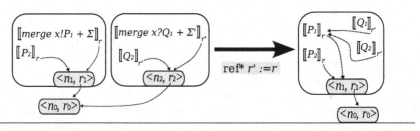

**Fig. 11.** Simplified diagrams illustrating enter, exit and merge rearrangements

potential of all elements in the compartment. For instance, for a compartment that contains $n$ copies of the same element $A(r)$, all updates can be done by a single inspection of the definition of $A(r)$, not $n$-times in contrast to priority-based encodings. Since dereferentiation might be required, this might still cost time $O(n)$ in the rare worst case, but will often be more efficient.

$\text{PARENT} =_{df} \lambda r. \text{ HERE } (\texttt{snd } (\texttt{val } (\texttt{ref}^\star \; r)))$       $\text{HERE} =_{df} \lambda r. \texttt{ fst } (\texttt{val } (\texttt{ref}^\star \; r))$

$\llbracket A(\tilde{x}) \triangleq P \rrbracket =_{df} A(\tilde{x}, r) \triangleq \llbracket P \rrbracket_r$       where $r \notin \{\tilde{x}\} \cup fn(P)$

$\llbracket A(\tilde{x}) \rrbracket_r =_{df} A(\tilde{x}, r)$       $\llbracket P \mid Q \rrbracket_r =_{df} \llbracket P \rrbracket_r \mid \llbracket Q \rrbracket_r$

$\llbracket \Sigma + \Sigma' \rrbracket_r =_{df} \llbracket \Sigma \rrbracket_r + \llbracket \Sigma' \rrbracket_r$       $\llbracket (\nu x) P \rrbracket_r =_{df} (\nu x : \texttt{unit}) \; \llbracket P \rrbracket_r$

$\llbracket [P] \rrbracket_r =_{df} (\nu n : \texttt{unit})(\nu r' : \langle n, r \rangle) \llbracket P \rrbracket_{r'}$       $\llbracket 0 \rrbracket_r =_{df} 0$

$\llbracket d \; x!(y).P \rrbracket_r =_{df} x[\langle d, r \rangle]!(y). \llbracket P \rrbracket_r$       for $d \in \{\texttt{local}, \texttt{s2s}, \texttt{p2c}, \texttt{c2p}\}$

$\llbracket \texttt{local} \; x?(y).P \rrbracket_r =_{df} x[\lambda \langle \texttt{local}, r' \rangle.(\text{HERE } r)\texttt{=}(\text{HERE } r')]?(y). \llbracket P \rrbracket_r$

$\llbracket \texttt{s2s} \; x?(y).P \rrbracket_r =_{df} x[\lambda \langle \texttt{s2s}, r' \rangle.(\text{PARENT } r)\texttt{=}(\text{PARENT } r')]?(y). \llbracket P \rrbracket_r$

$\llbracket \texttt{p2c} \; x?(y).P \rrbracket_r =_{df} x[\lambda \langle \texttt{c2p}, r' \rangle.(\text{HERE } r)\texttt{=}(\text{PARENT } r')]?(y). \llbracket P \rrbracket_r$

$\llbracket \texttt{c2p} \; x?(y).P \rrbracket_r =_{df} x[\lambda \langle \texttt{p2c}, r' \rangle.(\text{PARENT } r)\texttt{=}(\text{HERE } r')]?(y). \llbracket P \rrbracket_r$

$\llbracket c \; x!P \rrbracket_r =_{df} x[\langle c, r \rangle]!(). \llbracket P \rrbracket_r$       for $c \in \{\texttt{merge}, \texttt{in}, \texttt{out}\}$

$\llbracket \texttt{in} \; x?P \rrbracket_r =_{df} x \begin{bmatrix} \lambda \langle \texttt{in}, r' \rangle. \text{IF } (\text{PARENT } r)\texttt{=}(\text{PARENT } r') \\ \quad \text{THEN IF NOT } (\text{HERE } r)\texttt{=}(\text{HERE } r') \\ \quad \text{THEN } (\texttt{ref}^\star \; r' := \langle \text{HERE } r', r \rangle) \end{bmatrix} ?(). \llbracket P \rrbracket_r$

$\llbracket \texttt{out} \; x?P \rrbracket_r =_{df} x \begin{bmatrix} \lambda \langle \texttt{out}, r' \rangle. \text{IF } (\text{PARENT } r')\texttt{=}(\text{HERE } r) \\ \quad \text{THEN } (\texttt{ref}^\star \; r' := \langle \text{HERE } r', \texttt{snd}(\texttt{val}(\texttt{ref}^\star \; r)) \rangle) \end{bmatrix} ?(). \llbracket P \rrbracket_r$

$\llbracket \texttt{merge} \; x?P \rrbracket_r =_{df} x \begin{bmatrix} \lambda \langle \texttt{merge}, r' \rangle. \text{IF } (\text{PARENT } r)\texttt{=}(\text{PARENT } r') \\ \quad \text{THEN IF NOT } (\text{HERE } r)\texttt{=}(\text{HERE } r') \\ \quad \text{THEN } (\texttt{ref}^\star \; r' := r) \end{bmatrix} ?(). \llbracket P \rrbracket_r$

**Fig. 12.** Encoding BioAmbients

We encode BioAmbients into $\pi^{imp}(\mathcal{L}(cap, dir))$, which provides constants for all directions and capacities of BioAmbients. The encoding is given in Fig. 12. We first define two lambda expressions HERE and PARENT which map ambients $r$ to their name HERE $r = n$ and to the name of their parent PARENT $r = $ HERE $r'$.

For every BioAmbients process $P$ in ambient $r$, the encoding defines a unique process $\llbracket P \rrbracket_r$ in $\pi^{imp}(cap, dir)$. Encoding an ambient $[P]$ with parent $r$ consists in creating a new ambient name $n$ and a reference $r'$ to the CV $\langle n, r \rangle$, and proceed with the encoding $\llbracket P \rrbracket_{r'}$. In general, this is how one can dynamically create new ambients. Encodings of rearrangement prefixes are illustrated by the diagrams in Fig. 11. Dashed arrows link references to their CV's. The graphical boxes represent ambients $[P]$ and are annotated by the CV of the ambient.

In Diagram (a), ambient $r$ with CV $\langle n_1, r_1 \rangle$ enters ambient $r'$ with CV $\langle n_2, r_2 \rangle$. The translation has to specify that the first ambient becomes a child of the second. Therefore, we update the CV of $r$ to $\langle n_1, r' \rangle$, such that its parent is now $r'$. Note that the rearrangement is allowed only if the ambients are siblings. We thus have to perform the sibling test and the CV update in an atomic manner by a communication constraint on $x$ in $\llbracket \texttt{in} \; x?P \rrbracket_r$:

$\lambda \langle \texttt{in}, r' \rangle. \text{IF } (\text{PARENT } r)\texttt{=}(\text{PARENT } r') \text{ THEN}$
$\qquad \text{IF NOT } (\text{HERE } r)\texttt{=}(\text{HERE } r') \text{ THEN } (\texttt{ref}^\star \; r' := \langle \text{HERE } r', r \rangle)$

This function matches its argument against the pair $\langle \mathtt{in}, r' \rangle$, checks that the parent of receiver $r$ coincides with the parent of its communication partner $r'$, checks that both processes are not located within the same ambient and finally updates the CV of the sender accordingly. Note that an encoding of BioAmbients with stochastic aspects, as considered in [1,15], would simply make this function return the rate of $x$ (that is val $x$ assuming communication channels refer to their stochastic rate) in the sequence with the reference assignment. Diagram (b) describes the exiting ambients and Diagram (c) ambient merging. These used similar concepts as ambient entering in Diagram (a).

We define the top-level encoding $[\![P]\!]_{\nu r}$ by $(\nu r{:}\langle \mathtt{unit}, \mathtt{unit} \rangle) \, [\![P]\!]_r$ and call an environment $\rho$ for $P$ *ground* if $\rho(x) = \mathtt{unit}$ for all $x \in fn(P)$. We define $\simeq$ as the least congruence such that $\equiv \, \subseteq \, \simeq$ and $(\nu r'{:}v) \, (\nu r{:}r') \, P \simeq (\nu r'{:}v) \, P\{r'/r\}$. This equivalence is preserved by reduction of BioAmbients encodings, that is for any BioAmbients term $P$ and $\pi^{imp}(\mathcal{L}(cap, dir))$ processes $Q_1 = [\![P]\!]_{\nu r}$ and $Q_2$, such that $Q_1 \simeq Q_2$, then $(Q_1, \rho) \to (Q_1', \rho)$ iff $(Q_2, \rho) \to (Q_2', \rho)$ with $Q_1' \simeq Q_2'$.

**Theorem 1 (Soundness and completeness of BioAmbients encoding)**

1. *For all BioAmbients processes $P, P'$, if $P \to P'$ then there exists a process $Q' \simeq [\![P']\!]_{\nu r}$ of $\pi^{imp}(\mathcal{L}(cap, dir))$ such that $([\![P]\!]_{\nu r}, \rho) \to (Q', \rho)$ for every ground environment $\rho$ of $[\![P]\!]_{\nu r}$.*
2. *For all BioAmbients processes $P$, ground environment $\rho$ of $[\![P]\!]_{\nu r}$, and $\pi^{imp}(\mathcal{L}(cap, dir))$ process $Q'$, if $([\![P]\!]_{\nu r}, \rho) \to (Q', \rho)$ then there exists a BioAmbients process $P'$ of such that $P \to P'$ and $[\![P']\!]_{\nu r} \simeq Q'$.*

**BioAmbients with Variable Volumes.** Stochastic rates of reactions in compartmented systems depend on concentrations of reactants and thus on volumes of compartments. This was already illustrated by the osmosis example in Section 3. In this section, we discuss notions of volumes for ambients, and how to model them in the imperative $\pi$-calculus. Which logics for volumes to choose depends on the concrete geometry that is assumed.

When considering spatial systems where compartment nesting corresponds to geometrical nesting, we have to distinguish two notions of volumes: the *molecular volume* of a compartment, which sums up the volumes of all molecules that it contains, and the *geometric volume*, which adds the geometric volumes of all child compartments to the molecular volume. In the osmosis example, the geometric volume of the outer sphere shell (of which is outer radius R depends) does indeed include the volumes of all molecules of the inner sphere.

In order to model BioAmbients with molecular and geometric volumes in the imperative $\pi$-calculus, we can enrich the CV's of compartments by these volumes, and define lambda expressions MVOL $r$ and AVOL $r$ to access them when knowing the ambient's reference $r$. Furthermore, we have to update these volumes for all operations of the calculus, which can be expressed by using assignment operations and real arithmetics. These details need elaboration beyond 15 pages.

# 5  Conclusion and Outlook

We have shown that imperative assignments for the π-calculus yield global effects, that offer an alternative to priorities. These permit to express operations of compartment dissolution and merging in an efficient, simpler and stochastic manner. The imperative π-calculus thus answers the question for a better modeling language for dynamic compartments. In work, we would like to further investigate on the relation to Bigraphs.

**Acknowledgements.** Part of the research was financed by the DFG (Research Training School "dIEM oSiRiS") and the Agence Nationale de Recherche through a JEUNES CHERCHEURS grant (ANR BIOSPACE, 2009-2011).

# References

1. Brodo, L., Degano, P., Priami, C.: A stochastic semantics for bioambients. In: Malyshkin, V.E. (ed.) PaCT 2007. LNCS, vol. 4671, pp. 22–34. Springer, Heidelberg (2007)
2. Cardelli, L.: Brane calculi. In: Danos, V., Schachter, V. (eds.) CMSB 2004. LNCS (LNBI), vol. 3082, pp. 257–278. Springer, Heidelberg (2005)
3. Chabrier-Rivier, N., Fages, F., Soliman, S.: The biochemical abstract machine BIOCHAM. In: Danos, V., Schachter, V. (eds.) CMSB 2004. LNCS (LNBI), vol. 3082, pp. 172–191. Springer, Heidelberg (2005)
4. Ciocchetta, F., Hillston, J.: Bio-PEPA: An Extension of the Process Algebra PEPA for Biochemical Networks. ENTCS 194(3), 103–117 (2008)
5. Ciocchetta, F., Guerriero, M.L.: Modelling Biological Compartments in Bio-PEPA. ENTCS 227, 77–95 (2009)
6. Danos, V., Laneve, C.: Formal molecular biology. TCS 325(1), 69–110 (2004)
7. Dematté, L., Priami, C., Romanel, A.: Modelling and Simulation of Biological Processes in BlenX. SIGMETRICS Perf. Evaluation Review 35(4), 32–39 (2008)
8. Elf, J., Ehrenberg, M.: Spontaneous Separation of Bi-Stable Biochemical Systems into Spatial Domains of Opposite Phases. IEEE Proceedings Systems Biology 1(2), 230–236 (2004)
9. Gillespie, D.T.: Exact Stochastic Simulation of Coupled Chemical Reactions. Journal of Physical Chemistry 81, 2340–2361 (1977)
10. Himmelspach, J., Uhrmacher, A.M.: Plug'n Simulate. In: ANSS 2007, IEEE Proceedings, pp. 137–143 (2007)
11. John, M., Lhoussaine, C., Niehren, J., Uhrmacher, A.M.: The attributed pi calculus. In: Heiner, M., Uhrmacher, A.M. (eds.) CMSB 2008. LNCS (LNBI), vol. 5307, pp. 83–102. Springer, Heidelberg (2008)
12. Krivine, J., Milner, R., Troina, A.: Stochastic bigraphs. ENTCS 218, 73–96 (2008)
13. Kuttler, C., Lhoussaine, C., Niehren, J.: A Stochastic Pi-Calculus for Concurrent Objects. In: Anai, H., Horimoto, K., Kutsia, T. (eds.) Ab 2007. LNCS, vol. 4545, pp. 232–246. Springer, Heidelberg (2007)
14. Milner, R.: Pure bigraphs: Structure and dynamics. Information and Computation 204(1), 60–122 (2006)
15. Phillips, A.: An Abstract Machine for the Stochastic Bioambient Calculus. ENTCS 227, 143–159 (2009)

16. Phillips, A., Cardelli, L.: Efficient, correct simulation of biological processes in the stochastic pi-calculus. In: Calder, M., Gilmore, S. (eds.) CMSB 2007. LNCS (LNBI), vol. 4695, pp. 184–199. Springer, Heidelberg (2007)
17. Priami, C., Regev, A., Shapiro, E., Silverman, W.: Application of a Stochastic Name-Passing Calculus to Representation and Simulation of Molecular Processes. Information Processing Letters 80, 25–31 (2001)
18. Regev, A., Panina, E.M., Silverman, W., Cardelli, L., Shapiro, E.: BioAmbients: An Abstraction for Biological Compartments. TCS 325(1), 141–167 (2004)
19. Regev, A., Shapiro, E.: Cells as Computation. Nature 419, 343 (2002)
20. Versari, C.: A Core Calculus for a Comparative Analysis of Bio-Inspired Calculi. In: De Nicola, R. (ed.) ESOP 2007. LNCS, vol. 4421, pp. 411–425. Springer, Heidelberg (2007)
21. Versari, C., Busi, N.: Stochastic Biological Modelling in Presence of Multiple Compartments. TCS (to appear)

# Probabilistic Approximations of Signaling Pathway Dynamics

Bing Liu[1], P.S. Thiagarajan[1,2], and David Hsu[1,2]

[1] NUS Graduate School for Integrative Sciences and Engineering,
National University of Singapore
[2] Department of Computer Science, National University of Singapore
{liubing,dyhsu,thiagu}@comp.nus.edu.sg

**Abstract.** Systems of ordinary differential equations (ODEs) are often used to model the dynamics of complex biological pathways. We construct a discrete state model as a probabilistic approximation of the ODE dynamics by discretizing the value space and the time domain. We then sample a representative set of trajectories and exploit the discretization and the structure of the signaling pathway to encode these trajectories compactly as a dynamic Bayesian network. As a result, many interesting pathway properties can be analyzed efficiently through standard Bayesian inference techniques. We have tested our method on a model of EGF-NGF signaling pathway [1] and the results are very promising in terms of both accuracy and efficiency.

## 1 Introduction

Quantitative mathematical models are needed to understand the functioning of complex biological systems. In particular they are needed to capture the dynamics of various intra (and inter)-cellular processes. Here we focus on signaling pathways which typically sense extra-cellular or internal signals and in response, activate a cascade of intra-cellular reactions. A multitude of signaling pathways govern and coordinate the behavior of cells. As might be expected, many disease processes arise from defects in signaling pathways. Thus the study of signaling pathways via quantitative dynamic models is of critical importance.

A standard formalism used to model signaling pathways (and other bio-pathways) is a system of Ordinary Differential Equations (ODEs); the equations describe specific bio-chemical reactions while the variables typically represent concentration levels of molecular species (genes, RNAs, proteins). This formalism can be extended to include discrete aspects [2] and the techniques we develop here can be adapted to such extensions as well.

Signaling pathways usually involve a large number of molecular species and bio-chemical reactions. Hence the corresponding ODEs system will not admit closed form solutions. Instead, one will have to resort to numerically generated trajectories to study the dynamics. A second barrier is that the values of many of the parameters (rate constants) associated with the ODEs will be unknown. Even assuming all the parameters are known, the observables of the system will

P. Degano and R. Gorrieri (Eds.): CMSB 2009, LNBI 5688, pp. 251–265, 2009.

have very limited precision. Specifically, the initial concentration levels of the various proteins and rate constants will often be available only as *intervals* of values. Further, experimental data in the form of the measured concentration levels of a few proteins at a small number of time points will also be available only in terms of intervals of values. In addition, the data will often be gathered using a population of cells. Consequently, when numerically simulating the ODEs model, one must resort to Monte Carlo methods to ensure that sufficiently many point values from the relevant intervals of values are being sampled. As a result, analysis tasks such as model validation, parameter estimation and sensitivity analysis will require the generation of a large number of trajectories. This motivates our goal of probabilistically approximating the dynamics of ODEs via discretizations.

We start with a system of ODEs and a prior distribution of the initial states. Usually, this prior will consist of a uniform distribution over certain intervals of values of the variables and the rate constants. We then fix a suitable discretization of the value and time domains. This is followed by sampling the prior distribution of initial states to numerically pre-compute and store a representative subset of trajectories induced by the ODEs dynamics. The key idea is to exploit the dependencies/independecies in the pathway structure and the discretization, to compactly encode these trajectories as a time-variant dynamic Bayesian network [3]. The resulting approximation is called the Bayesian Dynamics Model (BDM). Since the trajectories are grouped together through the discretization, our method bridges the mismatch between the accuracy of the results obtained by ODE simulation and the limited precision of experimental data used for model construction and verification. Secondly, the BDM represents the global pathway dynamics more explicitly in the graph structure of the underlying dynamic Bayesian network (DBN). As a result, many interesting pathway properties can be analyzed efficiently through standard Bayesian inference techniques, instead of resorting to a large number of ODE simulations. There is a one-time computational cost incurred to construct the BDM but this cost can be amortized by performing multiple analysis tasks such as expected profiles estimation, parameter estimation, sensitivity analysis etc. using the BDM. We have tested our method on a model of EGF-NGF signaling pathway [1] and the results obtained are very promising in terms of both accuracy and efficiency.

In terms of related work, a variety of qualitative and quantitative computational models have been proposed in the recent years to study bio-pathways [2,4,5,6]. Among the quantitative models, one usually distinguishes between population-based models driven by stochastic simulations and ODEs based models driven by -deterministic- numerical simulations. Clearly, both approaches are needed to cover different contexts. Indeed, our work is, in spirit, related to the discretized approximations presented in [7,8,9] that can be applied to high level modeling formalism such as PEPA and PRISM. In these cited works, the dynamics of a process-algebra-based description of the bio-pathway is given in terms of a Continuous Time Markov Chain (CTMC) which is then discretized (using the notion of levels) to ease analysis. Apart from the fact that our starting point is a system of ODEs, a crucial additional step we take is to exploit the structure of the pathway to encode the dynamics more compactly as a dynamic Bayesian network and perform analysis tasks directly on this representation. In a similar vein, we feel that our model is a more compact discrete state model than than the graphical model of a network of non-homogenous Markov processes studied in [10]. We also believe that the techniques proposed in [11], as well as the verification techniques reported in [12,13] can be adapted to our setting. Interestingly, there have been recent attempts to synthesize ODEs from PEPA model [14], the motivation being that numerical simulations are faster than stochastic simulations. We note however, in our setting, though BDM is a probabilistic graphical model, we do not have to resort to stochastic simulations. The inferencing algorithm we use (the so called Factored Frontier Algorithm [15]), in one sweep, gathers information about the statistical properties of the family of trajectories encoded by the BDM.

In the next section, we describe our method for constructing our BDM approximation. In section 3, we present a basic inferencing technique and methods for performing tasks such as parameter estimation and global sensitivity analysis using the BDM. We also simultaneously use a realistic signaling pathway model to evaluate these techniques. In the final section, we summarize the paper and discuss future work. The interested reader can find additional technical material in the form an appendix and relevant supplementary material at [16].

## 2   The Bayesian Dynamics Model

Conceptually, our approximation technique consists of three steps:

1. We start with a system of ODEs; a discretization of the value space of each variable and rate constant into a finite set of intervals; and a digitalization of the temporal domain of interest into a finite set of time points $\{t_0, t_1, \ldots, t_{max}\}$. We also assume a prior distribution of the initial values (usually, a uniform distribution) over some of the intervals of the value space. These initial values will define an uncountably infinite family of trajectories $TRAJ_{ideal}$, which in turn, via the discretization, will induce a Markov chain $\mathcal{MC}_{ideal}$.

2. It is impossible to compute $\mathcal{MC}_{ideal}$ explicitly. However, it can be approximated by sampling the set of initial values according to the prior and using numerical integration to generate a representative subset $TRAJ_{approx} \subseteq TRAJ_{ideal}$ of trajectories. Then, using the discretization and simple counting, we can construct the Markov Chain $\mathcal{MC}_{approx}$ which will be an approximation of $\mathcal{MC}_{ideal}$.

3. However, $\mathcal{MC}_{approx}$ can be very large since the number of states that this Markov chain will be, in the worst case, exponential in the number of variables. To get around this, we exploit the pathway structure (i.e. the way the variables are coupled to each other in the system of ODEs) to represent $\mathcal{MC}_{approx}$ compactly as time-variant dynamic Bayesian network. This representation of $\mathcal{MC}_{approx}$ is called the Bayesian Dynamics Model (BDM).

We emphasize that this three step procedure is just a conceptual framework; we construct the BDM *directly* from the given system of ODEs. In what follows, we describe the main technical ideas. The interested reader can find background material and additional details in the appendix portion of the supplementary material.

## 2.1   ODEs and Flows

We assume a set of ODEs $\dot{x}_i(t) = f_i(\mathbf{x}(t), \mathbf{p})$ involving the continuous real-valued variables $\{x_1, x_2, \ldots, x_n\}$ and real-valued parameters $\{p_1, p_2, \ldots, p_m\}$. In our setting, we will often be interested in studying the dynamics for different combinations of values for the parameters. Hence it will be convenient to treat them also as variables. However they will be time-invariant in the sense once their values are fixed at $t = 0$, these values will not change through the passage of time. Consequently, we will implicitly assume the given system of ODEs to be augmented with $m$ additional differential equations of the form $\dot{p}_j(t) = 0$ with $j$ ranging over $\{1, 2, \ldots, m\}$. In what follows, we will often let $\mathbf{x}$, $\mathbf{v}$ range over $\mathbb{R}_+^n$, the values space of the variables and $\mathbf{k}$ range over $\mathbb{R}_+^m$, the values space of the parameters and $\mathbf{z}$ range over $\mathbb{R}_+^{n+m}$, the combined values space.

In vector form, our system of ODEs may be then represented as $\mathbf{Z}' = F(\mathbf{Z})$. The ODEs will be mainly modeling mass action kinetics or variants such as Michaelis-Menten kinetics. Hence we can assume $F : \mathbb{R}_+^{n+m} \to \mathbb{R}_+^{n+m}$ to be a $C^1$ (continuously differentiable) function. Furthermore, the variables representing the concentration level of a species within a single cell as well as the parameters capturing the reaction rates will take values from a bounded interval. Hence the domain of $F$ can be restricted to a bounded region $\mathcal{D}$ of $\mathbb{R}_+^{n+m}$.

Given $\mathbf{z}_0 = (\mathbf{v}_0, \mathbf{k})$ where $\mathbf{v}_0$ specifies the initial values of the variables and $\mathbf{k}$ specifies the parameters values, the system of ODEs will have a unique solution (due to $F \in C^1$) [17]. We shall denote this solution by $\mathbf{Z}(t)$ with $\mathbf{Z}(0) = \mathbf{z}_0$ and $\mathbf{Z}'(t) = F(\mathbf{Z}(t))$.

It will be convenient to define the flow $\Phi : \mathbb{R}_+ \times \mathcal{D} \to \mathcal{D}$ of $\mathbf{Z}' = F(\mathbf{Z})$ for arbitrary initial vectors $\mathbf{z}$. It will be a $C^0$ (continuous) function given by: $\Phi(t, \mathbf{z}) = \mathbf{Z}(t)$ with $\Phi(0, \mathbf{z}) = \mathbf{Z}(0) = \mathbf{z}$ and $\frac{d}{dt}(\Phi(t, \mathbf{z})) = F(\Phi(t, \mathbf{z}))$ for all $t$.

## 2.2   The Markov Chain $\mathcal{MC}_{ideal}$

Pathways models are usually validated by experimental data available only for a few time points with the concentrations measured at the last time point typically signifying the steady state value. Hence we assume the dynamics is of interest only for discrete time points and that too only up to a maximal time point. Consequently, we fix a time step $\Delta t > 0$ and the time points of interest is assumed to be the set $\{d \cdot \Delta t\}$ with $d$ ranging over $\{0, 1, \ldots, \hat{d}\}$ where $\hat{d} \cdot \Delta t$ is the maximal time point of interest.

Next we assume that the values of the variables can be observed with only finite precision and accordingly partition the range of each $x_i$ into $L^i$ intervals $[v_i^{min}, v_i^1), [v_i^1, v_i^2), \ldots, [v_i^{L_i-1}, v_i^{max}]$. We denote this set of intervals as $\mathcal{I}_i$. We also similarly discretize the range of each parameter $p_j$ into a set of intervals denoted as $\mathcal{I}_{n+j}$. The set $\mathcal{I} = \{\mathcal{I}_i\}_{1 \le i \le n} \cup \{\mathcal{I}_{n+j}\}_{1 \le j \le m}$ is called the **discretization**.

As pointed out earlier, the initial values vector as well as the rate constants (even when they are known) will be given not as point values but as distributions (usually uniform) over the intervals defined by the discretization. We correspondingly assume we are given a prior distribution in the form of a probability density function $\Upsilon^0$ capturing the distribution of initial values. For example, suppose we are given that the initial values are uniformly distributed within a hypercube $\hat{I}_1 \times \hat{I}_2 \times \ldots \times \hat{I}_{n+m}$, where $\hat{I}_i \in \mathcal{I}_i$ for each $i$. Let $\hat{I}_i = [l_i, u_i)$ and $\hat{w}_i = u_i - l_i$. Then the corresponding prior probability density function $\Upsilon^0$ will be given by:

$$\Upsilon^0(\mathbf{z}) = \begin{cases} \frac{1}{\hat{w}_1 \cdot \hat{w}_2 \cdot \ldots \cdot \hat{w}_{n+m}} & \text{if } \mathbf{z} \in \hat{I}_1 \times \hat{I}_2 \times \ldots \times \hat{I}_{n+m}, \\ 0 & \text{otherwise.} \end{cases}$$

The associated probability space we have in mind is $(\mathcal{D}, \mathcal{B}_{\mathcal{D}}, P^0)$ where $\mathcal{B}_{\mathcal{D}}$ is the Borel $\sigma$-algebra over $\mathcal{D}$; the minimal $\sigma$-algebra containing the open sets of $\mathcal{D}$ under the usual topology. $P^0$ is the probability distribution induced by $\Upsilon^0$ and is given by:

$$P^0(B) = \int_B \Upsilon^0(\mathbf{z}) d\mathbf{z}, \text{ for every } B \in \mathcal{B}_{\mathcal{D}}.$$

Further, $TRAJ_{ideal} = \{\mathbf{Z}(t)\}_{t \ge 0}$ with $\mathbf{Z}(0) \in \hat{I}_1 \times \hat{I}_2 \times \ldots \times \hat{I}_{n+m}$ is the family of trajectories starting from all the possible points in this hypercube. Since the flow is continuous and hence measurable we can associate a probability distribution $P^t$ over $\mathcal{B}_{\mathcal{D}}$ for every $t$. To define this, let $\Phi_t^{-1}(B) = \{\mathbf{z}' \mid \Phi(t, \mathbf{z}') \in B\}$ for $B \in \mathcal{B}_{\mathcal{D}}$. Since $\Phi(t, \cdot)$ is measurable, we have $\Phi_t^{-1}(B) \in \mathcal{B}_{\mathcal{D}}$ too. We can now define $P^t$ as:

$$P^t(B) = P^0(\Phi_t^{-1}(B)), \text{ for every } B \in \mathcal{B}_{\mathcal{D}}.$$

Let $v$ be in the range of $x_i$. We define $[v]$ as the interval in which $v$ falls. In other words, $[v] = I$ iff $v \in I$. Similarly, $[k] = J$ if $k \in J$ for a parameter value $k$ of $p_j$ with $J \in \mathcal{I}_{n+j}$.

Lifting this notation to the vector setting, if $\mathbf{z} = (v_1, v_2, \ldots, v_n, k_1, k_2, \ldots, k_m) \in \mathbb{R}_+^{n+m}$, we define $[\mathbf{z}] = ([v_1], [v_2], \ldots, [v_n], [k_1], \ldots, [k_m])$ and refer to it as a **discrete state**. An $\mathcal{MC}$-**state** is a pair $(\mathbf{s}, d)$, where $\mathbf{s}$ is a discrete state and $d \in \{1, 2, \ldots, \hat{d}\}$.

We next define $Pr(\mathbf{s}, d) = P^{d \cdot \Delta t}(\{\mathbf{z} \mid \mathbf{z} \in I_1 \times I_2 \times \ldots \times I_{n+m}\})$, where $\mathbf{s} = (I_1, I_2, \ldots, I_{n+m})$. We term the $\mathcal{MC}$-state $M$ to be *feasible* iff $Pr(M) > 0$.

The transition relation denoted as $\rightarrow$, between $\mathcal{MC}$-states is defined via: $M = (\mathbf{s}, d) \rightarrow M' = (\mathbf{s}', d')$ iff $d' = d + 1$ and both $M$ and $M'$ are feasible and there exist $\mathbf{z}_0$, $\mathbf{z}$, and $\mathbf{z}'$ such that $\Phi(d \cdot \Delta t, \mathbf{z}_0) = \mathbf{z}$ and $\Phi((d+1) \cdot \Delta t, \mathbf{z}_0) = \mathbf{z}'$. Furthermore, $[\mathbf{z}] = \mathbf{s}$ and $[\mathbf{z}'] = \mathbf{s}'$.

Let $E$, $F$ denote, respectively, the event that the system is in the discrete state $\mathbf{s}$ at time $d \cdot \Delta t$ and in the discrete state $\mathbf{s}'$ at time $(d+1) \cdot \Delta t$ for two feasible $\mathcal{MC}$-states $(\mathbf{s}, d \cdot \Delta t)$ and $(\mathbf{s}', (d+1) \cdot \Delta t)$. Let $EF = E \cap F$ denote joint event $\{\mathbf{z}_0 \mid \Phi(d \cdot \Delta t, \mathbf{z}_0) \in \mathbf{s}, \Phi((d+1) \cdot \Delta t, \mathbf{z}_0) \in \mathbf{s}'\}$. Consequently, we define the transition probability $Pr((\mathbf{s}, d) \to (\mathbf{s}', d')) = Pr(F|E) = Pr(EF)/Pr(E)$. Since $Pr(E) > 0$ this transition probability is well-defined.

Let $\mathcal{M} = \{M_1, M_2, \ldots, M_{\hat{n}}\}$ be the set of $\mathcal{M}$-states. We can now define the Markov chain $\mathcal{MC}_{ideal} = (\mathcal{M}, \{p_{ij}\})$ with transition probabilities $p_{ij} = Pr(M_i \to M_j)$ as above.

## 2.3   The Markov Chain $\mathcal{MC}_{approx}$

$\mathcal{MC}_{ideal}$ can not be explicitly computed. Hence we sample $\mathbf{z}_0$ a sufficiently larger number of times, say $N$, according to the prior distribution $P^0$ (we say more about $N$ below). For each sampled initial $\mathbf{z}_0$, we determine through numerical integration the $\mathcal{M}$-states $[\Phi(d \cdot \Delta t, \mathbf{z}_0)]$, with $d$ ranging over $\{0, 1, \ldots, \hat{d}\}$. We also determine the transitions along this trajectory. Then through a simple counting process involving these $N$ trajectories, we compute a Markov chain that we refer to as the $\mathcal{MC}_{approx}$.

Since $N$ is finite, there will be an error between the transition probabilities (also the $\mathcal{MC}$-state probabilities) computed using $\mathcal{MC}_{approx}$ and the ones defined by $\mathcal{MC}_{ideal}$. By the central limit theorem [18], this error can be probabilistically bounded. In other words, given an error bound $\epsilon$ and a confidence level $c$, we can compute $N$, the number of samples required to get an error less than or equal to $\epsilon$ with likelihood $c$ (the Appendix gives more details). Further, this error will tend to 0 with probability 1 as $N$ tends to $\infty$. There will be an additional error induced by the $p$th-order numerical integration method we use to compute the $N$ trajectories. This error will tend to 0 as $\Delta t$ tends to 0 or $p$ tends to $\infty$.

However, the number of states of this Markov chain will be exponential in $n$ and hence for many signaling pathways $\mathcal{MC}_{approx}$ will be too large a structure. Hence we shall construct a time-variant DBN called the BDM to compactly represent $\mathcal{MC}_{approx}$. We shall however compute the BDM directly from the $N$ sampled trajectories.

## 2.4   The BDM Representation

In what follows, we assume the basic background concerning Bayesian networks and dynamic Bayesian networks [3]. The graphical structure of the DBN used for our approximation can be derived from the differential equations. It will have $n + m$ random variables (corresponding to the variables and the parameters) as nodes for each time slice $d \cdot \Delta t$ with $d$ ranging over $\{0, 1, \ldots, \hat{d}\}$. For convenience, we will use the same name to denote a variable (parameter) and the corresponding random variable. From the context it should be clear which role is intended. The random variable $x_i$ ($p_j$) can assume as values, the finite set of intervals $\mathcal{I}_i$ ($\mathcal{I}_{n+j}$).

The variable (parameter) $x_i$ ($p_j$) in the time slice $d \cdot \Delta t$ will be written as $x_i^d$ ($p_j^d$). Edges connecting a node in the $d$-th slice to a node in the $(d+1)$-th slice will be determined by the dependencies of the variables and the parameters in the ODEs. Suppose $z_l^d$ is a (variable or parameter) node in the $d$-th time slice and $z_q^{d+1}$ is a node in the next time slice. Then there will be an edge from $z_l^d$ to $z_q^{d+1}$

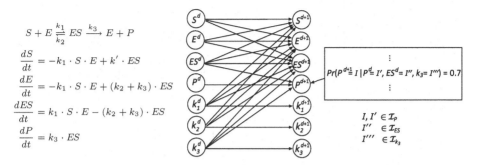

$$S + E \underset{k_2}{\overset{k_1}{\rightleftharpoons}} ES \xrightarrow{k_3} E + P$$

$$\frac{dS}{dt} = -k_1 \cdot S \cdot E + k' \cdot ES$$

$$\frac{dE}{dt} = -k_1 \cdot S \cdot E + (k_2 + k_3) \cdot ES$$

$$\frac{dES}{dt} = k_1 \cdot S \cdot E - (k_2 + k_3) \cdot ES$$

$$\frac{dP}{dt} = k_3 \cdot ES$$

**Fig. 1.** The ODE model of the enzyme-kinetic system and its BDM

iff $z_l = z_q$ or $z_q$ is a variable node and $z_l$ appears in the expression for $\dot{z}_q$ in the system of ODEs. As usual, the *parents* of the node $z_q^{d+1}$ will be the set of nodes of the form $z_l^d$ from which there is an edge into $z_q^{d+1}$. Suppose, $parents(x_i^{d+1}) = \{z_1^d, \ldots, z_l^d\}$. Then conditional probability table (CPT) associated with the node $x_i^{d+1}$ will have entries of the form $Pr(x_i^{d+1} = I \mid z_1^d = I^1, \ldots, z_l^d = I^l) = h$ with $I$ ranging over $\mathcal{I}_i$ and $I^k$ ranging over $\mathcal{I}_k$ and $h \in [0,1]$.

For instance, Figure 1 shows two adjacent slices of a enzyme-kinetic system. In this BDM, the parent nodes of $P^{d+1}$ are $P^d$, $ES^d$ and $k_3^d$. As mentioned earlier, the parameters are assumed to not change their values during a run and hence we denote $k_i^d$ as simply $k_i$ and there will be no CPTs associated with these nodes. As illustrated by the example, the connectivity between the nodes in successive slices will remain invariant. However, due to the fact that the CPTs associated with the nodes capture the transition probabilities of $\mathcal{MC}$-states, they will be time variant.

$\mathcal{MC}_{approx}$ will have, in the worst case, $O(\hat{d}K^n)$ states and $O((\hat{d}-1)K^{2n})$ transitions, where $K$ is the maximum of $|\mathcal{I}_i|$ with $1 \leq i \leq n+m$. In contrast, the number of nodes in the BDM representation is $O(\hat{d}(n+m))$ and the conditional probability table associated each node will have at most $O(K^{R+1})$ entries, where $R$ is the maximal number of parents a node can have. Usually, the reactants in pathway models will be sparsely coupled to each other and hence $R$ will be much smaller than $n$. For instance, in the case study to be presented, $n = 32$ and $R = 5$. Even in cases where $R$ is large, due to the nature of the ODEs we deal with, we can often break up the corresponding node into nodes with smaller fan-in degrees and thus reduce $R$ [16].

To fill up the entries of the CPTs associated with the nodes we randomly choose $N$ combinations of initial values for the variables and the parameters from their prior distribution as before. (If we want a coverage of $J$ samples per interval in an $n + m$ dimensional vector of intervals, then by exploiting the network structure, we can make do with $N = JK^{R+1}$ samples.) We then perform numerical integration to generate $N$ trajectories and discretize those trajectories by the predefined intervals and compute the conditional probabilities for each node by simple counting. For example, suppose $\alpha$ trajectories hit ($P^d = $

$I', ES^d = I'', k_3 = I''')$ and $\beta$ of them in turn hit $(P^{d+1} = I)$, then $Pr(P^{d+1} = I | P^d = I', ES^d = I'', k_3 = I''') = \frac{\beta}{\alpha}$.

Further, the $\mathcal{MC}_{approx}$ can be easily recovered from this DBN [19]. In this sense, our BDM representation is a principled probabilistic approximation of the dynamics induced by the system of ODEs. Various optimizations can be developed to reduce the practical complexity of the BDM construction. The details can be found in [16].

Though the construction of the BDM involves significant computational effort, it is a one time cost. Moreover, a substantial portion of the computation can be executed in parallel. Further, once the BDM has been constructed, many of the analysis tasks can be performed very efficiently and the one time cost of constructing the BDM can be easily amortized. We present some experimental results in support of this in the next section.

## 3   Analysis

We now present some of the analysis techniques that we have developed so far for the BDM representation. These techniques are based on the basic Bayesian inference method called the $FF$ (Factored Frontier) algorithm [15] and can be used to answer elementary probabilistic queries as well for performing parameter (rate constants) estimation and sensitivity analysis. Our goal here is not to develop new algorithms to solve these problems. Rather, we wish to demonstrate how standard techniques for tackling these problems can be adapted to BDM framework in a straightforward manner. We validate our techniques using a relatively large signaling pathway and show the relevant experimental results along with our techniques.

### 3.1   The EGF-NGF Signaling Pathway and Its BDM

PC12 cells are a valuable model system in neuroscience. They proliferate in response to EGF stimulation but differentiate into sympathetic neurons in response to NGF. This interesting phenomenon has been intensively studied [20]. It has been reported that the signal specificity is correlated with different Erk dynamics. Specifically, a transient activation of Erk1/2 has been associated with cell proliferation, while a sustained activity has been linked to differentiation. How EGF and NGF affect the dynamics of active Erk through a network of intermediate signaling proteins is shown schematically in Figure 2.

This model not only includes a common pathway to Erk through Ras shared by both the EGFR and NGFR, but also includes two important side branches through PI3K and C3G, which introduce multiple feedback loops thus complicating the dynamics. The ODE model of this pathway is available in the BioModels database[1]. It consists of 32 differential equations and 48 associated rate parameters (estimated from multiple sets of experimental data).

---

[1] http://www.ebi.ac.uk/biomodels/

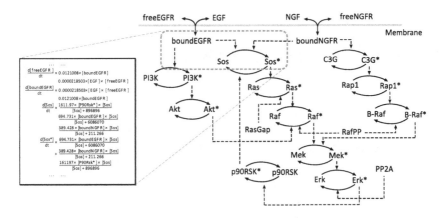

**Fig. 2.** EGF-NGF pathway [1]

To construct the BDM, we first derived its graph from its ODEs. We then discretized the ranges of each variable and parameter into 5 equal-size intervals and fixed the time step $\Delta t$ to be 1 minute. Our experimental data (western blot) is such that 5 uniform intervals seems an appropriate choice. However our construction can be easily extended to non-uniform values intervals and time points. To fill up the conditional probability tables associated with the nodes, $3 \times 10^6$ trajectories were generated up to 100 mins by sampling initial states and parameters from the prior which are assumed to be uniform distributions over certain intervals (see [16]). The computational workload was distributed on 10 Opteron 2.2GHz processors in a cluster. It took around 4 hours to construct the BDM. All the subsequent experiments reported below were done using an Intel Xeon 2.8GHz processor.

## 3.2   Probabilistic Inference

As pointed out earlier, although the dynamics defined by the ODEs is deterministic, to answer a basic query such as *"what will be the concentration of the protein $x_i$ at time $t$?"* one will have to numerically generate a representative sample of trajectories and compute the average of the values for $x_i$ at $t$ yielded by the individual trajectories.

Using our BDM approximation, we can answer such a basic query and other more sophisticated queries by Bayesian inference. Specifically, given a Bayesian network, some observed evidence and some knowledge about the distribution of values of a set of variables, Bayesian inference aims to compute posterior distribution for a set of query variables. In our setting, the observed evidence refers to the initial conditions, known parameters, and experimental data. Query variables potentially refer to all the random variables in the BDM. We adopt the approximate algorithm known as the Factored Frontier (FF) algorithm [15]. It approximates joint distributions over each time point as a product of marginal

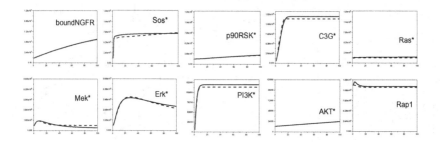

**Fig. 3.** Simulation results of EGF-NGF signaling pathway. Solid lines represent nominal profiles and dash lines represent BDM simulation profiles.

distributions and computes the posterior distribution according to:

$$Pr(x_i^d|D) = \sum_I (Pr(x_i^d|Pa(x_i^d)) = I) \prod_{u \in Pa(x_i^d)} Pr(u|D)). \qquad (1)$$

Here $Pr(u|D)$ are the marginal distributions over the parents, $D$ is the evidence, and $Pa(x_i)$ denotes the parents of $x_i$. The implementation of FF is straightforward. By storing $Pr(x_i^d|Pa(x_i^d))$ in the conditional probability tables and propagating $Pr(u|D)$ to the next time point, we can use equation 1 to compute $Pr(x_i^d|D)$. The time complexity of this algorithm is $O(\hat{d}(n+m)K^{R+1})$, where $K$ is the maximal number of intervals associated with a variable or rate constant's value domain. Further, $R$ is the maximal number of parents a node can have.

Using this algorithm, and with some additional simple computations, many queries can be answered. For instance, we identify each interval $I = [l, u)$ in $\mathcal{I}$ with its mid-point $\frac{l+u}{2}$. Then after inferring the probability distribution of $x_i$ over intervals, the expected value $E(x_i^d)$ at a time slice $d$ can be computed and used to validate the model by comparing it with the cell population based data that may be available for $x_i$ at $d \cdot \Delta t$.

To test the quality of our approximation, we implemented Monte Carlo integration for the ODE model to get good estimates by sampling. Specifically, we numerically generated a number of random trajectories -according to the prior- using ODEs, discretized them and computed the average values of the variables at the chosen time points. Our experiments show that the average values converge when the number of random trajectories generated is roughly $10^4$. The averaged trajectories projected to individual protein concentration time series values are termed to be the nominal simulation profiles. Using the implemented FF algorithm, the mean of each variable over time was computed. The resulting time profiles are termed to be the BDM-simulation profiles. As summarized in Figure 3, our BDM-simulation profiles fit the nominal simulation profiles quite well for most of the cases.

In terms of running time, a single execution of FF inference requires 0.29 seconds while generating a stable nominal profile requires 386.4 seconds. Thus, the total computation time will be sharply reduced by our approach when many such "queries" need to be answered. In the next subsection, we will further demonstrate this advantage by carrying out a simulation-intensive analysis task.

## 3.3   Parameter Estimation

Lack of knowledge about the parameters and hence the need to perform parameter estimation using limited data has long been identified as a major bottleneck of pathway modeling. Current approaches to parameter estimation formulate it as a non-linear optimization problem [21]. A typical procedure will involve searching in a high dimensional solution space, in which each point represents a vector of parameter values. Whether a point is good or not is measured by the objective function, which will capture the difference between experimental data and prediction generated by simulations using the corresponding parameters.

For a large pathway model, one often needs to evaluate a very large number of solution points involving a numerical integration for each evaluation. This makes the whole process computationally intensive. The BDM representation allows us to carry out the search for good parameter values in a hierarchical manner. Due to the discretized nature of the BDM, the solution space is transformed to a rectilinear grid consisting of a space-filling tessellation by hyperrectangles that we call *blocks*. An important observation is that kinetic parameters are often robust [22]. In other words, the points around the best solution in the search space will also have relatively small objective values. Thus, instead of searching point by point in the solution space, we can first search for a few promising blocks and then take a closer look within these small blocks. Therefore, the general scheme of our "grid search" algorithm will consist of two phases: (1) identify good blocks, (2) do local search within candidate blocks. We note that phase(2) is necessary only when we aim to estimate parameters with finer granularity than the granularity of the BDM's discretization. Otherwise, one can skip phase(2) and return a probabilistic estimate (typically a Maximal Likelihood Estimate) of a combination of intervals of parameter values. For executing phase(1), we can apply any standard search algorithms over the discretized search space. As the discretized search space is much smaller than the original one, simple direct search algorithm such as Hooke & Jeeves's search [23] can be adopted and the overall search process will only require a small number of executions of the FF algorithm.

In order to test the performance of the BDM-based parameter estimation method, we synthesized experimental time series data for 9 (out of 32) proteins {bounded EGFR, bounded NGFR, active Sos, active C3G, active Akt, active p90RSK, active Erk, active Mek, active PI3K}, measured at the time points {2, 5, 10, 20, 30, 40, 50, 60, 80, 100} (min).This data was synthesized using prior knowledge about initial conditions and parameters [16]. To mimic western blot data which is cell population based, we first averaged $10^4$ random trajectories generated by sampling initial states and rate constants, and then added observation noise with variance 5% to the simulated values. With the assumed measurement precision, those values were discretized into 5 intervals, which represent the concentration levels in western blot data. We reserved the data of 7

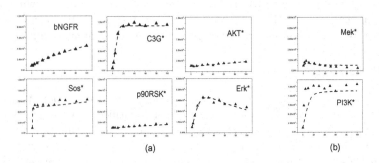

**Fig. 4.** Parameter estimation results. (a) BDM-simulation profiles vs. training data. (b) BDM-simulation profiles vs. test data.

proteins for training the parameters and reserved the rest data for testing the quality of the estimated parameter values.

Assuming that 20 of the 48 parameter values are unknown, a modified version of Hooke & Jeeves algorithm was implemented to search for in the discretized parameter space. The parameters obtained can be found in [16]. As shown in Figure 4, the BDM-simulation profiles generated using the estimated parameters obtained (with the match to training data as shown) has good agreement with the test data.

We compared the efficiency and quality of our results with the following ODEs based optimization algorithms: Levenberg-Marquardt (LM), Genetic Algorithm (GA), Stochastic Ranking Evolutionary Strategy (SRES), and Particle Swarm Optimization (PSO). These optimization algorithms were executed using the COPASI [24] tool. We scored the resulting parameters obtained from all the algorithms using the weighted sum-of-squares *difference* between the experimental data and the corresponding simulation profiles (i.e. low scores correspond to low errors). The results are summarized in Figure 5, which suggests that our method achieves a good balance between accuracy and performance. We also note that the cost of constructing the BDM representation gets rapidly amortized. In fact

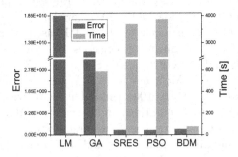

**Fig. 5.** Performance comparison of our parameter estimation method (BDM) and 4 other methods

the savings become even more significant when we perform additional analysis tasks such as sensitivity analysis.

## 3.4   Global Sensitivity Analysis

Sensitivity analysis has been used to identify the critical parameters in signal transduction. To overcome the limitations of traditional local sensitivity analysis methods global methods have been proposed recently, e.g. multi-parametric sensitivity analysis (MPSA) [25]. The MPSA procedure consists of: (1) draw samples from parameter space and for each combination of parameters, compute the weighted sum of squared error between experimental data and predictions generated by selected parameters; (2) classify the sampled parameter sets into two classes (good and bad) using a threshold error value; (3) plot the cumulative frequency of the parameter values associated with the two classes; (4) evaluate the sensitivities as the Kolmogorov-Smirnov statistic of cumulative frequency curves. To improve this process, [25] adopts Latin hypercube sampling (LHS) since it requires fewer samples while guaranteeing that individual parameter ranges are evenly covered. In our BDM setting, MPSA can be done in a similar manner using LHS since the parameter space is discretized into blocks. In addition, the number of samples used to reach convergence is reduced since we can quickly evaluate the whole block instead of having to draw samples from a block.

We modified and implemented the MPSA method for the BDM. Using the same experimental data set introduced in previous subsection, the global sensitivities (K-S statistics) of the parameters were computed. The results are shown in Figure 6. The cumulative frequency distributions for the acceptable and unacceptable cases of the rate constants can be found in [16]. Specifically, the reactions involved in the phosphorylation of Erk ($k_{23}$), Mek ($k_{17}$), Akt ($k_{34}$) and p90RSK ($k_{28}$) have the highest sensitivities, indicating that these reactions affect the system behavior most directly. These results are consistent with previous findings [20].

The MPSA method adopts Monte Carlo strategy for the ODE model. We recorded the running time of the algorithm till the K-S values converged. The total running time of the ODEs based MPSA method was about 22 *hours*, while the MPSA method based on the BDM required only 34 *minutes*.

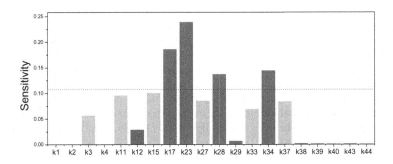

**Fig. 6.** Parameter sensitivities

## 4   Discussion

We have proposed a probabilistic approximation scheme for signaling pathway dynamics specified as a system of ODEs. Given a discretization and an initial distribution, it consists of pre-computing and storing a representative sample of trajectories induced by the system of ODEs. We use a dynamic Bayesian network representation, called the Bayesian Dynamics Model, to compactly represent these trajectories by exploiting the pathway structure. Basically, the underlying graph of the BDM captures the dependencies of the variables on other variables and rate constants as defined by the system of ODEs. Due to the probabilistic graphical representation, a variety of analysis questions concerning the pathway dynamics traditionally addressed using Monte Carlo simulations can be converted to Bayesian inference and solved more efficiently. Using the FF algorithm for doing basic Bayesian inference, we have adapted standard parameter estimation and sensitivity analysis algorithms to the BDM setting. We have demonstrated the applicability of our techniques with the help of the good sized EGF-NGF signaling pathway.

A number of further lines of work suggest themselves. Firstly, we need to apply our method to a variety of pathway models. We are currently doing so in collaboration with biologists. Secondly, it will be useful to augment the ODE model with some discrete features but this should be easy to achieve. A more challenging issue is to abstract the BDM representation to an input-output transducer so that one can efficiently model networks of pathways and inter-cellular interactions models. Finally, it will be important to develop formal verification techniques based on the BDM representation. In this context, it is worth noting that the FF algorithm can compute the marginal probabilities of the discretized values of variables at specific time points. Hence a good starting point will be to develop probabilistic bounded model checking methods for specifications based on the BDM model.

## References

1. Brown, K.S., Hill, C.C., Calero, G.A., Lee, K.H., Sethna, J.P., Cerione, R.A.: The statistical mechanics of complex signaling networks: nerve growth factor signaling. Phys. Biol. 1, 184–195 (2004)
2. Matsuno, H., Tanaka, Y., Aoshima, H., Doi, A., Matsui, M., Miyano, S.: Biopathways representation and simulation on hybrid functional Petri net. Silico Biol. 3(3), 389–404 (2003)
3. Murphy, K.P.: Dynamic Bayesian Networks: Representation, Inference and Learning. PhD thesis, University of California, Berkeley (2002)
4. Antoniotti, M., Policriti, A., Ugel, N., Mishra, B.: XS-systems: extended s-systems and algebraic differential automata for modeling cellular behavior. In: Sahni, S.K., Prasanna, V.K., Shukla, U. (eds.) HiPC 2002. LNCS, vol. 2552, pp. 431–442. Springer, Heidelberg (2002)
5. de Jong, H., Page, M.: Search for steady states of piecewise-linear differential equation models of genetic regulatory networks. IEEE/ACM T. Comput. Bi. 5(2), 208–223 (2008)

6. Ghosh, R., Tomlin, C.: Symbolic reachable set computation of piecewise affine hybrid automata and its application to biological modelling: Delta-notch protein signalling. Systems Biol. 1(1), 170–183 (2004)
7. Calder, M., Gilmore, S., Hillston, J.: Modelling the influence of RKIP on the ERK signalling pathway using the stochastic process algebra PEPA. In: Priami, C., Ingólfsdóttir, A., Mishra, B., Riis Nielson, H. (eds.) Transactions on Computational Systems Biology VII. LNCS (LNBI), vol. 4230, pp. 1–23. Springer, Heidelberg (2006)
8. Calder, M., Vyshemirsky, V., Gilbert, D., Orton, R.: Analysis of signalling pathways using continuous time Markov chains. In: Priami, C., Plotkin, G. (eds.) Transactions on Computational Systems Biology VI. LNCS (LNBI), vol. 4220, pp. 44–67. Springer, Heidelberg (2006)
9. Ciocchetta, F., Degasperi, A., Hillston, J., Calder, M.: CTMC with levels models for biochemical systems. Elsevier, Amsterdam (2009) (preprint submitted)
10. Nodelman, U., Shelton, C.R., Koller, D.: Continuous time Bayesian networks. In: Proceedings of the 18th Conference in Uncertainty in Artificial Intelligence, Alberta, Canada, pp. 378–387 (2002)
11. Langmead, C., Jha, S., Clarke, E.: Temporal logics as query languages for dynamic Bayesian networks: Application to D. Melanogaster embryo development. Technical report, Carnegie Mellon University (2006)
12. Clarke, E.M., Faeder, J.R., Langmead, C.J., Harris, L.A., Jha, S.K., Legay, A.: Statistical model checking in BioLab: Applications to the automated analysis of T-Cell receptor signaling pathway. In: Heiner, M., Uhrmacher, A.M. (eds.) CMSB 2008. LNCS (LNBI), vol. 5307, pp. 231–250. Springer, Heidelberg (2008)
13. Heath, J., Kwiatkowska, M., Norman, G., Parker, D., Tymchyshyn, O.: Probabilistic model checking of complex biological pathways. Theor. Comput. Sc. 319(3), 239–257 (2008)
14. Geisweiller, N., Hillston, J., Stenico, M.: Relating continuous and discrete PEPA models of signalling pathways. Theor. Comput. Sc. 404(2), 97–111 (2008)
15. Murphy, K.P., Weiss, Y.: The factored frontier algorithm for approximate inference in DBNs. In: Proceedings of the 17th Conference in Uncertainty in Artificial Intelligence, San Francisco, CA, USA, pp. 378–385 (2001)
16. Supplementary Materials, http://www.comp.nus.edu.sg/~rpsysbio/cmsb09
17. Ammann, H.: Ordinary Differential Equations: An Introduction to Nonlinear Analysis. Walter de Gruyter, Berlin (1990)
18. Durrett, R.: Probability: Theory and Examples. Duxbury Press (2004)
19. Nunez, L.M.: On the relationship between temporal Bayes networks and Markov chains. Master's thesis, Brown University (1989)
20. Kholodenko, B.N.: Untangling the signalling wires. Nat. Cell Biol. 9(3), 247–249 (2007)
21. Banga, J.R.: Optimization in computational systems biology. BMC Syst. Biol. 2(47), 1–7 (2008)
22. Gutenkunst, R.N., Waterfall, J.J., Casey, F.P., Brown, K.S., Myers, C.R., Sethna, J.P.: Universally sloppy parameter sensitivities in systems biology. PLoS Comput. Biol. 3(10), 1871–1878 (2007)
23. Hooke, R., Jeeves, T.A.: "Direct search" solution of numerical and statistical problems. J. ACM. 8(2), 212–229 (1961)
24. Hoops, S., Sahle, S., Gauges, R., Lee, C., Pahle, J., Simus, N., Singhal, M., Xu, L., Mendes, P., Kummer, U.: COPASI - a COmplex PAthway SImulator. Bioinformatics 22(24), 3067–3074 (2006)
25. Cho, K.H., Shin, S.Y., Kolch, W., Wolkenhauer, O.: Experimental design in systems biology, based on parameter sensitivity analysis using a monte carlo method: A case study for the TNFα-mediated NF-κB signal transduction pathway. Simulation 79(12), 726–739 (2003)

# A Reduction of Logical Regulatory Graphs Preserving Essential Dynamical Properties

Aurélien Naldi[1], Elisabeth Remy[2], Denis Thieffry[1,3], and Claudine Chaouiya[1,4]

[1] TAGC Inserm U928 - Université de la Méditerranée, Marseille, France
[2] IML UMR 6206, Marseille, France
[3] CONTRAINTES, INRIA Paris Rocquencourt, France
[4] IGC, Instituto Gulbenkian de Ciência, Oeiras, Portugal

**Abstract.** To cope with the increasing complexity of regulatory networks, we define a reduction method for multi-valued logical models.

Starting with a detailed model, this method enables the computation of a reduced model by iteratively "hiding" regulatory components. To keep a consistent behaviour, the logical rules associated with the targets of each hidden node are actualised to account for the (indirect) effects of its regulators.

The construction of reduced models ensures the preservation of a number of dynamical properties of the original model. In particular, stable states and more complex attractors are conserved. More generally, we focus on the relationship between the attractor configuration of the original model and that of the reduced model, along with the issue of attractor reachability.

The power of the reduction method is illustrated by its application to a multi-valued model of the segment-polarity network Controlling segmentation in the fly Drosophila melanogaster.

**Keywords:** Regulatory networks, logical modelling, model reduction, decision diagrams, regulatory circuits, stable states, complex attractors, Drosophila development, segmentation.

## 1 Introduction

Biological data generation and integration efforts result in the delineation of ever more comprehensive and complex regulatory networks involved in the control of numerous processes. Consequently, current modelling and analysis approaches are reaching their limits in terms of the number and variety of components and interactions that can be efficiently considered. This is true for quantitative frameworks (*e.g.*, differential or stochastic models), as well as for qualitative approaches. Indeed, although logical modelling enables to handle networks comprising relatively large numbers of components (see *e.g.* [1,2]), the size of the state space grows exponentially with the number of regulatory nodes.

P. Degano and R. Gorrieri (Eds.): CMSB 2009, LNBI 5688, pp. 266–280, 2009.
© Springer-Verlag Berlin Heidelberg 2009

One way to handle this problem consists in developing compositional approaches to compute the dynamical properties of comprehensive networks, relying on the knowledge of the properties of simpler sub-systems or modules. A complementary approach consists in reducing large systems, by focusing on the most relevant components and redefining their interactions in order to preserve relevant dynamical properties (*e.g.* stable states).

Most often, modellers intuitively and manually reduce regulatory networks to address specific questions. Such empirical reductions have several drawbacks: (i) the process is error prone and limited to relatively simple cases; (ii) the maintenance of different versions of a model (complete and reduced) is cumbersome; (iii) storing the sole reduced model leads to the loss of relevant biological information.

These considerations led us to develop a reliable, automated reduction method in the context of a logical modelling framework. In this respect, we lean on the software *GINsim*, which facilitates the definition of comprehensive logical regulatory graphs, as well as the analysis of their dynamical properties [3,4]. Established on firm mathematical bases, our reduction method allows the user to select nodes to be made implicit and to perform dynamical analyses on reduced model versions, which preserve relevant topological and dynamical properties.

The paper is organised as follows. Section 2 recalls the definitions of logical regulatory graphs and of the associated state transition graphs. Next, the reduction method is defined in Section 3. Relationships between the dynamical behaviour of the original model and that of the reduced model are delineated in Section 4. A multi-valued logical model of the segment-polarity network is then used to demonstrate the power of the proposed reduction method in Section 5. The paper ends with conclusions and further prospects.

All models presented in this paper can be opened, edited, simulated, and analysed with *GINsim*, which implements the logical formalism and the reduction method presented here.

## 2    Logical Modelling of Regulatory Networks

Our modelling approach leans on the generalised logical formalism initially developed by R. Thomas *et al.* [5,6,3]. In this context, a regulatory network and its dynamics are both represented in terms of oriented graphs.

### 2.1    Regulatory Graphs

**Definition 1.** A logical regulatory graph (LRG) *is a directed labelled multigraph* $\mathcal{R} = (\mathcal{G}, \mathcal{M}ax, \Gamma, \Theta, \mathcal{K})$ *where,*

- $\mathcal{G} = \{g_1, \ldots, g_N\}$ *is the set of nodes, representing regulatory components.*
- $\mathcal{M}ax : \mathcal{G} \rightarrow \mathbb{N}^*$ *associates a maximum level* $\mathcal{M}ax(g_i) = \mathcal{M}ax_i$ *to node* $g_i$. *The current level of* $g_i$, *denoted* $x_i$, *takes its values in* $\mathcal{D}_i = \{0, \ldots, \mathcal{M}ax_i\}$.
- $\Gamma$ *is the set of arcs, defined as a finite multiset of ordered pairs of elements of* $\mathcal{G}$ *representing regulatory interactions. If* $\mathcal{M}ax_i > 1$, $g_i$ *may have different*

*effects onto a component $g_j$, depending on level $x_i$. Hence, the arc connecting $g_i$ to $g_j$ may be a multi-arc encompassing different interactions. The multiplicity of the arc $(g_i, g_j)$ (i.e. the number of its constitutive interactions), is denoted $m_{i,j}$ ($1 \leq m_{i,j} \leq Max_i$). Loops (even multi-loops) are allowed: an arc $(g_i, g_i)$ denotes an autoregulation of $g_i$.*

*For each $g_j \in \mathcal{G}$, $Reg(j)$ denotes the set of its regulators: $g_i \in Reg(j)$ if and only if $(g_i, g_j) \in \Gamma$.*

– $\Theta$ *is a labelling function, which associates a threshold to each element of $\Gamma$. More precisely, $\theta_{i,j,k}$ is associated to the $k^{th}$ interaction between $g_i$ and $g_j$ (denoted $(g_i, g_j, \theta_{i,j,k})$, $k \in \{1, \ldots, m_{i,j}\}$), with $1 \leq \theta_{i,j,1} < \cdots < \theta_{i,j,m_{i,j}} \leq Max_i$. This interaction is* active, *when $x_i$, the level of its source $g_i$, lays between the threshold of this interaction and that of the next interaction: $\theta_{i,j,k} \leq x_i < \theta_{i,j,k+1}$ (by convention, $\theta_{i,j,m_{i,j}+1} = Max_i + 1$).*

– $\mathcal{K} = (\mathcal{K}_1, \ldots, \mathcal{K}_N)$ *defines the* logical rules *attached to the nodes specifying their behaviours: each $\mathcal{K}_i$ is a multi-valued logical function that gives the target value of $g_i$, depending on the levels of the regulators acting on $g_i$:*

$$\mathcal{K}_i : \left( \prod_{g_j \in \mathcal{G}} \mathcal{D}_j \right) \mapsto \{0, \ldots, Max_i\}.$$

The logical function $\mathcal{K}_i$ can be equivalently defined on the set $\prod_{g_j \in Reg(i)} \mathcal{D}_j$, giving the target value of $g_i$ depending on the current levels of its regulators. Figure 1 illustrates this definition of a logical regulatory graph. In the following, when no confusion is possible, we will use $i$ to denote $g_i$.

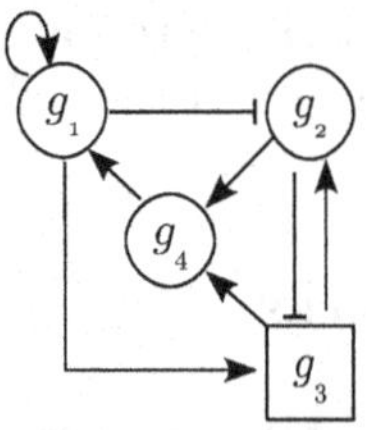

$\mathcal{G} = \{g_1, g_2, g_3, g_4\}$
$Max_1 = Max_2 = Max_4 = 1$
$Max_3 = 2$
$\mathcal{D}_1 = \mathcal{D}_2 = \mathcal{D}_4 = \{0, 1\}, \mathcal{D}_3 = \{0, 1, 2\}$
$Reg(2) = \{g_1, g_3\}$
$\theta_{3,2,1} = 2$

**Fig. 1.** Example of logical regulatory graph. Left: graphical representation of a LRG. Blunt arrows depict inhibitions while normal arrows depict activations (this is only a graphical convention, since the logical functions encode the regulatory effects). The rectangular node $g_3$ is ternary, whereas the others nodes are Boolean. The thresholds of all interactions are set to 1, except that of $(g_3, g_2)$, which is set to 2. Right: illustration of the notations of Definition 1. Examples of logical functions $\mathcal{K}_i$ are displayed in Figure 2 for the same model.

## 2.2   State Transition Graphs

We represent the dynamical behaviour of a LRG in terms of a state transition graph, defined as follows.

**Definition 2.** *Given a LRG* $\mathcal{R} = (\mathcal{G}, \mathcal{M}ax, \Gamma, \Theta, \mathcal{K})$, *its associated full state transition graph (STG)* $\mathcal{E} = (\mathcal{S}, \mathcal{T})$ *is a directed graph, where:*

- $\mathcal{S} = \Pi_{i \in \mathcal{G}} \mathcal{D}_i$ *is the state space, a state of the system being a vector* $x = (x_i)_{i=1,\ldots,N}$, *with* $x_i \in \mathcal{D}_i$, $\forall i \in \mathcal{G}$,
- $\mathcal{T} \subset \mathcal{S}^2$ *is the set of transitions defined as follows:* $(x, y) \in \mathcal{T}$ *if and only if* $\exists i \in \mathcal{G}$ *such that:*

  $x_i \neq \mathcal{K}_i(x)$,

  $y = x + \Delta_i(x).e^i$, *where* $\Delta_i(x) = \frac{\mathcal{K}_i(x) - x_i}{|\mathcal{K}_i(x) - x_i|}$ *and* $e^i$ *is the canonical vector in* $\mathcal{S}$ ($e_i^i = 1$ *and* $e_j^i = 0$, $\forall j \in \mathcal{G}, j \neq i$).

Here $\Delta_i(x)$ gives the sign of the update of $i$ (increase or decrease). One can also consider a state transition graph related to an initial (set of) condition(s). It is then a subgraph of the full STG.

When analysing the behaviour of a LRG, we mainly focus on attractors, which represent asymptotic dynamical properties. Given a STG, attractors are its terminal strongly connected components, classified as:

- *stable states*: reduced to a unique terminal node,
- *cyclic attractors*: terminal elementary (oriented) cycles,
- *complex attractors*: other terminal strongly connected components (*i.e.* involving intertwined cycles).

Cyclic and complex attractors will be called *non-trivial* attractors.

In what follows, LRGs are assumed consistent, *i.e.* all interactions are effective and autoregulations functional, meaning that all interactions have a dynamical role and could be recovered from the logical functions $\mathcal{K}$ (see further explanations in the Appendix A).

# 3   Logical Regulatory Graph Reduction

This section presents the principles underlying the reduction of a regulatory graph and then defines the new model, called *reduced model*. In what follows, we consider a reduction consisting of the *removal* of a single regulatory component (making it implicit). The generalisation to a reduction encompassing a set of nodes is obtained by iterating the corresponding one-node reductions. However, the ordering of a sequence of one-node reductions may have an impact on the resulting reduced model (see Appendix B).

Here, we aim at defining a reduction method, which preserves, as much as possible, the dynamical properties of the original model. The underlying principle is already intuitively applied by modellers when they make regulatory nodes implicit in their networks.

The removal of a node $r$ basically consists in connecting directly its regulators to its targets, which logical functions are thus revised. In the revised logical functions, the effect of $r$ at a given value $x_r$ is conveyed by the values of the regulators leading $r$ to $x_r$. In other words, we consider the update of the removed component as a *fast process*, which is performed before anything else.

Following this principle, it is impossible to remove an autoregulated component since fixed values of its other regulators may not lead to a unique target value. Thus, the removal of an autoregulated component implies additional decisions, impeding the definition of a systematic procedure. In the following, we will require that autoregulated components should not be removed.

To properly implement an algorithm producing the reduced model, we need further notations to manipulate the logical functions. Given a regulatory graph $\mathcal{R} = (\mathcal{G}, \mathcal{M}ax, \Gamma, \Theta, \mathcal{K})$ and a node $i \in \mathcal{G}$, we denote:

- $x_i^{\{l\}}$ ($l \in \mathcal{D}_i$) the Boolean variable with value 1 when $x_i = l$, 0 otherwise.
- $x_i^S$ the Boolean variable that is true if $x_i \in S$, false otherwise. Hence $x_i^S$ is defined by,

$$x_i^S \triangleq \bigvee_{l \in S} x_i^{\{l\}}, \; S \subseteq \mathcal{D}_i.$$

Note that $x_i^\emptyset$ is always false and $x_i^{\mathcal{D}_i}$ always true.

- For all $v \in \mathcal{D}_i$, the logical function $\mathcal{K}_i^v$ that gives the conditions under which the target value of node $i$ is $v$. This function is defined as follows:

$$\mathcal{K}_i^v = \bigvee_{n=1,\dots p} \mathcal{C}_i^n, \tag{1}$$

where $\mathcal{C}_i^n$ are conjunctive clauses $\mathcal{C}_i^n = \bigwedge_{j \in Reg(i)} x_j^{S_{j,i,n}}$, where $S_{j,i,n} \subseteq \mathcal{D}_j$. Each clause $\mathcal{C}_i^n$ defines a situation (*i.e.* sets of combinations of incoming interactions acting upon $i$) for which the target value of $i$ is $v$.

In Equation (1), each clause $\mathcal{C}_i^n$ defines a subset of $\mathcal{S}$, $D = \Pi_{j \in \mathcal{G}} S_{j,i,n}$ (with $S_{j,i,n} = \mathcal{D}_j, \forall j \notin Reg(i)$), such that for all $x \in D$, $\mathcal{K}_i(x) = v$. Hence, Equation (1) defines a set of cubes in the state space $\mathcal{S}$, where the target value of $i$ is $v$.

**Definition 3.** *Given a LRG $\mathcal{R} = (\mathcal{G}, \mathcal{M}ax, \Gamma, \Theta, \mathcal{K})$, the reduced LRG $\mathcal{R}^r = (\mathcal{G}^r, \mathcal{M}ax^r, \Gamma^r, \Theta^r, \mathcal{K}^r)$ obtained by removing a non-autoregulated component $r \in \mathcal{G}$ is defined as follows:*

- $\mathcal{G}^r = \mathcal{G} \setminus \{r\}$.
- $\mathcal{M}ax^r : \mathcal{G}^r \rightarrow \mathbb{N}^*$, *s.t.* $\forall i \in \mathcal{G}^r \; \mathcal{M}ax^r(i) = \mathcal{M}ax_i$.
- *For all $i \in \mathcal{G}^r$, and for all $v \in \mathcal{D}_i$, the logical function $\mathcal{K}_i^{rv}$ is defined as follows. Consider $\mathcal{K}_i^v = \bigvee_{n=1,\dots p} \mathcal{C}_i^n$, the disjunctive form of $\mathcal{K}_i^v$, as defined previously. For all $n = 1, \dots p$ (i.e. for each clause $\mathcal{C}_i^n$), let define $\mathcal{F}_i^{rn}$ as:*

$$\mathcal{F}_i^{rn} = \left( \bigvee_{w \in S_{r,i,n}} \mathcal{K}_r^w \right) \wedge \left( \bigwedge_{j \in Reg(i) \setminus \{r\}} x_j^{S_{j,i,n}} \right)$$

*Then $\mathcal{K}_i^{rv} = \bigvee_{n=1,\dots p} \mathcal{F}_i^{rn}$.*
- *$\Gamma^r$ and $\Theta^r$ are deduced from $\mathcal{K}^r$; for all $i \in \mathcal{G}^r$, $j \in \mathcal{G}^r$,*

$$m_{i,j}^r = \sum_{v \in [1, \mathcal{M}ax_i]} \mathbb{1}_{i,j,v},$$

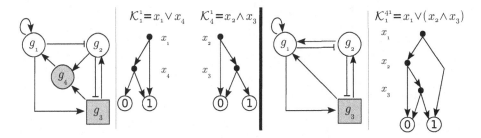

**Fig. 2.** Reduction in terms of MDDs. Left: the same LRG as in Figure 1, where $g_4$ (greyed-out) is selected for removal. Logical functions for $g_1$ and $g_4$ are shown on the right, along with their MDD representations. Right: the reduced LRG after removal of $g_4$, along with the resulting logical function for $g_1$. In the MDDs, internal nodes are labelled with the associated variable ($x_i$), whereas leaves represent the value of the logical functions. Children of internal nodes are ordered from left to right: the leftmost (resp. rightmost) child is the root of the sub-diagram corresponding to the case $x_i = 0$ (resp. $x_i = \mathcal{M}ax_i$).

*where $\mathbb{1}_{i,j,v} = 1$ if it exists $x \in \mathcal{S}$ such that $x_i = v-1$ and $\mathcal{K}_j^r(x) \neq \mathcal{K}_j^r(x+e^i)$. Then $(i,j) \in \Gamma^r$ if $m_{i,j}^r > 0$ (and the multiplicity of $(i,j)$ in $\Gamma^r$ is given by $m_{i,j}^r$). Finally, the ordered set of values $v$ such that $\mathbb{1}_{i,j,v} = 1$ defines the thresholds $\theta_{i,j,k}^r$ ($k = 1, \ldots, m_{i,j}^r$).*

The logical function $\mathcal{K}_i^{rv}$ is deduced from the logical function $\mathcal{K}_i^v$ by replacing, in each clause, literals $x_r^S$ by the formulae giving the conditions under which the target value of $r$ is in $S$ (remark that this definition may not give $\mathcal{K}_i^{rv}$ in a proper disjunctive form). Note that if $\mathcal{C}_i^n$ does not depend on $r$ (*i.e.* $r \notin Reg(i)$) then $S_{r,i,n} = \mathcal{D}_r$ and $\mathcal{F}_i^{rn} = \mathcal{C}_i^n$ for all $n$, therefore $\mathcal{K}_i^{rv} = \mathcal{K}_i^v$.

The set of arcs verifies:

$$\Gamma^r \subseteq \{(i,j) \in \mathcal{G}^r \times \mathcal{G}^r, \text{ s.t. } (i,r), (r,j) \in \Gamma \text{ or } (i,j) \in \Gamma\}.$$

In practice, the construction of the new logical function is performed using Reduced Ordered Multivalued Decision Diagrams (ROMDDs or MDDs for short). Decision diagrams are rooted directed acyclic graphs, widely used to represent logical functions (see *e.g.* [7,8]). In these diagrams, internal nodes are labelled with *decision variables* and have one child per value, while leaves represent the values of the function. Decision variables are ordered: each internal node has a rank and the sub-diagrams rooted by the children of a node of rank $i$ do not contain internal nodes of rank $j \leq i$. In [9] we used MDDs to represent the logical functions $\mathcal{K}_i$. In this context, decision variables are the levels of the components of the model. For the sake of simplicity, we consider that the ordering of the MDD variables is the same as that of the LRG components. Given the MDD representation of $\mathcal{K}_i$ and a state $x$, a unique path from the root of the MDD to one of its leaves is defined. Along this path, the child chosen for each non-terminal node is labelled with the value of the corresponding variable in state

$x$. The terminal node reached through this path gives the value of $\mathcal{K}_i(x)$. Each clause of $\mathcal{K}_i^v$ corresponds to a path leading to a leaf valued $v$.

To compute the MDD representing $\mathcal{K}_i^r$, we define the recursive algorithm given in Appendix C and illustrated in Figure 2.

## 4   Dynamics of the Reduced Model

In this section, the dynamical behaviour of a reduced LRG (as specified in Definition 3) is compared to that of the original LRG. In particular, we show that the reduction preserves existing attractors and does not add any spurious path.

Let $\mathcal{E} = (\mathcal{S}, \mathcal{T})$ be the full state transition graph of $\mathcal{R} = (\mathcal{G}, Max, \Gamma, \Theta, \mathcal{K})$ and $r \in \mathcal{G}$ a node not autoregulated. Let $\mathcal{E}^r = (\mathcal{S}^r, \mathcal{T}^r)$ be the full STG of $\mathcal{R}^r = (\mathcal{G}^r, Max^r, \Gamma^r, \Theta^r, \mathcal{K}^r)$, the LRG obtained after the removal of $r$ from $\mathcal{G}$.

Consider the projection $\pi_r : \mathcal{S} \to \mathcal{S}^r$ such that, $\forall i \in \mathcal{G}^r$, $\forall x \in \mathcal{S}$, $(\pi_r(x))_i = x_i$, and the equivalence relation on $\mathcal{S}$: $\forall x, y \in \mathcal{S}$, $x \sim_r y$ iff $\pi_r(x) = \pi_r(y)$.

We denote $[x]_{\sim_r}$ the equivalence class: $[x]_{\sim_r} = \{y \in \mathcal{S} \text{ s.t. } y \sim_r x\}$. The class $[x]_{\sim_r}$ contains all states of $\mathcal{S}$ that differ only by their $r^{th}$ component, i.e. the $(Max_r + 1)$ states $\{x^i \in \mathcal{S}, i = 0, \ldots, Max_r\}$, such that $x^i \sim_r x$ and $x_r^i = i$. Because $r$ is not autoregulated, $\forall x^i \in [x]_{\sim_r}$, $\mathcal{K}_r(x^i) = \mathcal{K}_r(x)$. This implies that:

- $(x^i, x^{i+1}) \in \mathcal{T}$, for all $i < \mathcal{K}_r(x)$,
- $(x^i, x^{i-1}) \in \mathcal{T}$, for all $i > \mathcal{K}_r(x)$,
- $(x^{\mathcal{K}_r(x)}, x^i) \notin \mathcal{T}$, for all $i$.

Hence, for all $x \in \mathcal{S}$, there exists a path in $\mathcal{S}$ from $x$ to $x^{\mathcal{K}_r(x)}$, which is the representative state of $[x]_{\sim_r}$.

**Definition 4.** $x \in \mathcal{S}$ is the representative state *of an equivalence class for* $\sim_r$ *iff* $x_r = \mathcal{K}_r(x)$.

We can then define the *retrieval* function $s_r : \mathcal{S}^r \to \mathcal{S}$ such that, $\forall z \in \mathcal{S}^r$,

$$(s_r(z))_i = z_i, \text{ for all } i \in \mathcal{G} \setminus \{r\},$$

$$(s_r(z))_r = \mathcal{K}_r(x), \text{ with } x \text{ such that } \pi_r(x) = z.$$

In other words, $s_r(z)$ is the representative state of the equivalence class projected on $z$ (see Figure 3). Relying on this, we can introduce an alternative definition of the logical functions in the reduced LRG: $\forall i \in \mathcal{G}^r$, $\mathcal{K}_i^r : \mathcal{S}^r \mapsto \mathcal{D}_i$ is defined as $\mathcal{K}_i^r(z) = \mathcal{K}_i(s_r(z))$. Note that if $(r, i) \notin \Gamma$ (i.e. $r$ is not a regulator of $i$), $\mathcal{K}_i^r(\pi_r(x)) = \mathcal{K}_i(x)$.

*Remark 1.* It follows from their definitions that functions $\pi_r$ and $s_r$ verify:

1. $\pi_r \circ s_r$ is the identity function.
2. For any $x \in \mathcal{S}$, $(s_r \circ \pi_r(x)) \sim_r x$.
3. If $x \in \mathcal{S}$ is a representative state, then, $s_r \circ \pi_r(x) = x$.
4. For any $z \in \mathcal{S}^r$, $\mathcal{K}^r(z) = \pi_r(\mathcal{K}(s_r(z)))$; indeed, $\forall x \in \mathcal{S}$, $\forall i \in \mathcal{G}^r$, $\mathcal{K}_i^r(\pi_r(x)) = \mathcal{K}_i(s_r \circ \pi_r(x))$.

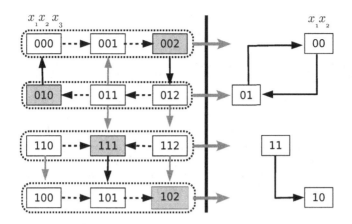

**Fig. 3.** Dynamical behaviour of the reduced model given in Figure 2, before and after removal of the ternary node $g_3$. Left: State transition graph (STG), partitioned into four equivalence classes for $g_3$. Each equivalence class contains 3 states; its representative state is greyed out and internal transitions are dashed. Right: STG of the reduced model, each state corresponding to an equivalence class of the original STG. After the reduction, the stable state 102 is projected on 10 and all transitions are preserved except the one from the second equivalence class to the third one. This results in the isolation of the non-terminal strongly connected component involving the first two equivalence classes of the original STG, hence generating the attractor $(01, 00)$.

The following lemma establishes the relationships between transitions in $\mathcal{E}$ and $\mathcal{E}^r$.

**Lemma 1.**  *1. Let $z, z' \in \mathcal{S}^r$.*

$$(z, z') \in \mathcal{T}^r \implies \exists x \in \mathcal{S} \ s.t. \ \pi_r(x) = z' \ and \ (s_r(z), x) \in \mathcal{T}.$$

*2. Let $x, y \in \mathcal{S}$. If $x$ is a representative state, then*

$$(x, y) \in \mathcal{T} \implies (\pi_r(x), \pi_r(y)) \in \mathcal{T}^r.$$

*Proof.* Recall that $\Delta_i(x) \triangleq \frac{\mathcal{K}_i(x) - x_i}{|\mathcal{K}_i(x) - x_i|}$. For $z \in \mathcal{S}^r$ s.t. $z_i \neq \mathcal{K}_i^r(z)$, we similarly denote:

$$\Delta_i^r(z) \triangleq \frac{\mathcal{K}_i^r(z) - z_i}{|\mathcal{K}_i^r(z) - z_i|} = \frac{\mathcal{K}_i(s_r(z)) - (s_r(z))_i}{|\mathcal{K}_i(s_r(z)) - (s_r(z))_i|} = \Delta_i(s_r(z)).$$

1. Consider $z, z' \in \mathcal{S}^r$ such that $(z, z') \in \mathcal{T}^r$. Then $\exists i \neq r$ s.t. $\mathcal{K}_i^r(z) \neq z_i$, and $z' = z + \Delta_i^r(z) e^i$. By definition, $\mathcal{K}_i^r(z) = \mathcal{K}_i(s_r(z)) \neq (s_r(z))_i = z_i$. This implies that $(s_r(z), x) \in \mathcal{T}$ with $x \in \mathcal{S}$ and $x = s_r(z) + \Delta_i(s_r(z)) e^i$, and then $\pi_r(x) = z'$.
2. Consider $x, y \in \mathcal{S}$ such that $\mathcal{K}_r(x) = x_r$. The hypothesis $(x, y) \in \mathcal{T}$ implies that $\exists i \in \mathcal{G}, i \neq r$ s.t. $\mathcal{K}_i(x) \neq x_i$, and $y = x + \Delta_i(x) e^i$.

We have $\mathcal{K}_i^r(\pi_r(x)) = \mathcal{K}_i(x)$ (since $x$ is a representative state), and $x_i = (\pi_r(x))_i$, since $i \neq r$. So, $\mathcal{K}_i^r(\pi_r(x)) \neq (\pi_r(x))_i$, and then $\exists z \in \mathcal{S}^r$ s.t. $(\pi_r(x), z) \in \mathcal{T}^r$, with

$$z = \pi_r(x) + \Delta_i^r(\pi_r(x)) e^i = \pi_r(x) + \Delta_i(s_r \circ \pi_r(x)) e^i = \pi_r(y) . \qquad \square$$

The first item of Lemma 1 states that any transition in $\mathcal{T}^r$ corresponds to at least one transition in $\mathcal{T}$. Clearly, the reverse is not true. The second item of the lemma gives a condition under which transitions are preserved from $\mathcal{T}$ to $\mathcal{T}^r$. Of course, it is important to know which transitions are lost through the reduction.

**Definition 5.** *The reduction preserves a transition* $(x, y) \in \mathcal{T}$ *if* $(\pi_r(x), \pi_r(y)) \in \mathcal{T}^r$, *or* $\pi_r(x) = \pi_r(y)$. *The reduction preserves a path* $(s_1, \ldots, s_n) \in \mathcal{E}$ *if all its transitions are preserved.*

In other words, a path $(s_1, \ldots, s_n)$ in $\mathcal{E}$ is preserved if the reduction preserves the transitions between equivalence classes, in the required order.

The following property characterises the transitions that are not preserved by the reduction.

*Property 1.* A transition $(x, y) \in \mathcal{T}$ is *not* preserved by the reduction if and only if the three following conditions are satisfied:

1. $x$ is not a representative state,
2. $y \notin [x]_{\sim r}$ ($\Rightarrow \exists i \neq r$ s.t. $y_i \neq x_i$),
3. $\Delta_i(x) \neq \Delta_i(s_r \circ \pi_r(x))$.

The last condition means that there is no call for updating $i$ in the same direction in state $s_r \circ \pi_r(x)$.

*Proof.* Consider a transition $(x, y) \in \mathcal{T}$, which satisfies the three conditions. Suppose that $(x, y)$ is preserved by the reduction, then $(\pi_r(x), \pi_r(y)) \in \mathcal{T}^r$ (the case $\pi_r(x) = \pi_r(y)$ is not possible because of the second condition). This means that there exists $j \neq r$ s.t. $(\pi_r(x))_j \neq (\pi_r(y))_j$, and $(\pi_r(x))_k = (\pi_r(y))_k$ for any $k \neq j$. With Condition 2 and by definition of $\pi_r$, we deduce that $j = i$. Moreover, we know that:

$$\pi_r(y) = \pi_r(x) + \Delta_i^r(\pi_r(x)) e^i = \pi_r(x) + \Delta_i(s_r \circ \pi_r(x)) e^i .$$

Finally, $y = x + \Delta_i(x) e^i$, and, as $y_i = (\pi_r(y))_i$, we have $\Delta_i(x) = \Delta_i(s_r \circ \pi_r(x))$. This contradicts Condition 3. Hence, $(x, y)$ is not preserved by the reduction.

Conversely, let $(x, y) \in \mathcal{T}$ be a transition not preserved by the reduction.

- Condition 1 is satisfied by the second item of Lemma 1.
- Condition 2 is satisfied because $y \in [x]_{\sim r} \Rightarrow \pi_r(x) = \pi_r(y) \Rightarrow (x, y)$ preserved, hence a contradiction.
- We know that $y = x + \Delta_i(x) e^i$. As $\mathcal{K}^r(\pi_r(x)) = \pi_r(\mathcal{K}(s_r \circ \pi_r(x)))$ (cf. Remark 1),

$$\mathcal{K}_i^r(\pi_r(x)) = (\pi_r(\mathcal{K}(s_r \circ \pi_r(x))))_i = \mathcal{K}_i(s_r \circ \pi_r(x))$$
$$\neq x_i = (\pi_r(x))_i .$$

Hence, there exists $z \in \mathcal{S}^r$ s.t. $(\pi_r(x), z) \in \mathcal{T}^r$ with

$$z_i = \pi_r(x) + \Delta_i^r(\pi_r(x))\, e^i$$
$$= \pi_r(x) + \Delta_i(s_r \circ \pi_r(x))\, e^i = \pi_r(x) + \Delta_i(x)\, e^i\,.$$

Consequently, $\pi_r(y) = z$ and $(x, y)$ is preserved, hence a contradiction. □

Given $C$, a set of states in $\mathcal{S}$, we denote $\pi_r(C) \triangleq \{\pi_r(x),\ x \in C\}$. Given $C'$, a set of states in $\mathcal{S}^r$, we denote $s_r(C') \triangleq \{s_r(z),\ z \in C'\}$. Note that $\pi_r(C)$ may contain less elements than $C$, and that $s_r(C')$ contains only representative states. The following results relate attractors in $\mathcal{E}$ and $\mathcal{E}^r$. Proofs are provided in Appendix D and E.

**Theorem 1.** *Consider a LRG $\mathcal{R} = (\mathcal{G}, \mathcal{M}ax, \Gamma, \Theta, \mathcal{K})$ and $\mathcal{R}^r$ the reduced LRG. Let $\mathcal{E}$ (resp. $\mathcal{E}^r$) be the full STG of $\mathcal{R}$ (resp. of $\mathcal{R}^r$), then:*

1. *Stable states in $\mathcal{E}$ and $\mathcal{E}^r$ verify:*
   - *$x$ stable state in $\mathcal{E} \implies \pi_r(x)$ stable state in $\mathcal{E}^r$. Furthermore no other stable state is projected on $\pi_r(x)$,*
   - *$z$ stable state in $\mathcal{E}^r \implies s_r(z)$ stable state in $\mathcal{E}$.*
   *Hence, the number of stable states is conserved by the reduction.*
2. *If $(s_1, \ldots s_n)$ is a cyclic attractor in $\mathcal{E}$, then $(\pi_r(s_1), \ldots \pi_r(s_n))$ is a cyclic attractor in $\mathcal{E}^r$.*
3. *If $C$ is a complex attractor in $\mathcal{E}$, $z \in \pi_r(C)$ and $(z, z') \in \mathcal{T}^r$, then $z' \in \pi_r(C)$. As a consequence, $\pi_r(C)$ contains at least one non-trivial attractor in $\mathcal{E}^r$.*

Theorem 1 characterises the dynamical properties conserved by the reduction. Going further, it is possible to identify the situations leading to the generation of additional non-trivial attractors. A non-trivial attractor in the reduced STG corresponds to a (part of a) strongly connected component of the original STG. This SCC is itself a non-trivial attractor or involves outgoing transitions all in conflict with transitions concerning the removed component. In other words, we can fully characterise the set of states in the original STG giving rise to a non-trivial attractor in the reduced dynamics. Interestingly, this set corresponds to transient oscillatory behaviour from which the system cannot escape provided that updates of the removed component are always faster than other concurrent changes. This is formalised by Theorem 2 in Appendix E.

## 5 Application: Segment Polarity

We demonstrate the power and flexibility of our reduction method through its application to the segment-polarity network, which plays a key role in the segmentation of the fly embryo. This system has been thoroughly analysed by developmental geneticists and has been already modelled using continuous [10,11,12] and logical approaches [13,14,15]. However, all these studies involved important simplifications of the network, particularly so as a proper modelling of

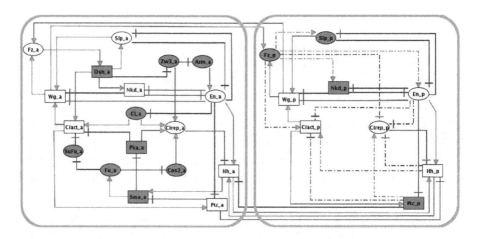

**Fig. 4.** Logical model of the segment polarity network for two cells, based on [15]. Ellipsoid and rectangular nodes denote Boolean and ternary components, respectively. The two cellular networks have been properly connected to take into account Wg and Hh diffusion, as well as Hh sequestration by Ptc, as in [16]. The anterior cell contains the extended version of the model, where greyed-out components will be removed, leading to the model on the right. Dashed arrows denote indirect interactions resulting from this reduction. Greyed-out components in the posterior cell are candidates for further reduction.

its behaviour requires the chaining of several identical networks to account for inter-cellular interactions through Wingless (Wg) and Hedgehog (Hh) signalling. Describing the most complete model to date, [15] had to discard various components known to play important roles in Wg and Hh signalling to keep dynamical simulations and analyses computationally tractable for up to six cells. Here, we propose a logical model based on their full description of the segment polarity network. The resulting regulatory graph encompasses 18 components and 31 regulatory interactions (left part of Figure 4).

In order to model the intercellular interactions involved in the formation of segment boundaries, we have to connect neighbouring cells (along the anterior-posterior axis) through Wg and Hh signalling. Wg is known to bind its receptor, Frizzled (Fz), only at very short range, amounting here to neighbouring cells. This can be represented by positive arcs linking each Wg node to Fz nodes of neighbouring cells. In contrast, Hh is able to reach more distant cells, but can be sequestered by its receptor Patched (Ptc). Similar interactions have been modelled in [16] in terms of positive arcs between Hh nodes in neighbouring cells (diffusion) and negative arcs from Ptc onto the Hh node of neighbouring cells (sequestering). Figure 4 illustrates the intercellular network obtained after coupling two cells and reducing one of the cellular sub-networks down to nine components.

The reduction method described above can be advantageously applied to ease the identification of all attractors of such intercellular models (Sánchez *et al.*

**Table 1.** Dynamical characteristics of different reduced models derived from that of Figure 4 (involving 2 x 9 nodes after applying the same reduction to both cells). The number of reachable states decreases drastically with the number of considered nodes. Note that the three stable states remain reachable for all reductions listed, but the last one (removal of Slp).

| LRG size | Removed components | Number of reached states | Reached stable states |
|----------|--------------------|--------------------------|-----------------------|
| 2x9      | –                  | $> 10^6$                 | TT, WE, EW            |
| 2x7      | Fz,Ptc             | 12476                    | TT, WE, EW            |
| 2x6      | Fz,Ptc,Nkd         | 1625                     | TT, WE, EW            |
| 2x8      | Slp                | 11350                    | TT, WE                |

considered six cells). The modeller can select the sets of nodes to discard from the network, depending on biological considerations (*e.g.* different time scales, specific mutations, etc.). In a first step, it is reasonable to conserve transcription factors and components involved in intercellular communications: Wg, Hh and their receptors (Fz and Ptc). However, since the transcription factor Cubitus interruptus is represented by three nodes here (full length immature Ci protein, activator Ci-act and repressor Ci-rep forms), we choose to retain only the two nodes corresponding to active regulatory forms. These choices correspond to the removal of the greyed-out components in the left part of Figure 4.

The reduced model involves half of the nodes of the original one, which amounts to a much higher reduction of the number of possible states, as this grows exponentially with the number of regulatory nodes. The resulting regulatory graph (Figure 4, right) remains easy to grasp as it reasonably unfolds most intra-cellular and inter-cellular regulatory pathways. As we shall see, this logical model can be further reduced to facilitate analyses encompassing more cells.

For proper logical rules (cf. [15] and supplementary material), one can check that the detailed and the reduced two-cells models have exactly three stables states (as predicted by Theorem 1). These multi-cellular stable states combine three types of cellular states: a Wg expressing state (denoted W), an En expressing state (E), and a *trivial* state (T) expressing neither Wg, nor En. The three stable states for the two connected cells correspond to TT, WE and EW cell combinations reported by Sánchez *et al.* All three stable states are reachable from biologically relevant initial conditions (significant amounts of Wg and Slp in the anterior cell, significant amount of En in the posterior cell), provided as an outcome of the activity of the pair-rule system, cf. [17,15]. However, the size of the corresponding state transition graph still impedes detailed dynamical analyses (see Table 1).

As shown in Table 1, the removal of Fz, Ptc and Nkd drastically reduces the number of reached states without changing the reachability of the three stable states from the considered initial state. However, the sole removal of Slp impedes the reachability of the stable state with inverted Wg and En expressing cells. It also suppresses the functionality contexts of all negative circuits [9,15], implying that the state transition graph does not contain any cyclic attractor. Indeed, after further reduction to three nodes per cell (Wg, En and Hh), we were able to check the absence of non-trivial attractors in the full STG. As the reduction cannot delete existing non-trivial attractors (see Theorem 1), this implies that all attractors of the original model are stable states.

# 6 Conclusions and Prospects

We have defined a reduction method that can be applied to multi-valued logical models while preserving important dynamical properties. In particular, all attractors of the original dynamics have a counterpart in the dynamics of the reduced model. Furthermore, trajectories in the reduced model can be formally related to trajectories in the original one. This enables to infer the existence of paths in the dynamics of a detailed model whenever it is possible to show (by simulation and graph analysis) that paths exist between the corresponding states in a reduced version of the model. However, the reverse is not true. Indeed, a reduction can lead to the loss of reachability properties. Whenever several asynchronous component updates are possible at a given state, the elimination of one of the updated components amounts to consider it as "faster" than the concurrent ones, leading to the possible exclusion of some transitions in the reduced STG. Such reductions relate to the delineation of specific priority class configurations [18].

One particular feature of the reduction method defined here is that the removal of (functional) autoregulated components is forbidden. This rule is related to previous work on the dynamical roles of the regulatory circuits. Indeed, it has been recently proven in the discrete framework that positive regulatory circuits are necessary to generated multiple attractors, whereas negative circuits are necessary to generate cyclic attractors (cf. [19] and references therein). At least in the discrete framework, these properties depend only on the sign of the regulatory circuit, *i.e.* on the product of the sign of the involved interactions and not on their number. From a qualitative dynamical point of view, it is thus possible to reduce the number of components of a circuit down to a single autoregulated component, while keeping the corresponding property, as long as we conserve the sign of the circuit (along with some functionality constraints).

Our formal presentation of the reduction method mainly focuses on the removal of a single component. However, iterating this process enables the removal of several components. This raises the question of the impact of the order in which reductions are performed. As shown in Appendix B, the removal of a component may be possible only after the prior removal of others. If we aim at removing as many components as possible, the ordering of removals may thus be crucial. Further work is needed to properly define optimal or maximal reductions for the general case. When the removal of a set of components is possible in several orders, we suspect that the dynamics of the resulting model does not depend on the order (work in progress).

The worst case complexity of the algorithm for the reduction of a node $r$ that regulates $k$ targets is in $O(m^d)$, where $m$ is the highest number of levels of the involved components and $d$ is the depth of the MDDs representing the revised logical functions associated to the target nodes. In most cases, $m \leq 3$ and $d \leq 5$.

Applying our reduction method to a detailed model of the segment-polarity network, we were able to show the absence of non-trivial attractors in a state transition graph too large to be stored. As indicated for this application, the reduction method offers a great flexibility to the modeller. Biological arguments (*e.g.* information on relative reaction speeds) can be used to select sets of nodes for consistent model reduction. In the course of the dynamical analysis of complex networks (*e.g.* multicellular networks), further reduction can be performed to identify all attractors and check their reachability from specific initial states.

To ease the maintenance of a detailed model along with its reduced versions, the *GINsim* implementation enables the user to define and record various reductions for the same reference model. In order to handle still larger and more complex networks, such reduction could be combined with algorithmic methods enabling the analysis of large state transition graphs ([20] and references therein), or yet with model checking techniques ([21] and references therein).

**Supplementary Materials.** *GINsim* can be downloaded from `http:// gin.univ-mrs.fr/GINsim`. The Appendix, and the models, are available at the following URL: `http://gin.univ-mrs.fr/GINsim/publications/ naldi2009.html`.

**Acknowledgments.** A.N. has been supported by PhD grant from the French Ministry of Research and Technology. C.C. acknowledges the support provided by the Calouste Gulbenkian Foundation. This work was further supported by research grants from the French National Agency (project ANR-08-SYSC-003), from EU FP7 (APO-SYS Project), and by the Belgian Science Policy Office (IAP BioMaGNet).

# References

1. Saez-Rodriguez, J., Simeoni, L., Lindquist, J., Hemenway, R., Bommhardt, U., Arndt, B., Haus, U., Weismantel, R., Gilles, E., Klamt, S., Schraven, B.: A logical model provides insights into t cell receptor signaling. PLoS Comput. Biol. 3(8), e163 (2007)
2. Franke, R., Müller, M., Wundrack, N., Gilles, E.D., Klamt, S., Kähne, T., Naumann, M.: Host-pathogen systems biology: logical modelling of hepatocyte growth factor and helicobacter pylori induced c-met signal transduction. BMC Syst. Biol. 2, 4 (2008)
3. Chaouiya, C., Remy, E., Mossé, B., Thieffry, D.: Qualitative analysis of regulatory graphs: a computational tool based on a discrete formal framework. LNCIS, vol. 294, pp. 119–126 (2003)
4. Naldi, A., Berenguier, D., Fauré, A., Lopez, F., Thieffry, D., Chaouiya, C.: Logical modelling of regulatory networks with GINsim 2.3. BioSystems (in press)
5. Thomas, R.: Regulatory networks seen as asynchronous automata: A logical description. J. Theor. Biol. 153, 1–23 (1991)
6. Thomas, R., Thieffry, D., Kaufman, M.: Dynamical behaviour of biological regulatory networks–i. biological role of feedback loops and practical use of the concept of the loop-characteristic state. Bull. Math. Biol. 57(2), 247–276 (1995)

7. Bryant, R.E.: Graph-based algorithms for boolean function manipulation. IEEE Trans. Comput. 35, 677–691 (1986)
8. Kam, T., Villa, T., Brayton, R.K., Sangiovanni-Vincentelli, A.L.: Multi-valued decision diagrams: Theory and applications. Int. J. Multi. Logic 4, 9–12 (1998)
9. Naldi, A., Thieffry, D., Chaouiya, C.: Decision diagrams for the representation and analysis of logical models of genetic networks. In: Calder, M., Gilmore, S. (eds.) CMSB 2007. LNCS (LNBI), vol. 4695, pp. 233–247. Springer, Heidelberg (2007)
10. Meinhardt, H.: Hierarchical inductions of cell states: a model for segmentation in drosophila. J. Cell Sci. Suppl. 4, 357–381 (1986)
11. von Dassow, G., Meir, E., Munro, E.M., Odell, G.M.: The segment polarity network is a robust developmental module. Nature 406(6792), 188–192 (2000)
12. Ingolia, N.T.: Topology and robustness in the drosophila segment polarity network. PLoS Biol. 2(6), e123 (2004)
13. Albert, R., Othmer, H.G.: The topology of the regulatory interactions predicts the expression pattern of the segment polarity genes in drosophila melanogaster. J. Theor. Biol. 223(1), 1–18 (2003)
14. Chaves, M., Albert, R., Sontag, E.D.: Robustness and fragility of boolean models for genetic regulatory networks. J. Theor. Biol. 235(3), 431–449 (2005)
15. Sánchez, L., Chaouiya, C., Thieffry, D.: Segmenting the fly embryo: logical analysis of the role of the segment polarity cross-regulatory module. Int. J. Dev. Biol. 52(8), 1059–1075 (2008)
16. González, A., Chaouiya, C., Thieffry, D.: Logical modelling of the role of the hh pathway in the patterning of the drosophila wing disc. Bioinformatics 24(16), i234–i240 (2008)
17. Sánchez, L., Thieffry, D.: Segmenting the fly embryo: a logical analysis of the pair-rule cross-regulatory module. J. Theor. Biol. 224(4), 517–537 (2003)
18. Fauré, A., Naldi, A., Chaouiya, C., Thieffry, D.: Dynamical analysis of a generic boolean model for the control of the mammalian cell cycle. Bioinformatics 22(14), e124–e131 (2006)
19. Remy, E., Ruet, P.: From minimal signed circuits to the dynamics of boolean regulatory networks. Bioinformatics 24(16), i220–i226 (2008)
20. Garg, A., Di Cara, A., Xenarios, I., Mendoza, L., De Micheli, G.: Synchronous versus asynchronous modeling of gene regulatory networks. Bioinformatics 24(17), 1917–1925 (2008)
21. Monteiro, P.T., Ropers, D., Mateescu, R., Freitas, A.T., de Jong, H.: Temporal logic patterns for querying dynamic models of cellular interaction networks. Bioinformatics 24(16), i227–i233 (2008)

# On the Use of Stochastic Petri Nets in the Analysis of Signal Transduction Pathways for Angiogenesis Process

Lucia Napione, Daniele Manini, Francesca Cordero, András Horváth, Andrea Picco, Massimiliano De Pierro, Simona Pavan, Matteo Sereno, Andrea Veglio, Federico Bussolino, and Gianfranco Balbo

[1] Department of Computer Science, University of Torino, Torino, Italy
[2] Institute for Cancer Research and Treatment, Candiolo (TO), Italy
[3] Department of Clinical and Biological Sciences, University of Torino, Torino, Italy
[4] Department of Oncological Sciences, University of Torino, Torino, Italy

**Abstract.** In this paper we consider the modeling of a selected portion of signal transduction events involved in the angiogenesis process. The detailed model of this process contains a large number of parameters and the data available from wet-lab experiments are not sufficient to obtain reliable estimates for all of them. To overcome this problem, we suggest ways to simplify the detailed representation that result in models with a smaller number of parameters still capturing the overall behaviour of the detailed one.

Starting from a detailed stochastic Petri net (SPN) model that accounts for all the reactions of the signal transduction cascade, using structural properties combined with the knowledge of the biological phenomena, we propose a set of model reductions.

## 1 Introduction

Formal modeling is a central theme in systems biology in which mathematical modeling and simulation can play an important role. The Petri net (PN) formalism [18] is a framework that allows the construction of a precise and clear representation of biological systems based on solid mathematical foundations. This formalism permits the study of qualitative properties related to the structure of the model (e.g., the structure of a biological pathway). The variant of PNs, called Stochastic Petri Nets (SPNs) [16,15,2] and characterized by the addition of timing and/or stochastic information, can be used for quantitative analysis (e.g., analysis that involve the rates in biochemical reactions). PNs have been first proposed for the representation of biological pathways by Reddy et al [17]. Since their introduction, many other researchers constructed PN models of biological pathways [11] with the aim of using their representations to obtain qualitative information about the behavior of these systems, mostly via simulation [12,9]. The interaction of qualitative and quantitative analysis is necessary to check a model for consistency and correctness; following this idea, Heiner et al [10] proposed a methodology to develop and analyze large biological models in a step-wise manner.

In this paper we present our experience in modeling signal transduction pathways for the angiogenesis process using SPNs. The general goal is to analyze the temporal dynamics of a few relevant biological products and this requires to build and parameterize

P. Degano and R. Gorrieri (Eds.): CMSB 2009, LNBI 5688, pp. 281–295, 2009.

the model of the phenomenon under study. A detailed model is built by biologists and
then the parameters are estimated on the basis of data obtained by wet-lab experiments.
It is often the case however that the amount of available wet-lab data is not sufficient to
have reliable estimates of the many parameters involved in the model. The key contribution of this paper aims at alleviating this problem by providing a simplification process which transforms the detailed model into a simpler one with less parameters. The
proposed simplification process is guided by qualitative properties together with knowledge on the phenomenon under study and it is validated by comparing the quantitative
properties of the detailed and simplified models. Moreover, this process represents the
basis of the development of arguments useful for identifying both critical complexes
and interactions that play a crucial role in the biochemical system under study. With
respect to the framework proposed by Gilbert et al [8], where the main idea is to illustrate the complementarity among the three different ways of modeling biochemical
network , i.e. qualitative, stochastic and continuous, here we focus our attention on
the definition and robustness of the simplification process to limit the complexity of
the model.

Techniques that can be seen as simplification procedures have already been published
in the literature. See, for example, [4,3] where approximate analysis methods based on
aggregation of states are proposed. The goal of these techniques is however different
from ours, since they aim at reducing the complexity of the analysis of the model and
not the difficulty of its parametrization. Indeed, they result in simpler models in which
the number of parameters is identical to that of the original one.

The paper is organized as follows. Section 2 provides an overview of PNs and SPNs
and of their use in biochemical systems. Section 3.1 describes the angiogenesis case
study. Section 3.2 presents the approach we followed to build the SPNs and Section 3.3
shows the formal and biological rules used in the simplification process as well as the
resulting SPNs. The quantitative analysis performed in order to verify the mathematical
robustness of the simplified model is proposed in Section 3.4. We conclude with a
discussion and an outlook of future works in Section 4.

## 2   Modeling Formalism and Solution Techniques

The descriptions commonly applied in biology, where the relations among components
are expressed by biochemical reactions, or by interactions of genes as well as by cell
population interactions, are easy to transform into PNs in which places correspond to
genes/proteins/compounds (substrates) and transitions to their interactions.

### 2.1   Petri Net Representation for Biochemical Entities Interactions

PNs are a graphical language for the formal description of distributed systems with
concurrency and synchronization. PNs are bipartite graphs with two types of nodes,
namely places and transitions, connected by directed arcs. The state of the system is
given by the distribution of tokens over the places of the net. The dynamics of the
model (starting from an initial marking) is captured by state changes due to firing of
transitions and by the consequent movement of tokens over the places.

**Definition 1.** *[PN - syntax ]*A Petri Net graph is a tuple $(P, T, W, \mathbf{m}_0)$, where

- $P$ is a finite set of *places*;
- $T$ is a finite set of *transitions*;
- $P$ and $T$ are such that $P \cap T = \emptyset$;
- $W : (P \times T) \cup (T \times P) \rightarrow \mathbb{N}$ defines the arcs of the net and assigns to each of them a multiplicity;
- $\mathbf{m}_0$ is the *initial* marking which associates with each place a number of *tokens*.

When applied in systems biology, places represent biochemical entities (enzymes, compounds, etc.) and transitions represent the interactions among entities [17]. The quantities of the entities are represented by tokens in the places. The biological system we consider consists of biochemical reactions similar to those reported in Fig. 1-a where we show the PN representation scheme we adopted to describe all reactions of this type. Fig. 1-b represents the state evolution due to firing of transition $k_{53}$.

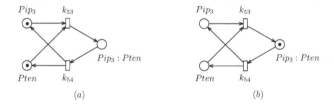

**Fig. 1.** PN representation of reactions $Pip3 + Pten \underset{k_{54}}{\overset{k_{53}}{\rightleftarrows}} Pip3 : Pten$

## 2.2  Analysis Techniques Based on Structural Properties

The PN graph inspection can provide several functional properties of the model, whose validity is true independently of the initial state of the system: such properties are, for instance, the boundedness and the existence of structural deadlocks and traps [18,2]. In deriving such kinds of information an important role is played by the so called net's *invariants*. There exist two kinds of invariants: place invariants (P-invariants) and transition invariants (T-invariants) [18]. In this paper we deal with P-invariants only. A P-invariant is a weighted sum of tokens contained in a subset of places of the net that remains constant through the entire evolution of the model, starting from an initial marking. The subset of places used for computing the P-invariant is the support (i.e., the set of nonzero components) of a P-semiflow $\mathbf{f}$ [14], which is a vector of nonnegative weights assigned to all the places of the net. A P-semiflow $\mathbf{f}$ is an integer and nonnegative solution of the matrix equation $\mathbf{f}\mathbf{C} = 0$, where $\mathbf{C}$ is the incidence matrix of the net, obtained by properly using the information provided by the flow relation $\mathbf{W}$ .

The interpretation of a P-invariant in a biological context, where tokens represent compounds, enzymes etc., is relatively simple: the places that support the semiflow $\mathbf{f}$ represent a portion of the PN where a given kind of correlated matter is preserved.

## 2.3   Quantitative Temporal Analysis

To study the temporal dynamics of a biological system it is natural to apply an extension of PNs that allows to introduce in the model temporal specifications. SPNs are time extensions of PNs in which exponentially distributed random delays (interpreted as durations of certain activities) are associated with the firings of the transitions. SPNs are qualitatively equivalent to PNs, meaning that for their structural analysis it is sufficient to disregard their time specifications. The temporal stochastic behaviour of an SPN is isomorphic to that of a continuous time Markov chain (CTMC). This CTMC can be built automatically from the description of the SPN and corresponds to the behaviour of the biological system described by the Master Chemical Equations [7]. This "stochastic approach" based on SPNs, adopts a discrete view of the quantity of the entities and sees their temporal behaviour as a random process.

Another possibility is to adopt a "deterministic approach" in which the temporal behaviour of the entities is seen as a continuous and completely predictable process. In our context we make use also of the deterministic approach because it allows for faster and simpler evaluation of the simplification process we propose for SPNs.

The deterministic approach translates the interactions into a set of coupled, first-order, ordinary differential equations (ODE) with one equation per entity. These equations describe how the quantities of the entities change based on the speed and the structure of the interactions among reactants. Referring again to the reactions considered in Figure 1, the corresponding ODEs are

$$\frac{dX_{Pip3}(t)}{dt} = -k_{53}X_{Pip3}(t)X_{Pten}(t) + k_{54}X_{Pip3:Pten}(t),$$

$$\frac{dX_{Pten}(t)}{dt} = -k_{53}X_{Pip3}(t)X_{Pten}(t) + k_{54}X_{Pip3:Pten}(t),$$

$$\frac{dX_{Pip3:Pten}(t)}{dt} = k_{53}X_{Pip3}(t)X_{Pten}(t) - k_{54}X_{Pip3:Pten}(t)$$

where $X_i(t)$ denotes the quantity of reactant $i$ at time $t$. Having the ODEs and information on the initial amount of the different entities, numerical integration of the ODEs is applied to calculate the quantities at a given time instant.

# 3   A Stochastic Petri Nets Based Approach Applied to Signal Transduction Pathways for the Angiogenesis Process

One main objective in systems biology is to model and analyze temporal dynamics of the phenomenon under study. By using SPNs as the formalism for the construction of the model, the analysis is performed in two steps: the first provides qualitative information on the structure of the model and the second investigates quantitative properties including statistical indices describing the temporal behavior of the system. Here we use this approach to study the angiogenesis process.

## 3.1   Biological Case Study Definition

Angiogenesis, defined as the formation of new vessels from the existing ones, is a topic of great interest in all areas of human biology, particularly to scientists studying vascular development, vascular malformation and cancer biology. Angiogenesis is a complex process involving the activities of many growth factors and relative receptors, which trigger several signaling pathways resulting in different cellular responses. The Vascular Endothelial Growth Factor (VEGF) family proteins are widely regarded as the most important growth factors involved in angiogenesis. VEGF-A, a member of VEGF family, has been most carefully studied and is thought to be of singular importance. VEGF receptor-2 ($KDR$ in humans) is thought to mediate most of VEGF-A's angiogenic functions, including cell proliferation, survival, and migration. Although the core components of the main $KDR$-induced pathways have been identified, the molecular mechanisms involved need to be characterized in fine details in order to better understand the flow of information. Indeed, a strong body of evidences indicates the presence of common adaptor/effector proteins involved in the survival and proliferation pathways induced by VEGF-A/$KDR$ axis, pointing out the difficulty to isolate a specific pathway and suggesting the presence of common nodes which contribute to create an intricate signaling network. In particular, the phosphorylated active receptor, indicated as $KDR^*$, catalyzes phosphorylation of several intracellular substrates including the adaptor protein $Gab1$ [13,5]. The main pathway through which VEGF-A induces cell proliferation involves the activation of $PLC\gamma$ [19]. Activation of $PLC\gamma$ promotes phosphatidylinositol 4,5-bisphosphate ($Pip_2$) hydrolysis giving rise to 1,2-diacylglycerol ($DAG$). VEGF-A-induced cell survival is dependent on the activity of $Pi3K$ [6]. The activated $Pi3K$ phosphorylates $Pip_2$ generating phosphatidylinositol-3,4,5-triphosphate ($Pip_3$). This recruits $Akt$ to the membrane where it is activated trough phosphorylation. Activated $Akt$ induces cell survival. Taking into account these notions, we wrote a system of biochemical reactions based on the available biological information together with further supposed mechanisms which could contribute to underline the presence of additional molecular nodes in the context of VEGF-A-induced proliferation and survival pathways.

### 3.2   Model Construction

In this section we discuss the approach we followed to represent the signal transduction cascade by SPN. Consider the detailed biological model depicted in Fig. 2.

These reactions describe $KDR$-proximal signaling events in the context of the survival and proliferation signal modules induced by receptor activation. In particular, reactions are split into four blocks. The First Block represents the earliest signaling events which include $KDR^*$ (we use the star to denote that proteins are active), $Gab1$, and $Pip_3$. The Second Block concerns the regeneration of $Pip_2$, a common substrate for the two signal modules that we are considering. In this block $Pip_2$ recovery was considered to result from the contribution of Pten-dependent dephosphorylation of $Pip_3$ in combination with $DAG$ catabolism (here recapitulated in the pseudo-enzyme E). The Third Block includes the reactions describing the survival pathway triggered by the $PI3K/Akt$ axis. The Fourth Block represents the proliferation pathway involving $PLC\gamma$ activation. Using the reaction representations outlined in Section 2.1 and the GreatSPN tool [1] the SPN model of the angiogenesis process was built as illustrated in Fig. 3. Exploiting the block organization and the structure of the model we analyzed the biochemical reactions in order to identify possible pathways and sub-pathways that describe embedded behaviors of the complete model. We denoted the reactions by means of their kinetic constants.

| KDR-Receptor (First Block) | Survival (Third Block) |
|---|---|
| $Kdr^* + Gab1 \underset{k_1}{\overset{k_0}{\rightleftarrows}} Kdr^*{:}Gab1$ <br><br> $Kdr^*{:}Gab1 \overset{k_2}{\rightarrow} Kdr^*{:}Gab1^*$ <br><br> $Gab1 + Pip3 \underset{k_4}{\overset{k_3}{\rightleftarrows}} Gab1{:}Pip3$ <br><br> $Kdr^* + Gab1{:}Pip3 \underset{k_6}{\overset{k_5}{\rightleftarrows}} Kdr^*{:}Gab1{:}Pip3$ <br><br> $Kdr^*{:}Gab1{:}Pip3 \overset{k_7}{\rightarrow} Kdr^*{:}Gab1^*{:}Pip3$ <br><br> $Kdr^*{:}Gab1^*{:}Pip3 \underset{k_9}{\overset{k_8}{\rightleftarrows}} Gab1^*{:}Pip3 + Kdr^*$ <br><br> $Kdr^*{:}Gab1^* + Pip3 \underset{k_{11}}{\overset{k_{10}}{\rightleftarrows}} Kdr^*{:}Gab1^*{:}Pip3$ | $Gab1^*{:}Pip3 + Pi3k \underset{k_{13}}{\overset{k_{12}}{\rightleftarrows}} Gab1^*{:}Pip3{:}Pi3k$ <br><br> $Gab1^*{:}Pip3{:}Pi3k + Kdr^* \underset{k_{15}}{\overset{k_{14}}{\rightleftarrows}} Kdr^*{:}Gab1^*{:}Pip3{:}Pi3k$ <br><hr> $Kdr^*{:}Gab1^* + Pi3k \underset{k_{17}}{\overset{k_{16}}{\rightleftarrows}} Kdr^*{:}Gab1^*{:}Pi3k$ <br><br> $Kdr^*{:}Gab1^*{:}Pi3k \overset{k_{18}}{\rightarrow} Kdr^*{:}Gab1^*{:}Pi3k^*$ <br><br> $Kdr^*{:}Gab1^*{:}Pi3k^* + Pip2 \underset{k_{20}}{\overset{k_{19}}{\rightleftarrows}} Kdr^*{:}Gab1^*{:}Pi3k^*{:}Pip2$ <br><br> $Kdr^*{:}Gab1^*{:}Pi3k^*{:}Pip2 \overset{k_{21}}{\rightarrow} Kdr^*{:}Gab1^*{:}Pi3k + Pip3$ <br><hr> $Kdr^*{:}Gab1^*{:}Pip3 + Pi3k \underset{k_{23}}{\overset{k_{22}}{\rightleftarrows}} Kdr^*{:}Gab1^*{:}Pip3{:}Pi3k$ <br><br> $Kdr^*{:}Gab1^*{:}Pip3{:}Pi3k \overset{k_{24}}{\rightarrow} Kdr^*{:}Gab1^*{:}Pip3{:}Pi3k^*$ <br><br> $Kdr^*{:}Gab1^*{:}Pip3{:}Pi3k^* + Pip2 \underset{k_{26}}{\overset{k_{25}}{\rightleftarrows}} Kdr^*{:}Gab1^*{:}Pip3{:}Pi3k^*{:}Pip2$ <br><br> $Kdr^*{:}Gab1^*{:}Pip3{:}Pi3k^*{:}Pip2 \overset{k_{27}}{\rightarrow} Kdr^*{:}Gab1^*{:}Pip3{:}Pi3k + Pip3$ <br><hr> $Pip3 + Akt \underset{k_{29}}{\overset{k_{28}}{\rightleftarrows}} Pip3{:}Akt$ <br><br> $Pip3{:}Akt \overset{k_{30}}{\rightarrow} Pip3 + Akt^*$ |
| **Pip2 Regeneration (Second Block)** | **Proliferation (Fourth Block)** |
| $Pip3 + Pten \underset{k_{54}}{\overset{k_{53}}{\rightleftarrows}} Pip3{:}Pten$ <br><br> $Pip3{:}Pten \overset{k_{55}}{\rightarrow} Pip2 + Pten$ <br><hr> $Pten + Pip2 \underset{k_{57}}{\overset{k_{56}}{\rightleftarrows}} Pten{:}Pip2$ <br><br> $Pten{:}Pip2 + Pip3 \underset{k_{59}}{\overset{k_{58}}{\rightleftarrows}} Pten{:}Pip2{:}Pip3$ <br><br> $Pten{:}Pip2{:}Pip3 \overset{k_{60}}{\rightarrow} Pten{:}Pip2 + Pip2$ <br><hr> $Dag + E \underset{k_{62}}{\overset{k_{61}}{\rightleftarrows}} Dag{:}E$ <br><br> $Dag{:}E \overset{k_{63}}{\rightarrow} Pip2 + E$ | $Kdr^* + Plc_\gamma \underset{k_{32}}{\overset{k_{31}}{\rightleftarrows}} Kdr^*{:}Plc_\gamma$ <br><br> $Kdr^*{:}Plc_\gamma \overset{k_{33}}{\rightarrow} Kdr^*{:}Plc_\gamma^*$ <br><br> $Kdr^*{:}Plc_\gamma^* + Pip2 \underset{k_{35}}{\overset{k_{34}}{\rightleftarrows}} Kdr^*{:}Plc_\gamma^*{:}Pip2$ <br><br> $Kdr^*{:}Plc_\gamma^*{:}Pip2 \overset{k_{36}}{\rightarrow} Kdr^*{:}Plc_\gamma + Dag$ <br><hr> $Kdr^*{:}Gab1^* + Plc_\gamma \underset{k_{38}}{\overset{k_{37}}{\rightleftarrows}} Kdr^*{:}Gab1^*{:}Plc_\gamma$ <br><br> $Kdr^*{:}Gab1^*{:}Plc_\gamma \overset{k_{39}}{\rightarrow} Kdr^*{:}Gab1^*{:}Plc_\gamma^*$ <br><br> $Kdr^*{:}Gab1^*{:}Plc_\gamma^* + Pip2 \underset{k_{41}}{\overset{k_{40}}{\rightleftarrows}} Kdr^*{:}Gab1^*{:}Plc_\gamma^*{:}Pip2$ <br><br> $Kdr^*{:}Gab1^*{:}Plc_\gamma^*{:}Pip2 \overset{k_{42}}{\rightarrow} Kdr^*{:}Gab1^*{:}Plc_\gamma + Dag$ <br><hr> $Kdr^*{:}Gab1^*{:}Pip3 + Plc_\gamma \underset{k_{44}}{\overset{k_{43}}{\rightleftarrows}} Kdr^*{:}Gab1^*{:}Pip3{:}Plc_\gamma$ <br><br> $Kdr^*{:}Gab1^*{:}Pip3{:}Plc_\gamma \overset{k_{45}}{\rightarrow} Kdr^*{:}Gab1^*{:}Pip3{:}Plc_\gamma^*$ <br><br> $Kdr^*{:}Gab1^*{:}Pip3{:}Plc_\gamma^* + Pip2 \underset{k_{47}}{\overset{k_{46}}{\rightleftarrows}} Kdr^*{:}Gab1^*{:}Pip3{:}Plc_\gamma^*{:}Pip2$ <br><br> $Kdr^*{:}Gab1^*{:}Pip3{:}Plc_\gamma^*{:}Pip2 \overset{k_{48}}{\rightarrow} Kdr^*{:}Gab1^*{:}Pip3{:}Plc_\gamma + Dag$ <br><hr> $Gab1^*{:}Pip3 + Plc_\gamma \underset{k_{50}}{\overset{k_{49}}{\rightleftarrows}} Gab1^*{:}Pip3{:}Plc_\gamma$ <br><br> $Gab1^*{:}Pip3 : Plc_\gamma + Kdr^* \underset{k_{52}}{\overset{k_{51}}{\rightleftarrows}} Kdr^*{:}Gab1^*{:}Pip3{:}Plc_\gamma^*$ |

**Fig. 2.** Reactions of the detailed model

In the model $Akt$ and $DAG$ have been considered as the end points of the survival and proliferation pathways, respectively. Taking into account these end points in combination with the notion that $Akt$ activation is strictly $Pip_3$-dependent, we examined the signal transduction cascade focusing our attention mainly on the reactions that lead to the production of $Pip_3$ (i.e. $k_{21}$ and $k_{27}$) and $DAG$ (i.e. $k_{36}$, $k_{42}$, and $k_{48}$).

This analysis (supported also by a careful drawing of the SPN) allowed us to recognize different sub-pathways that lead to the survival or proliferation effects. In the context of the survival signal module we identified two sub-pathways, each one characterized by the presence of a distinguishing complex, $KDR^*{:}Gab1^*$ or $KDR^*{:}Gab1^*{:}$ $Pip_3$, belonging to the First Block. Actually, the sub-pathways that determine the survival behavior are three since an additional element, $Gab1^*{:}Pip_3$, also contributes to the formation of $KDR^*{:}Gab1^*{:}Pi3k{:}Pip_3$ complex already involved in one of the identified sub-pathways. Summarizing there are three sub-pathways that lead to survival effect starting from: $KDR^*{:}Gab1^*$, $KDR^*{:}Gab1^*{:}Pip_3$ and $Gab1^*{:}Pip_3$. Considering the proliferation module, we identified four different sub-pathways that are distinguished by the compounds belonging to the First Block, i.e.: $KDR^*$, $KDR^*{:}Gab1^*$, $KDR^*{:}Gab1^*{:}Pip_3$ or $Gab1^*{:}Pip_3$. Notice that the distinguishing elements of the detected sub-pathways are the same within the survival and proliferation modules, with the exception of the compound $KDR^*$.

Referring again to the SPN of Fig. 3, we can notice that the time evolution of this SPN is intuitively portrayed by a top-down view. On the top is depicted the place $KDR^*$ that represents the starting point of the signal cascade induced by its ligand. From the $KDR^*$ cascade start all the sub-pathways that characterize the proliferation and survival pathways. The places describing $DAG$ and $Pip_3$ are aligned on the bottom of the net. It is interesting to note that the sub-pathways we identified in the detailed model are represented in the SPN with structures, such as that outlined by a dashed box in Fig. 3, which correspond to the reaction groups separated by continuous lines in Fig. 2. We denote these Sub-Components by SC. Each SC involves:

- the binding between an enzyme and other species present in the cascade (e.g. transitions $k_{31}k_{32}$);
- the enzyme activation (e.g. transition $k_{33}$);
- the recruitment of the $Pip_2$ (e.g. transitions $k_{34}k_{35}$);
- the production of the molecules representing the pathway end point and the enzyme deactivation (e.g. transition $k_{36}$).

### 3.3 Model Simplification

The SPN we obtained requires a simplification process to take place in order to limit the complexity of the parameterization and analysis of the model as we pointed out before. The computation of the P-semiflows of this SPN show that the net is bounded (the net is covered by P-semiflows, i.e., every place of the net is member of the support of one P-semiflow, at least). Interpreting the P-semiflows in biological terms, we can recall again that this means that all the compounds associated with the places of the net, independently of their original amounts, cannot grow indefinitely during the evolution of the model out of its initial state

The presence of repeated structures in the SPN corresponds to the fact that the biological model is characterized by the existence of several similar reaction groups, and this observation can be used to identify simplification steps to be applied to the detailed

288    L. Napione et al.

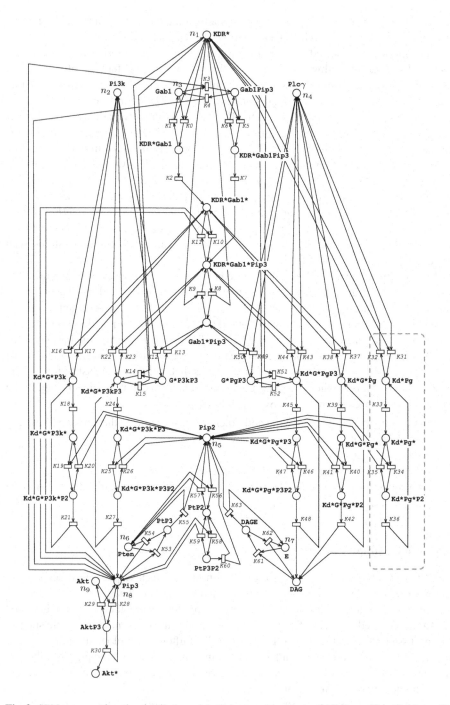

**Fig. 3.** SPN representing the detailed model. Compound symbols: $KDR \equiv Kd$, $Gab1 \equiv G$, $Pi3k \equiv P3k$, $Plc_\gamma \equiv Pg$, $Pip_3 \equiv P3$, $Pip_2 \equiv P2$, $Pten \equiv Pt$.

model. In particular, the SCs shown in the previous section are enzymatic kinetics re-action groups. Each of these groups can be written as the reactions set (1) where the enzyme $E$ binds reversibly to the compound $C$.

$$E + C \rightleftarrows EC \rightarrow E^*C \qquad (1)$$
$$E^*C + S \rightleftarrows E^*CS$$
$$E^*CS \rightarrow EC + P$$

This complex $EC$ irreversibly becomes $E^*C$, which means that the enzyme is activated. $E^*C$ binds reversibly to the substrate $S$ (forming $E^*CS$), before converting it into a product $P$ and releasing the complex $EC$. In order to simplify the detailed model we can represent each SC by the following couple of "merging" pseudo-reactions:

$$E + S \rightarrow E^*S \qquad (2)$$
$$E^*S + S' \rightleftarrows E + S + P$$

| KDR-Receptor (First Block) | Survival (Third Block) |
|---|---|
| $Kdr^* + Gab1 \overset{k_0}{\underset{k_1}{\rightleftarrows}} Kdr^*{:}Gab1$ $Kdr^*{:}Gab1 \overset{k_2}{\rightarrow} Kdr^*{:}Gab1^*$ $Gab1 + Pip3 \overset{k_3}{\underset{k_4}{\rightleftarrows}} Gab1{:}Pip3$ $Kdr^* + Gab1{:}Pip3 \overset{k_5}{\underset{k_6}{\rightleftarrows}} Kdr^*{:}Gab1{:}Pip3$ $Kdr^*{:}Gab1{:}Pip3 \overset{k_7}{\rightarrow} Kdr^*{:}Gab1^*{:}Pip3$ $Kdr^*{:}Gab1^*{:}Pip3 \overset{k_8}{\rightarrow} Gab1^*{:}Pip3 + Kdr^*$ $Kdr^*{:}Gab1^* + Pip3 \overset{k_9}{\underset{k_{11}}{\overset{k_{10}}{\rightleftarrows}}} Kdr^*{:}Gab1{:}Pip3$ | $Gab1^*{:}Pip3 + Kdr^* + Pi3k \overset{k_{12}}{\rightarrow} Kdr^*{:}Gab1^*{:}Pip3{:}Pi3k^*$ $Kdr^*{:}Gab1^*{:}Pip3{:}Pi3k^* + Pip2 \overset{k_{13}}{\rightarrow} Gab1^*{:}Pip3 + Kdr^* + Pi3k + Pip3$ $\overline{\phantom{xxxxxxxxxxxxxxxxxxxxx}}$ $Kdr^*{:}Gab1^* + Pi3k \overset{k_{14}}{\rightarrow} Kdr^*{:}Gab1^*{:}Pi3k^*$ $Kdr^*{:}Gab1^*{:}Pi3k^* + Pip2 \overset{k_{15}}{\rightarrow} Kdr^*{:}Gab1^* + Pi3k + Pip3$ $\overline{\phantom{xxxxxxxxxxxxxxxxxxxxx}}$ $Kdr^*{:}Gab1^*{:}Pip3 + Pi3k \overset{k_{16}}{\rightarrow} Kdr^*{:}Gab1^*{:}Pip3{:}Pi3k^*$ $Kdr^*{:}Gab1^*{:}Pip3{:}Pi3k^* + Pip2 \overset{k_{17}}{\rightarrow} Kdr^*{:}Gab1^*{:}Pip3 + Pi3k + Pip3$ $\overline{\phantom{xxxxxxxxxxxxxxxxxxxxx}}$ $Pip3 + Akt \overset{k_{18}}{\underset{k_{19}}{\rightleftarrows}} Pip3{:}Akt$ $Pip3{:}Akt \overset{k_{20}}{\rightarrow} Pip3 + Akt^*$ |
| **Pip2 Regeneration (Second Block)** | **Proliferation (Fourth Block)** |
| $Pip3 + Pten \overset{k_{29}}{\underset{k_{30}}{\rightleftarrows}} Pip3{:}Pten$ $Pip3{:}Pten \overset{k_{31}}{\rightarrow} Pip2 + Pten$ $\overline{\phantom{xxxxxxxxxx}}$ $Pten + Pip2 \overset{k_{32}}{\underset{k_{33}}{\rightleftarrows}} Pten{:}Pip2$ $Pten{:}Pip2 + Pip3 \overset{k_{34}}{\underset{k_{35}}{\rightleftarrows}} Pten{:}Pip2{:}Pip3$ $Pten{:}Pip2{:}Pip3 \overset{k_{36}}{\rightarrow} Pten{:}Pip2 + Pip2$ $\overline{\phantom{xxxxxxxxxx}}$ $Dag + E \overset{k_{37}}{\underset{k_{38}}{\rightleftarrows}} Dag{:}E$ $Dag{:}E \overset{k_{39}}{\rightarrow} Pip2 + E$ | $Kdr^* + Plc_\gamma \overset{k_{21}}{\underset{k_{22}}{\rightleftarrows}} Kdr^*{:}Plc_\gamma$ $Kdr^*{:}Plc_\gamma^* + Pip2 \overset{k_{22}}{\rightarrow} Kdr^* + Plc_\gamma^* + DAG$ $\overline{\phantom{xxxxxxxxxxxxxxxxxx}}$ $Kdr^*{:}Gab1^* + Plc_\gamma \overset{k_{23}}{\rightleftarrows} Kdr^*{:}Gab1^*{:}Plc_\gamma^*$ $Kdr^*{:}Gab1^*{:}Plc_\gamma^* + Pip2 \overset{k_{24}}{\rightarrow} Kdr^*{:}Gab1^* + Plc_\gamma + DAG$ $\overline{\phantom{xxxxxxxxxxxxxxxxxx}}$ $Kdr^*{:}Gab1^*{:}Pip3 + Plc_\gamma \overset{k_{25}}{\rightleftarrows} Kdr^*{:}Gab1^*{:}Pip3{:}Plc_\gamma^*$ $Kdr^*{:}Gab1^*{:}Pip3{:}Plc_\gamma^* + Pip2 \overset{k_{26}}{\rightarrow} Kdr^*{:}Gab1^*{:}Pip3 + Plc_\gamma + DAG$ $\overline{\phantom{xxxxxxxxxxxxxxxxxx}}$ $Gab1^*{:}Pip3 + Kdr^* + Plc_\gamma \overset{k_{27}}{\rightleftarrows} Kdr^*{:}Gab1^*{:}Pip3{:}Plc_\gamma^*$ $Kdr^*{:}Gab1^*{:}Pip3{:}Plc_\gamma^* + Pip2 \overset{k_{28}}{\rightarrow} Gab1^*{:}Pip3 + Kdr^* + Plc_\gamma + DAG$ |

**Fig. 4.** Reactions after first step of simplification

By exploiting this representation, we rewrite the reactions of the Third and the Fourth Blocks as shown in Fig. 4, and we use them to simplify the original SPN obtaining the net depicted in Fig. 5, that is still covered by P-semiflows, meaning that these trans-formations are acceptable also from a qualitative point of view. Note that in this new

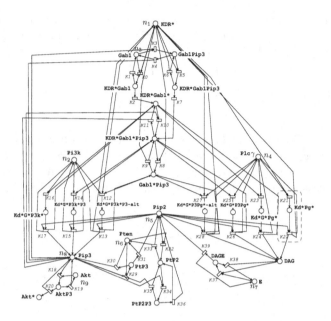

**Fig. 5.** SPN obtained after the first step of simplification (Simpl1)

SPN any SC is represented by a sequence of transition-place-transition, such as the one outlined by the dashed box, again. These SCs are defined by reaction groups separated by continuous line in Fig. 4. A further simplification step can be performed on the basis of the following reaction:

$$E + S + S' \rightarrow E + S + P \tag{3}$$

The application of this observation to reactions of the Third and Fourth Blocks provides the reactions illustrated in Fig. 6, used to define the new simplified net illustrated in Fig. 7 (which is again covered by P-semiflows). Note that in this last SPN any SC is represented by a singular transition such as the one outlined again by the dashed box. Moreover reactions $k_{29}k_{30}$, $k_{31}$ and $k_{37}k_{38}$, $k_{39}$ of the Second Block are simplified following the scheme: $E + S \rightarrow E + P$. The structural criteria that we used to guide the simplification process helped us to verify that the reduced models maintain the biological significance of the original one and provide a good approximation of its behavior. Notice that the reaction substitutions represented by Eq. (2 and 3) can be seen as patterns (or net substructures) that can be replaced every time they occur in the detailed and intermediate nets of the simplification process. In reality, it has not been possible to perform these transformations "mechanically". Instead, the invariant conditions had to be checked for each simplification substitution. At the end of this process, considering that the P-semiflows of the simplified SPNs have smaller supports, we found the same eight P-semiflows in all the three nets:

– one P-semiflow including $KDR^*$ and the complexes containing it that are present in the sub-pathways (both in proliferation and in survival) that bring to the recruiting of substrate $Pip_2$;

| KDR-Receptor (First Block) | Survival (Third Block) |
|---|---|
| $Kdr^* + Gab1 \overset{k_0}{\underset{k_1}{\rightleftarrows}} Kdr^*{:}Gab1$ | $Gab1^*{:}Pip3 + Kdr^* + Pi3k + Pip2 \overset{k_{12}}{\rightarrow} Gab1^*{:}Pip3 + Kdr^* + Pi3k + Pip3$ |
| $Kdr^*{:}Gab1 \overset{k_2}{\rightarrow} Kdr^*{:}Gab1^*$ | $Kdr^*{:}Gab1^*{:}Pip3 + Pi3k + Pip2 \overset{k_{13}}{\rightarrow} Kdr^*{:}Gab1^*{:}Pip3 + Pi3k + Pip3$ |
| $Gab1 + Pip3 \overset{k_3}{\underset{k_4}{\rightleftarrows}} Gab1{:}Pip3$ | $Kdr^*{:}Gab1^* + Pi3k + Pip2 \overset{k_{14}}{\rightarrow} Kdr^*{:}Gab1^* + Pi3k + Pip3$ |
| $Kdr^* + Gab1{:}Pip3 \overset{k_5}{\underset{k_6}{\rightleftarrows}} Kdr^*{:}Gab1{:}Pip3$ | |
| $Kdr^*{:}Gab1{:}Pip3 \overset{k_7}{\rightarrow} Kdr^*{:}Gab1^*{:}Pip3$ | $Pip3 + Akt \overset{k_{15}}{\underset{k_{16}}{\rightleftarrows}} Pip3{:}Akt$ |
| $Kdr^*{:}Gab1^*{:}Pip3 \overset{k_8}{\rightleftarrows} Gab1^*{:}Pip3 + Kdr^*$ | $Pip3{:}Akt \overset{k_{17}}{\rightarrow} Pip3 + Akt^*$ |
| $Kdr^*{:}Gab1^* + Pip3 \overset{k_{10}}{\underset{k_{11}}{\rightleftarrows}} Kdr^*{:}Gab1^*{:}Pip3$ | |

| Pip2 Regeneration (Second Block) | Proliferation (Fourth Block) |
|---|---|
| $Pip3 + Pten \overset{k_{22}}{\rightarrow} Pten + Pip2$ | $Kdr^* + Plc_\gamma + Pip2 \overset{k_{18}}{\rightarrow} Kdr^* + Plc_\gamma + DAG$ |
| $Pten + Pip2 \overset{k_{23}}{\underset{k_{24}}{\rightleftarrows}} Pten{:}Pip2$ | $Kdr^*{:}Gab1^* + Plc_\gamma + Pip2 \overset{k_{19}}{\rightarrow} Kdr^*{:}Gab1^* + Plc_\gamma + DAG$ |
| $Pten{:}Pip2 + Pip3 \overset{k_{25}}{\underset{k_{26}}{\rightleftarrows}} Pten{:}Pip2{:}Pip3$ | $Kdr^*{:}Gab1^*{:}Pip3 + Plc_\gamma + Pip2 \overset{k_{20}}{\rightarrow} Kdr^*{:}Gab1^*{:}Pip3 + Plc_\gamma + DAG$ |
| $Pten{:}Pip2{:}Pip3 \overset{k_{27}}{\rightarrow} Pten{:}Pip2 + Pip2$ | $Gab1^*{:}Pip3 + Kdr^* + Plc_\gamma + Pip2 \overset{k_{21}}{\rightarrow} Gab1^*{:}Pip3 + Kdr^* + Plc_\gamma + DAG$ |
| $Dag + E \overset{k_{28}}{\rightarrow} Pip2 + E$ | |

**Fig. 6.** Reactions after second step of simplification

- one P-semiflow including $Gab1$ and the complexes containing it that are present in the sub-pathways (both in proliferation and in survival) that bring to the recruiting of substrate $Pip_2$;
- one P-semiflow including $Akt$ and the complexes that lead to its activation;
- one P-semiflow including both $Pip_3$ and $Dag$, and $Pip_2$ that is the common substrate in both pathways. This semiflow includes also the cascade complexes containing $Pip_3$;
- each enzyme present in the model ($Pi3k$, $Plc\gamma$, $Pten$, $E$) has a semiflow including the complexes containing it.

The consistency among the structural properties of all the nets allowed us to consider the simplified models valid from a qualitative point of view.

### 3.4 Model Analysis for Accuracy Assessment

The simplification process proposed in Section 3.3 results in SPNs which maintain the qualitative properties of the original SPN, but are approximations of the detailed model from a quantitative point of view. In this section we report in silico experiments that were performed in order to check the validity of the simplifications from the point of view of quantitative properties. Indeed, before using the simplified models in a parameter identification experiment which uses real data coming from wet-lab experiments, it is necessary to make sure that an overall agreement exists between the quantitative temporal behaviours of the detailed and the simplified models for a wide range of model parameters. This test allows to build confidence on the fact that the reduced model is

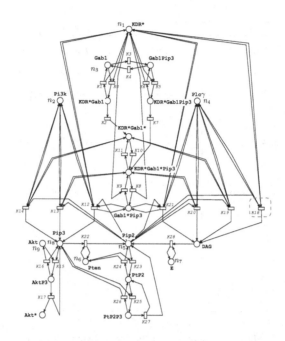

**Fig. 7.** SPN obtained after the second step of simplification (Simpl2)

suited for a preliminary analysis of the angiogenesis process as it can be done with in silico experiments.

This accuracy assessment was performed applying the "deterministic approach" described in Section 2.3. We compared the temporal behaviours of the detailed SPN with those of the simplified SPNs obtained for several different initial markings and several sets of transition rate parameters. Throughout the comparisons we concentrated on three important entities, in particular $Pip_3$ and $Dag$, and the substrate $Pip_2$ which is common to both the survival and the proliferation pathways. Hereafter we report on two cases which illustrate the obtained results. The initial condition is identical for the two cases: for all three SPNs (see Figures 3, 5 and 7) we use the initial marking $n_1 = 2, n_2 = n_3 = n_4 = 1, n_5 = 20, n_6 = n_7 = n_8 = n_9 = 1$ which reflects the concentration differences that are likely to exist in wet-lab experiments. Different sets of transition rates are used in the two cases to push the behaviours of the models in opposite directions. In the first case the rates are such that the transitions along the survival pathway are ten times faster than all the others. Figure 8A depicts the temporal behaviour for the detailed model. With these parameters the concentration of $Pip_3$ increases, the concentration of $Pip_2$ decreases and the concentration of $DAG$ remains low. The temporal behaviour of the simplified models, depicted in Figures 8B-C, shows the same major characteristics. In the second case the transitions along the proliferation pathway are ten times faster than all the others. Figures 8E-F-G depict the temporal behaviour for the three models. Also in this case, the major characteristic, i.e. the fact that the concentration of $DAG$ prevails over the concentration of $Pip_2$ and $Pip_3$, is maintained. The general agreement among the results of all these models was also tested by

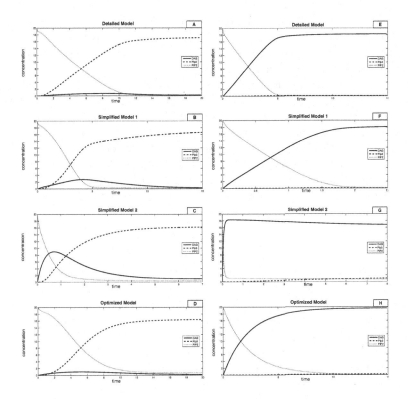

**Fig. 8.** The left (right) column shows the ODE results from the first (second) set of parameters

computing the steady state distribution of tokens in the places corresponding to compounds $Pip_3$ and $DAG$ and to the substrate $Pip_2$, obtained from the solutions of the CTMCs corresponding to all the SPN models we have constructed (that we do not report in detail in this paper, due to space constraints).

For the first set of experiments reported here, the rates of corresponding transitions in the simplified and detailed models were set equal independently of the fact that in the simplified SPNs they often represent the compound effect of a few transitions of the detailed one. As a result, even if the overall characteristics are maintained, the shape of the curves can be rather different and the dynamics take place on different time scales. Focusing our attention on the curves corresponding to the first set of parameters (Figures 8A-B-C), we can notice that the crossing of the two concentration curves for $Pip_2$, which decreases, and $Pip_3$ that grows takes place at time instants that are not of the same order of magnitude in all the three cases. A more important difference, however, is observed when we concentrate on the dynamics of $DAG$ that shows a small initial growth followed by a descent to a value next to zero. For this case, the most simplified model predicts an important initial climb that makes the shape of the curve quite different from the others (see Figure 8C). Turning our attention to the curves corresponding to the second set of parameters (Figures 8E-F-G), we can notice that all the models predict a crossing between the concentration of $Pip_2$ which decreases and the concentration of $DAG$ that instead grows. With these parameters the concentration of $Pip_3$ remains always extremely small. In this case the predictions of the more compact model are quantitatively quite different since the crossing is reported to happen much sooner and the shapes of the curves are very different.

In order to mimic better the temporal behaviour of the detailed SPN, we applied an optimization technique to determine the transition rates for the simplified SPNs. We used the nonlinear optimization function of MatLab to find these transition rates with which the curves resulting from the simplified SPNs became closer to the curves resulting from the detailed SPN. The results are illustrated for the SPN obtained after the second step of simplification (Simpl2) in Figure 8D and 8H. It can be seen that by properly setting the transition rates, even the simplest of our SPNs (Simpl2) can mimic quite precisely the behaviour of the detailed model (compare the diagrams A with D and E with H of Figure 8).

## 4   Discussion

In this paper we showed how to use structural properties of SPNs and biochemical properties of the system in guiding the simplification process. The procedure was presented through a case study, namely, the model of signaling transduction pathways involved in the angiogenesis process. We showed that the procedure results in simplified SPNs that are able to mimic precisely the temporal behavior of the detailed SPN.

One non trivial step is to determine the transition rates in the simplified SPNs in such a way that the resulting temporal behavior is a good approximation of that of the detailed SPN. In this work we faced this problem by applying optimization.

In the future, on the basis of the simplification schemes presented in this paper, we plan to work on generalizations of these reduction steps which will allow to operate on other portions of the detailed model and on the identification of rules concerning the relations existing among the corresponding rates of the detailed and simplified models. In addition, we will study the possibility of defining formally the quantitative characteristics that have to be maintained by the simplification process. In particular, temporal logics will be considered to this purpose.

Furthermore, we will consider the study of the whole VEGF-induced intracellular network, including signal modules that were not considered here, such as the migration pathway. This could contribute to a better understanding of the intricate signaling induced by VEGF-A during the angiogenesis.

### Acknowledgment

Cordero is the recipient of research fellowship supported by grants from Italian Association for Cancer Research and Regione Piemonte.

### References

1. Baarir, S., Beccuti, M., Cerotti, D., De Pierro, M., Donatelli, S., Franceschinis, G.: The Great-SPN tool: recent enhancements. SIGMETRICS Perform. Eval. Rev. 36(4), 4–9 (2009)
2. Balbo, G.: Introduction to stochastic petri nets. In: Brinksma, E., Hermanns, H., Katoen, J.-P. (eds.) EEF School 2000 and FMPA 2000. LNCS, vol. 2090, p. 84. Springer, Heidelberg (2001)

3. Busch, H., Sandmann, W., Wolf, V.: A numerical aggregation algorithm for the enzyme-catalyzed substrate conversion. In: Priami, C. (ed.) CMSB 2006. LNCS (LNBI), vol. 4210, pp. 298–311. Springer, Heidelberg (2006)

4. Calder, M., Vyshemirsky, V., Gilbert, D., Orton, R.: Analysis of Signalling Pathways Using Continuous Time Markov Chains. In: Priami, C., Plotkin, G. (eds.) Transactions on Computational Systems Biology VI. LNCS (LNBI), vol. 4220, pp. 44–67. Springer, Heidelberg (2006)

5. Dance, M., Montagner, A., Yart, A., Masri, B., Audigier, Y., Perret, B., Salles, J.P., Raynal, P.: The adaptor protein Gab1 couples the stimulation of vascular endothelial growth factor receptor-2 to the activation of phosphoinositide 3-kinase. Journal of Biological Chemistry 281, 23285–23295 (2006)

6. Gerber, H.P., McMurtrey, A., Kowalski, J., Yan, M., Keyt, B.A., Dixit, V., Ferrara, N.: Vascular endothelial growth factor regulates endothelial cell survival through the phosphatidylinositol 3'-kinase/Akt signal transduction pathway. requirement for Flk-1/KDR activation. Journal of Biological Chemistry 273, 30336–30343 (1998)

7. Gillespie, D.T.: A rigorous derivation of the master chemical equation. Physica 188, 404–425 (1992)

8. Glbert, D., Heiner, M., Lehrack, S.: A unifying framework for modelling and analysing biochemical pathways using Petri nets. In: Proc. Int. Conf. Computational Methods in System Biology, pp. 200–216 (2007)

9. Goss, P., Pecoud, J.: Quantitative modeling of stochastic systems in molecular biology by using stochastic Petri nets. Proc. Natl. Acad. Sci. 95(12), 6750–6755 (1998)

10. Heiner, M., Koch, I., Will, J.: Model validation of biological pathways using Petri nets demonstrated for apoptosis. BioSystems 75, 10–28 (2004)

11. Hofestädt, R.: A Petri net application of metabolic processes. Journal of System Analysis, Modeling and Simulation 16, 113–122 (1994)

12. Hofestädt, R., Thelen, S.: Quantitative modeling of biochemical networks. Silico Biology 1(6) (1998)

13. Laramée, M., Chabot, C., Cloutier, M., Stenne, R., Holgado-Madruga, M., Wong, A.J., Royal, I.: The scaffolding adapter Gab1 mediates Vascular Endothelial Growth factor signaling and is required for endothelial cell migration and capillary formation. Journal of Biological Chemistry 282, 7758–7769 (2007)

14. Memmi, G., Vautherin, J.: Advanced algebraic techniques. In: Brawer, W., Reisig, W., Rozenberg, G. (eds.) APN 1986, Part I. LNCS, vol. 254. Springer, Heidelberg (1987)

15. Molloy, M.K.: On the integration of delay and throughput measures in distributed processing models. Ph.D. Thesis, UCLA (1981)

16. Natkin, S.: Les réseaux de Petri stochastiques et leur application à l'évaluation des systèmes informatiques. Thèse de Docteur Ingégneur, CNAM (1980)

17. Reddy, V., Mavrovouniotis, M., Liebman, M.: Petri net representation in metabolic pathways. In: Proc. Int. Conf. Intelligent Systems for Molecular Biology, pp. 328–336 (1993)

18. Reisig, W.: A Primer in Petri Net Design. Springer Compass International, Heidelberg (1992)

19. Takahashi, T., Ueno, H., Shibuya, M.: VEGF activates protein kinase C-dependent, but Ras-independent Raf-MEK-MAP kinase pathway for DNA synthesis in primary endothelial cells. Oncogene 18, 2221–2230 (1999)

# CSL Model Checking of Biochemical Networks with Interval Decision Diagrams

Martin Schwarick and Monika Heiner

Department of Computer Science, Brandenburg University of Technology
Postbox 10 13 44, 03013 Cottbus, Germany
ms@informatik.tu-cottbus.de,
monika.heiner@informatik.tu-cottbus.de

**Abstract.** This paper presents an Interval Decision Diagram based approach to symbolic CSL model checking of Continuous Time Markov Chains which are derived from stochastic Petri nets. Matrix-vector and vector-matrix multiplication are the major tasks of exact analysis. We introduce a simple, but powerful algorithm which uses explicitly the Petri net structure and allows for parallelisation. We present results demonstrating the efficiency of our first prototype implementation when applied to biochemical network models, specifically with increasing token numbers. Our tool currently supports CSL model checking of time-bounded operators and the Next operator for ordinary stochastic Petri nets.

## 1 Motivation

Stochastic Petri nets are a natural way to model biochemical networks, where token values may be interpreted as molecules or concentration levels [GHL07], [HGD08]. Petri nets reflect explicitly the network structure, which contributes to a better understanding of the network behaviour, and – as we are going to see – supports efficiency gains otherwise not possible.

A stochastic Petri net's semantics is a Continuous Time Markov Chain (CTMC) which can be investigated by simulative approaches, or analysed analytically by transient and steady-state analysis [Ste94], or model checking of Continuous-time Stochastic Logic (CSL) [ASSB00]. In this paper we concentrate on (analytic) CSL model checking, which has been proven to be particularly useful for model validation and model-based experiment design in systems and synthetic biology: special behavioural properties are expressed in CSL, a flexible and powerful query language, and then checked exhaustively against all behaviour the model can exhibit.

The tool of choice when applying CSL model checking of CTMCs is often the probabilistic model checker PRISM [PNK06], which seems to represent the current state of the art [JKO+08]. Stochastic Petri nets can be easily translated into the PRISM input language as it has been done in [CDDS06], [GHL07], [HGD08]. However, computational experiments reach pretty fast their limits, as they always do if the famous state space explosion problem is one of the game players.

P. Degano and R. Gorrieri (Eds.): CMSB 2009, LNBI 5688, pp. 296–312, 2009.

PRISM's approach to cope with the problem is symbolic analysis based on Multi Terminal Binary Decision Diagrams (MTBDD), which are basically Binary Decision Diagrams (BDD) allowing more than two terminal nodes, each standing for a different value. While this often works fine for technical systems resulting into 1-bounded networks, it does not smoothly scale to the generalised bounded case. First of all, prior knowledge of the boundedness degree of each place is required. A place with an upper bound of $k$ tokens is represented by $\lceil ld(k) \rceil$ MTBDD variables. This may result in an overhead in computation time and memory. Since tokens may represent concentration levels, increasing the analysis accuracy implies an increase of the possible number of tokens on places. Secondly, PRISM creates an MTBDD which represents the entire CTMC with states and transitions encoded in a matrix. Therefore it is necessary to double the number of MTBDD variables to index rows and columns. Finally, a further drawback occurs if the CTMC contains many different rate values, since the number of terminal nodes in the MTBDD equals this amount. These lessons learnt from the PRISM approach made us elaborate a new technique for symbolic CSL model checking, specifically designed for biochemical networks with increasing token numbers.

The efficient analysis of qualitative Petri nets, provided they are bounded, but not necessarily 1-bounded, is discussed by A. Tovchigrechko in [Tov08]. He deploys Interval Decision Diagrams (IDD), which generalise BDDs by allowing more than two outgoing arcs for each node, but keeping the idea of two terminal nodes only. The developed data structures and algorithms support state space based analysis, including model checking of Computational Tree Logic (CTL). They do neither require a priori knowledge of the boundedness degree nor a suitable network partitioning as Kronecker-based approaches do, see e.g. [CJMS06]. The IDDs' inherent compression effect often yields compact representations of very large state spaces [HST09], see also caption of Table 2 in Section 4.

In this paper we are going to demonstrate how these IDD techniques can be transfered and adapted to CSL model checking, which basically requires to incorporate matrix-vector multiplication. In doing so we always bear in mind the option of parallelised processing on nowadays standard workstations. It goes without saying, the application of our results is not restricted to stochastic Petri nets. Specifically we will demonstrate how PRISM's efficiency may take advantage of our pre-analysis of a network's inherent structure.

## 2   Preliminaries

**Stochastic Petri Net.** An ordinary stochastic Petri net $SPN$ is a tuple $(P, T, F, V, s_0)$. As usual, $P$ denotes the set of places, $T$ the set of transitions, $F : ((P \times T) \cup (T \times P)) \rightarrow \{0, 1\}$ the arc weight function, and $s_0$ the initial state (marking). The mapping $V : T \rightarrow H$, where $H$ is the set of *hazard* functions, associates to each transition a function $h_t$ from $H$, defining a generally state-dependent, but always exponentially distributed firing rate. We deal with biologically interpreted stochastic Petri nets; thus we consider besides arbitrary arithmetic functions specifically functions representing *biomass action semantics (BMA)* and *biolevel interpretation semantics (BLI)*. All these functions have in common that the domain is restricted to the preplaces of the corresponding transition. For more details see [GHL07].

**Continuous Time Markov Chain.** The semantics of a stochastic Petri net is a CTMC which is isomorphic to the reachability graph of the underlying qualitative Petri net, but state transitions are labelled with firing rates. Without loss of generality we assume, if $s \xrightarrow{t} s'$ and $s \xrightarrow{t'} s'$ are state transitions in the CMTC, then $t = t'$. A CTMC is a tuple $(S, \mathbf{R}, L, s_0)$, with $S$ denoting the set of reachable states of the underlying net, $\mathbf{R} : S \times S \to \mathbb{R}_{\geq 0}$ the rate function, usually represented as matrix, $L : S \to 2^{AP}$ the labelling function, and $s_0$ the initial state. The set $AP := \{p \circ n | p \in P, \circ \in \{<, \leq, =, \neq, \geq, >\}, n \in \mathbb{N}_0\} \cup \{true, false\}$ of atomic propositions is defined over the set of places, which serve as integer variables. The entry $\mathbf{R}(s, s')$ is defined as:

$$\mathbf{R}(s, s') = \begin{cases} h_t(s) & \text{if } \exists t \in T : s \xrightarrow{t} s' \\ 0 & \text{otherwise .} \end{cases}$$

The total rate $E(s) = \Sigma_{s' \in S} \mathbf{R}(s, s')$ is the sum of entries of the matrix row indexed with $s$. A state $s$ with $E(s) = 0$ is called an absorbing state, since there is no way to leave it when reached. The probability of a transition $t$ enabled in state $s$ to fire (which results in state $s'$) within $n$ time units is $1 - e^{-\mathbf{R}(s,s') \cdot n}$. The transient probability $\pi(\alpha, s, \tau)$ is the probability to be in state $s$ at time $\tau$ starting from a certain probability distribution $\alpha$, with $\alpha : S \to [0, 1]$ and $\Sigma_{s \in S} \alpha(s) = 1$. The vector of transient probabilities for all states at time $\tau$ with the initial distribution $\alpha$ is denoted by $\underline{\pi}(\alpha, \tau)$. An established technique to compute the transient probabilities (transient analysis) of CTMCs is the uniformisation method. Its basic operation is vector-matrix multiplication which must be done for a certain number of iterations. For more details see [Ste94].

**Continuous time Stochastic Logic.** CSL is the stochastic counterpart to Computation Tree Logic (CTL). We consider CSL without the steady state operator and time-unbounded path formulae, and define state formulae

$$\phi ::= a \mid \neg\phi \mid \phi \wedge \phi \mid \phi \vee \phi \mid \mathcal{P}_{\bowtie p}[\varphi] ,$$

and path formulae

$$\varphi ::= X\phi \mid \phi U_{[\tau_1, \tau_2]}\phi \mid F_{[\tau_1, \tau_2]}\phi \mid G_{[\tau_1, \tau_2]}\phi ,$$

with $a \in AP$, $\bowtie \in \{<, \leq, \geq, >\}$, $p \in [0, 1]$, and $\tau_1, \tau_2 \in \mathbb{R}_{\geq 0} \wedge \tau_1 \leq \tau_2 \wedge \tau_2 < \infty$. For convenience we introduce the operators $F\phi$ and $G\phi$ as short-hand notations for the frequently used patterns $true\,U\phi$, and $\neg(true\,U\neg\phi)$.

CSL model checking of a CTMC $M$ can be realised by transient analysis. The basic concept is to do transient analysis for a CTMC $M'$ which has been derived from $M$ by making certain states absorbing, depending on the formula to be checked. For more details, e.g. formal semantics definition, see [BHHK00].

**Interval Decision Diagrams.** An IDD is a rooted, directed and acyclic graph with nodes having an arbitrary number of outgoing edges. Each edge is labelled with a left-closed and right-open interval on $\mathbb{N}_0$. The intervals of the outgoing edges of each IDD node define a partition of $\mathbb{N}_0$, inducing a total order of the edges. There are two nodes without outgoing edges: the terminal nodes, labelled with ONE and ZERO.

Each IDD node gets associated a variable, in our context a place of the stochastic Petri net. We assume that the variables occur in the same order on each path from the root to a terminal node – we get ordered IDDs. Furthermore we assume that an IDD does not contain isomorphic subgraphs – we get reduced ordered IDDs. As for BDDs, the variable ordering may influence the IDD size.

We use IDDs to encode sets of states of stochastic Petri nets, see Figure 1. The height of an IDD always equals the number of places, independently of the places' boundedness degree. IDD's grow in the breadth: a large variety of tokens on a given place may increase the number of outgoing edges of the corresponding IDD nodes, depending on the IDD-inherent compression effect. We consider bounded Petri nets; thus, each IDD node for a k-bounded place has at least two outgoing edges: [0, k+1), and [k+1, ∞).

A path (sequence of IDD nodes connected by edges) reaching the terminal node ONE represents generally a set of states. We get one state by choosing exactly one value from each of the intervals of all edges occurring along a path. For the efficient manipulation of state sets we assume operations like $\cap, \cup, \backslash$. Further we assume operations for the manipulation of state sets by the firing of transitions. $Fire(S,t) := \{s' | s \in S \wedge s \xrightarrow{t} s'\}$ represents the set of states

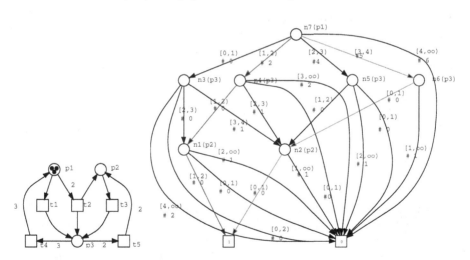

**Fig. 1.** A Petri net and the IDD, encoding its six reachable states. The path $n7 \xrightarrow{3} n6 \xrightarrow{0} n2 \xrightarrow{0} 1$ represents the initial state $m \equiv (p1 : 3, p2 : 0, p3 : 0)$. The path $n7 \xrightarrow{1} n4 \xrightarrow{1} n1 \xrightarrow{1} 1$ represents the state $m' \equiv (p1 : 1, p2 : 1, p3 : 1)$ which is reached from $m$ by firing transition $t2$. Edges are labelled with intervals and additional index data, see Section 3.1.

obtained by firing the transition $t$ for each state in $S$, and $Img(S)$ represents
the set of all direct successor states of $S$. Analogously, we define $RevFire(S, t)$
and $PreImg(S)$ for backward firing. For more details see [Tov08].

While CTL model checking can be completely reduced to the manipulation of
sets of integer states, CSL model checking by transient analysis requires the (repeated) multiplication of a real-valued matrix with a real-valued vector (or vice
versa). On account of the state space explosion problem it is not worth thinking
about implementing this vector-matrix multiplication explicitly with matrix and
vector indexed by states. There is no way to avoid the explicit representation of
the vector $\pi$. Actually, we need at least three (four) copies of it. Thus, the whole
problem boils down to the question: How to multiply with a matrix without
having (explicitly represented) the matrix?

In the following we present a matrix-free on-the-fly approach to realise CSL
model checking more efficiently than other tools available so far. Our tool IDD-
CSL computes all required data at each iteration anew from one augmented IDD
representing the reachable states of an SPN. Thus, our technique does not care
about the number of different matrix entries in the rate matrix. This is – in
terms of data structures – the main difference to PRISM's approach, where the
CTMC's state space $S$ and its rate matrix $\mathbf{R}$ are represented symbolically by a
BDD and an MTBDD.

# 3    Multiplication with IDDs

## 3.1    Basic Algorithm

We do not use dedicated data structures to represent the CTMC as other symbolic model checkers do. All necessary information is derived from the set of
reachable states encoded as an IDD and the Petri net structure itself. However,
we do need the lexicographic index for each state in the state set, which will be
determined by each depth first search traversal of the decision diagrams. One
slight extension of the IDD is required to get these indices, which brings us to
the index-labelled IDD, LIDD for short.

Now the basic idea of our approach is simply explained. The traversal of an
IDD representing a state set $S' \subseteq S$ drives the traversal of the LIDD, representing
$S$. This indirect, partial traversal of the LIDD $S$ allows to compute the index
for each state $s' \in S'$. Additionally we keep track of the index of the state
$s''$, reached by firing a given transition $t \in T$ in $s'$, assuming $s'$ enables $t$. We
compute the enabling states for a transition $t$ by $ES_t := S \cap RevFire(S, t)$. When
traversing an IDD, we always consider the pre- and post-conditions for the firing
of a transition $t$ of the underlying Petri net to determine the LIDD paths of
the related target states. This idea is inspired by the fire algorithm proposed
in [Tov08]. Each traversal extracts the indices of all state transitions (matrix
entries) of the CTMC induced by the firing of a transition $t$. Traversing the
LIDD for all transitions controlled by their enabling states eventually extracts all
non-zero matrix entries. Thus, each iteration required for the transient analysis
means the LIDD traversal for all transitions of the Petri net.

**Algorithm 1.** LIDD index labelling

```
procedure AUGMENTIDD(states : LIdd)
    AUGMENTNODE(states.root);
end procedure

function AUGMENTNODE(node : LIddNode) int :
    if node = ONE then
        return 1;
    end if
    if node = ZERO then
        return 0;
    end if
    count, reachable : int;
    count := 0;
    for 0 ≤ i < node.edges() do
        edge : LIddEdge;
        edge := node.edge(i);
        edge.smaller = count;
        reachable :=AUGMENTNODE(edge.node());
        count := count + reachable * edge.intervalWidth();
    end for
    return count;
end function
```

**Augmentation of an IDD with index information.** The required state indexing calls for an augmentation of the IDD, representing the reachable states, by some information which allows the necessary computation. Inspired by [MC99] we store the amount of lexicographic smaller states for each edge, which can be reached by all its previous sibling edges. This style to organise the index information allows to keep several sets within one and the same LIDD with different sets having different nodes as root. Algorithm 1 sketches how to derive recursively the LIDD representing $S$. The generated additional index data are labelled with a pound (#) in Figure 1.

**Determining a matrix entry.** We need the value (rate) of the current matrix entry $\mathbf{R}(i, j)$ to multiply the rate matrix with a vector or vice versa. This rate is determined by the hazard function of the Petri net transition which is responsible for the state transition from the state with index $i$ to the state with index $j$. Consequently, while computing the index pair for each state transition, we have to compute this function value, too. See Algorithm 2.

**Manipulating the matrix.** Please recall, model checking of time-bounded CSL operators for a CTMC $M$ can be reduced to the problem of applying transient analysis to $M'$ which has been derived from $M$ by making certain states absorbing. For this purpose, PRISM creates a new MTBDD representing the rate matrix of $M'$. In our approach this means only to call the procedure *traverse* for the non-absorbing subset $NES_t$ of the enabling states $ES_t$ of a transition $t$. When $A$ is an IDD representing the set of absorbing states, $NES_t$ can be computed efficiently by $ES_t \setminus A$.

Model checking the **X** operator involves the so-called Embedded Markov Chain (EMC). The EMC is a Discrete-time Markov Chain (DTMC), i.e. transitions are labelled with probabilities and the rate matrix $\mathbf{R}$ is replaced by the

## Algorithm 2. LIDD traversal

**procedure** TRAVERSEALLTRANSITIONS
  //the following for loop can be parallelised
  $t : Transition$;
  $ES_t : Idd$;
  **for** $0 <= j < SPN.transitions()$ **do**
    $t := SPN.getTransition(j)$;
    $fa : FunctionArgumentSet$;
    $ES_t := S \cap RevFire(S, t)$;
    TRAVERSE$(t, ES_t.root, S.root, S.root, 0, 0, fa)$;
  **end for**
**end procedure**

**procedure** TRAVERSE$(t \quad : \quad Transition, \quad root \quad : \quad IddNode, \quad src, dest \quad : \quad LIddNode,$
$srcIndex, destIndex : $ **int**, $fa : FunctionArgumentSet)$
  **if** $root = ONE$ **then**
    //e.g. vector-matrix r=v*M :
    //r[srcIndex] = v[destIndex]*rf.compute(fa)
    //rate is M[srcIndex][destIndex]
    $rate : $ **double**;
    $rate := t.rateFunction.compute(fa)$;
    $processData(srcIndex, destIndex, rate)$;
    **return** ;
  **end if**
  $p : Place$;
  $value, value2, srcIndex2, destIndex2 : int$;
  $edgeIndexSrc, edgeIndexDest : $ **int**;
  $edgeIndexSrc, edgeIndexDest := 0$;
  $src2, dest2 : LIddNode$;
  $edge : Edge$;
  $p := src.correspondingPlace()$;
  **for** $0 \leq i < root.edges()$ **do**
    $edge := root.edge(i)$;
    **if** $edge.node() \neq ZERO$ **then**
      $value := edge.lowerBound()$;
      **while** $value < edge.upperBound()$ **do**
        $value2 := value$;
        **if** $isPrePlace(p, t)$ **then**
          $fa.setArgument(p, value)$;
        **end if**
        $value2 := value + getWeight(p, t)$;
        $edgeIndexSrc := nextEdgeIndex(src, edgeIndexSrc, value)$;
        $edgeIndexDest := nextEdgeIndex(src, edgeIndexDest, value2)$;
        $src2 := src.edge(edgeIndexSrc).node()$;
        $dest2 := dest.edge(edgeIndexDest).node()$;
        $srcIndex2 := srcIndex+$SMALLERSTATES$(src, edgeIndexSrc, value)$;
        $destIndex2 := destIndex+$SMALLERSTATES$(dest, edgeIndexDest, value2)$;
        TRAVERSE$(edge.node(), src2, dest2, srcIndex2, destIndex2, fa)$;
        $value := value + 1$;
      **end while**
    **end if**
  **end for**
**end procedure**

**function** SMALLERSTATES$(node : LIddNode, edgeIndex, val : $ **int**$)$ **int** :
  $smaller : $ **int**;
  $edge : LIddEdge$;
  $edge : node.edge(edgeIndex)$;
  $smaller := 0$;
  **if** $edgeIndex > 0$ **then**
    $smaller := node.edge(edgeIndex - 1).smaller$;
  **end if**
  **return** $smaller + (val - edge.lowerBound()) * edge.node().lastEdge().smaller$;
**end function**

probability matrix $\mathbf{P}$, where each entry $(s, s')$ represents the probability of a state transition from $s$ to $s'$. The sum of each row of $\mathbf{P}$ is 1. The EMC $M_e$ is derived from a CTMC $M$ by defining $\mathbf{P}$ as $\mathbf{P}(s, s') := \mathbf{R}(s, s')/E(s)$. We assume that the values $E(s)$ for all states are stored in a vector. Multiplying the EMC with a vector means to adapt Algorithm 2 by the code given in Algorithm 3.

This approach should also work for PRISM's traversal algorithm. PRISM uses a further MTBDD to represent the EMC, for which the number of terminal nodes and thus the overall number of nodes can explode (see Section 4).

---

**Algorithm 3.** Adaptation of Algorithm 2 to handle Embedded Markov Chains

---

```
if root = ONE then
    // e.g. vector-matrix r=v*M :
    // r[srcIndex] = v[destIndex]*rf.compute(fa)
    // rate is M[srcIndex][destIndex]
    rateEmbedded : double;
    rateEmbedded := t.rateFunction.compute(fa)/E[srcIndex];
    processData(srcIndex, destIndex, rateEmbedded);
    return ;
end if
```

---

### 3.2   Optimization Techniques

In this section we sketch some optimization techniques contributing to the efficiency of our approach.

**Variable Ordering.** It is well known that the chosen variable order is crucial for the size of decision diagrams and thus for the efficiency of related algorithms. [Noa99] suggests a greedy algorithm to obtain a static variable order for Zero Suppressed Binary Decision Diagrams which is based on heuristics exploiting the Petri net structure. The basic idea is to create an order where related variables are close together. Related variables are in our case places, which are directly connected by a Petri net transition. The heuristic algorithm creates step-wise an order $\omega$, starting at the lowest IDD level and using the weight function $W(p)$ to determine the next place from the set of unprocessed places to be inserted in $\omega$ based on the set of already processed places $Q$. Using the standard dot notation to specify the set of pre- or postnodes of a given node we define $W(p)$ by:

$$W(p) := \frac{\Sigma_{t\in^\bullet p}\frac{|^\bullet t\cap Q|}{|^\bullet t|} + \Sigma_{t\in p^\bullet}\frac{|t^\bullet\cap Q|}{|t^\bullet|}}{|\,^\bullet p \cup p^\bullet\,|} \ . \tag{1}$$

Our approach benefits from the observation that the variable orders obtained by this algorithm usually yield small IDDs, see [Tov08], [HST09], and Section 4.

A prominent heuristics to represent matrices as MTBDDs relies on variable orders with alternating row and column variables. Additionally, it is worthwhile to find a good overall variable order, as we will see in Section 4. PRISM reads models as they are, i.e. it will not change the order of modules and the variables therein contained. Using the sketched ordering algorithm when specifying

PRISM models generally speeds up the state space construction and the model checking significantly.

**Caching.** As for every implementation of decision diagrams, efficiency depends on considering redundancies. Generally, nodes on lower IDD levels will be visited many times. Subpaths beginning in these nodes will be traversed each time anew. Following [Par02] we set a certain layer of the LIDD and cache index and rate information for each of its nodes for all paths to reach this node. Each time a node of the cache layer is reached, the cache data must be retrieved only.

Our algorithm traverses the LIDD transition-wise driven by an unlabelled IDD representing the enabling states of the transition or a non-absorbing subset of it. Thus it is necessary to store transition-specific information for all LIDD nodes of the cache layer. The cache data contain for each transition the index pair and the rate function of all possible path extensions. Each visit of a cache layer node comes with a unique index pair and a unique set of function arguments which allow to compute all related matrix entries using the cache data. A cache datum consists of an index pair and a rate function. Often many index pairs refer to the same function. Thus we associate to a function a set of index pairs. In general, a cache layer node keeps several rate function instances and their assigned index pairs for each transition. Changing the enabling states or the rates of the CTMC, e.g. by uniformisation, needs to reinitialise all cache data.

To use cache data requires a modification of Algorithm 2. When visiting a cache layer node, the transition-related cache data will be processed and the procedure *traverse* returns, compare Algorithm 4.

A crucial point for an implementation of this approach is to store the cache data, in particular the index sets, as memory-saving as possible. A naive way of doing this is to store lists of index pairs. But a closer look to the possible values reveals that there are often consecutive pairs with a fixed step size. This is a consequence of the fact that we obtain a huge state space by filling a Petri net with tokens, but without changing its structure. If such sequences of consecutive index pairs exceed a critical length it is worthwhile to represent them by a tuple $(first\_rowIndex, first\_colIndex, row\_stepSize, col\_stepSize, steps)$. Then, the sequence $(0,0); (5,10); (10,20); \ldots; (100,200)$ can be encoded by the tuple $(0,0,5,10,20)$. An issue here is to find a suitable critical length.

**Traversal for transition sets and arbitrary state sets.** The basic algorithm sketched so far requires a separate traversal for each transition of the stochastic Petri net. An improvement is to generalise the algorithm such that it controls the traversal of the LIDD $S$ for a set of transitions and an IDD encoding an arbitrary set of states $S' \subseteq S$. Then the algorithm must treat lists of source and target indices. The lists contain for each transition an entry holding the current traversal data. The basic algorithm ensures that the traversal-controlling IDD contains enabling states of a transition only. A generalization of the algorithm must deal with disabling states, too. Our prototype tool IDD-CSL implements the generalised algorithm.

**Algorithm 4.** Adaptation of Algorithm 2 in order to use cache data

```
if cacheLayerReached() then
    rate : double;
    rf : RateFunction;
    indices : IndexSet;
    actSrcIndex, actDestIndex : int;
    cd : CacheData;
    cd := src.cacheData(t);
    for 0 ≤ i < cd.entries() do
        rf := cd.getFunction(i);
        indices := cd.getIndices(i);
        rate := rf.compute(fa);
        for 0 ≤ j < indices.size() do
            actSrcIndex := srcIndex + indices.getSrcIndex(j);
            actDestIndex := destIndex + indices.getDestIndex(j);
            processData(actSrcIndex, actDestIndex, rate);
        end for
    end for
    return ;
end if
```

**Parallelisation.** Today's workstations or even standard personal computers in an everyday secretary office tend to possess two or more processors. Thus it is appealing to take advantage of the available multiple processors. There are basically two approaches to divide the problem of a matrix-vector multiplication or vice versa into smaller tasks, which can be solved concurrently.

On the one hand one could divide the Petri net's transition set and apply the algorithm concurrently to each subset. Doing so obviously requires some kind of synchronisation techniques. On the other hand one could partition the state space. Applying our algorithm concurrently with forward (backward) firing transitions with a partitioned state space means to devide the matrix row-wise (column-wise) into submatrices and requires no synchronisation when doing a matrix-vector (vector-matrix) multiplication, because each row (column) is considered for all transition by only one thread. When synchronisation is required all threads get their own complete result vector and collect the results of all other threads after each computation phase.

Although parallelisation is not the focus of this paper, we are going to indicate its potential by presenting some related results in the following section.

## 4   Benchmarks

In this section we present results comparing our prototype implementation IDD-CSL with PRISM, and by transitivity with a couple of CSL model checking tools on the market [JKO+08]. As benchmarks we consider stochastic Petri nets of the following popular biochemical networks.

- The mitogen-activated protein kinase (MAPK) cascade published in [LBS00] and discussed as three related Petri net models in [GHL07], [HGD08]. All initial states considered in our paper are multiples of level 4. This is the minimal initial (integer) state respecting the ratio in the initial (real-valued) concentrations as given in [LBS00]. Our model is structurally identical with the MAPK cascade given on the PRISM website. The models only differ in the names of variables, the initial state and the specified rate constants.

- The RKIP inhibited ERK pathway (ERK) published in [CSK+03], analysed with PRISM in [CVOG05], discussed as qualitative and continuous Petri nets in [GH06], and as three related Petri net models in [HDG10].
- The circadian clock model (CC) published in [BL00] and available as PRISM model on the PRISM website.

For the comparison with PRISM we either use the export feature of our modelling tool Snoopy [Sno08] (MAPK, ERK) or an available PRISM model (CC). The latter example needs capacities to enforce boundedness, which we simulate in the Petri nets by complementary places. All models have scalable initial states. The experiments consider *biomass action (resp. biolevel interpretation) semantics*, for which IDD-CSL offers the predefined *BioMassAction* (resp. *BioLevelInterpretation*) function.

Our implementation makes use of Intel's instruction set extension $SSE2$ which could also speed up PRISM. In some cases the efficiency gain is about 10 percent. Our test system is a Dell Precision workstation with 4 GB main memory and an Intel Xeon with $4 \times 2.83$GHz running a 64bit Linux. In our computational experiments we focus on runtime. All related figures are given in seconds.

**The influence of variable order.** In contrast to the modelling style in [KNP08], our generated monolithic PRISM models consist of one module only, with a module variable for each place. The value range of the variables (boundedness degree) and the variable order were computed by our IDD-based tool box. Table 1 illustrates the impact of the chosen variable order on PRISM's efficiency for different levels of the MAPK cascade, and Table 2 the CTMC size for different levels computed with PRISM using a good variable order.

**Table 1.** Comparison of two variable orders. The table shows the time and the number of MTBDD nodes, which PRISM needs to construct the rate matrix of the CTMC for a good variable order, computed using formula (1), and for the plain order of the original PRISM model, specified according to [KNP08].

| levels | terminal nodes[a] | good order | | original order | |
|---|---|---|---|---|---|
| | | time | nodes | time | nodes |
| 4 | 30 | 0.12 | 8,672 | 2.47 | 123,730 |
| 8 | 76 | 1.56 | 60,452 | 401.68 | 3,881,914 |
| 12 | 140 | 22.99 | 199,496 | - | - |
| 16 | 219 | 71.25 | 542,339 | - | - |
| 20 | 320 | 296.87 | 953,146 | - | - |
| 24 | 453 | 635.92 | 2,029,598 | - | - |
| 28 | 697 | 928.45 | 3,771,617 | - | - |
| 32 | 770 | 1847.60 | 6,015,521 | - | - |

[a] *i.e., number of different entries in rate matrix; '-' exceeds the available memory;*

**Table 2.** CTMC size for different levels computed with PRISM using a good variable order, see Section 3.2. In principle our tool IDD-CTL allows to compute the state space up to level 320 (2.627e+27 states ) in about 2 minutes on a standard personal computer [HST09]. However, the transient analysis it limited to the available memory to store the required vectors $\pi(\alpha, \tau)$ in the size of the state space.

| levels | number of states | number of edges[a] |
|---|---|---|
| 4 | 24,065 | 206,007 |
| 8 | 6,110,643 | 78,948,888 |
| 12 | 315,647,600 | 4,958,809,056 |
| 16 | 6,920,337,880 | 122,381,517,819 |
| 20 | 88,125,763,956 | 1,689,018,298,500 |
| 24 | 769,371,342,640 | 15,635,976,824,982 |
| 28 | 5,084,605,436,988 | 108,065,356,604,208 |
| 32 | 27,124,071,792,125 | 597,236,499,605,178 |

[a] *i.e., number of non-zero entries in rate matrix;*

**Table 3.** The formula $\mathcal{P}_{>0.0}[\mathbf{F}_{[0,1]} RafP = 2]$ is true for all states for which the probability is not zero to reach a state within one time unit which satisfies $RafP = 2$. For model checking of this formula all states satisfying $RafP = 2$ become absorbing; there are $1,083,102$ of them. The derived CTMC $M'$ comprises $64,368,742$ state transitions.

| PRISM[a] | | | IDD-CSL | | | | |
|---|---|---|---|---|---|---|---|
| cl[b] | total[c] | iter[d] | cl | 1 thread | | 2 threads | |
| | | | | total | iter | total | iter |
| 65 | 208.35 | **140.82** | 3 | 440.23 | 170.06 | 432.77 | 157.08 |
| 60 | 222.67 | 169.76 | 5 | 158.65 | 110.02 | 158.71 | 99.75 |
| 55 | 201.23 | 154.94 | 7 | 93.84 | 79.48 | 72.71 | 55.01 |
| 50 | 200.13 | 158.99 | 9 | 84.62 | 75.51 | 62.57 | 50.55 |
| 45 | **195.83** | 159.90 | 10 | **84.49** | **75.04** | **60.81** | **49.17** |
| 40 | 198.43 | 163.03 | 11 | 90.65 | 81.73 | 64.08 | 52.18 |
| 35 | 214.64 | 179.90 | 13 | 100.40 | 91.39 | 67.47 | 55.97 |
| 30 | 226.16 | 191.22 | 15 | 127.72 | 118.09 | 81.40 | 69.71 |
| 25 | 218.51 | 184.06 | 17 | 253.60 | 243.05 | 147.85 | 134.09 |
| 20 | 230.66 | 195.92 | 19 | 692.84 | 676.05 | 387.62 | 368.08 |
| 1 | 2318.86 | 2275.71 | 21 | 1808.66 | 1771.00 | 957.26 | 917.07 |

[a] *using a good variable order, determined by the network structure, see Table 1;*
[b] *cache layers;* [c] *includes time for state space construction, initialisation, computation and determining the satisfying states;* [d] *effective probability computation time;*

**The influence of caching.** Tables 3 and 4 compare the runtime of IDD-CSL and PRISM with different cache layers[1] for the eight level version of the MAPK cascade with *biolevel interpretation semanctics*. We use the flat PRISM model with a good variable order, compare Table 1, for these experiments. We take formulae which differ only in the specified time intervals. The interval $[0, 1]$ – in

---

[1] In PRISM the highest cache layer is the root node layer, in IDD-CSL it is the terminal node layer. PRISM's hybrid engine is used.

**Table 4.** The formula $\mathcal{P}_{>0.0}[\mathbf{F}_{[1,1]}RafP = 2]$ is true for all states for which the probability is not zero to be at time 1 in a state which satisfies $RafP = 2$ . Any path formula $\mathbf{F}_{[\tau,\tau]}\phi$ is suitable to trigger transient analysis up to time point $\tau$ using CSL model checking.

| PRISM[a] | | | IDD-CSL | | | | |
|---|---|---|---|---|---|---|---|
| | | | | 1 thread | | 2 threads | |
| cl | total | iter | cl | total | iter | total | iter |
| 65 | 242.80 | **163.05** | 3 | 382.00 | 148.05 | 365.91 | 129.08 |
| 60 | 231.65 | 170.69 | 5 | 135.06 | 94.01 | 124.50 | 74.27 |
| 55 | 275.53 | 219.89 | 7 | 88.74 | 75.84 | 67.32 | 50.43 |
| 50 | 222.23 | 173.32 | 9 | 82.22 | 73.47 | 60.25 | 48.34 |
| 45 | 210.13 | 167.48 | 10 | **81.58** | **72.82** | **58.55** | **47.21** |
| 40 | **209.91** | 168.83 | 11 | 87.67 | 78.98 | 62.42 | 51.11 |
| 35 | 212.23 | 171.38 | 13 | 97.53 | 88.98 | 65.55 | 54.34 |
| 30 | 211.43 | 170.39 | 15 | 124.67 | 116.00 | 78.85 | 67.56 |
| 25 | 221.78 | 181.28 | 17 | 260.95 | 250.09 | 150.66 | 137.09 |
| 20 | 231.90 | 192.46 | 19 | 782.22 | 766.00 | 428.38 | 409.02 |
| 1 | 2745.47 | 2691.70 | 21 | 2128.20 | 2087.00 | 1150.41 | 1106.00 |

[a] *using a good variable order, determined by the network structure, see Table 1;*

contrast to the interval $[1,1]$ – generally results into a set of absorbing states due to the CSL model checking algorithm [BHHK00]. A high amount of absorbing states reduces memory consumption and run time.

The MTBDD needs 66 row and 66 column Boolean variables. The IDD needs – independently of the number of levels – 22 integer variables, i.e. as many as there are places in the Petri net model. Thus, the MTBDD hight is 132, and the IDD hight is 22. The tables show the total processing time and the effective iteration time for the computation of the probability vector $\pi(\alpha, 1)$, which requires 218 iterations for each experiment of the used formulae. The best results in terms of total time and iteration time, depending on the used cache layer are highlighted in bold. The last line in each table represents the case where caching is disabled.

**Further results.** Table 5 and 6 present results for the transient analysis using CSL model checking for the ERK model and the circadian clock model with *biomass action semantics*. We give the total run time for both tools. In general PRISM's explicit sparse matrix engine is faster then its hybrid MTBDD engine at the expense of a higher memory usage [JKO+08]. The tables show that our tool outperforms also the sparse engine. The CTMC size affects the model checking performance. Thus, a high amount of absorbing states (which depends on the CSL formula) may significantly speed up the model checking. Except from choosing the engine or the number of threads, we run the tools with their default settings. Please note that changing, e.g., the cache layer would affect the run time and the memory consumption.

We also performed experiments with the **X** operator; we report here of one of them, which relates to the MAPK cascade. The formula $\mathcal{P}_{<0.1}[\mathbf{X}\,RafP = 2]$ is true for all states for which the probability is less than 0.1 to reach in one step

**Table 5.** CSL-based transient analysis of ERK for several initial markings. The formula $\mathcal{P}_{>0.0}[\mathbf{F}_{[1,1]}Raf1Star = 1]$ is true for all states for which the probability is not zero to reach a state within one time unit which satisfies $Raf1Star = 1$.

| level | CTMC size | | PRISM | | IDD-CSL | |
|---|---|---|---|---|---|---|
| | states | edges | hybrid | sparse | 1 thread | 2 threads |
| 5 | 1,974 | 12,236 | 0.73 | 0.65 | 0.20 | 0.17 |
| 10 | 47,047 | 372,372 | 5.52 | 3.97 | 1.27 | 1.05 |
| 15 | 368,220 | 3,213,408 | † | † | 20.73 | 16.67 |
| 20 | 1,696,618 | 15,609,594 | † | † | 148.92 | 118.59 |
| 25 | 5,723,991 | 54,438,930 | † | † | 740.28 | 581.39 |
| 30 | 15,721,464 | 152,964,146 | † | † | 3,005.62 | 2,455.57 |

† *time for initialisation exceeds 24 hours;*

**Table 6.** CSL-based transient analysis of the circadian clock for several initial markings. The formula $\mathcal{P}_{>0.0}[\mathbf{F}_{[1,1]}a = 1]$ is true for all states for which the probability is not zero to reach a state within one time unit which satisfies $a = 1$.

| level | CTMC size | | PRISM | | IDD-CSL | |
|---|---|---|---|---|---|---|
| | states | edges | hybrid | sparse | 1 thread | 2 threads |
| 5 | 31,104 | 290,160 | 3.90 | 2.36 | 1.76 | 1.16 |
| 10 | 644,204 | 6,766,320 | 122.55 | 64.94 | 44.65 | 26.10 |
| 15 | 4,194,304 | 45,972,480 | 1,090.44 | 570.43 | 466.47 | 312.83 |
| 20 | 16,336,404 | 183,032,640 | 5,569.65 | 2,835.89 | 2,471.70 | 1684.96 |
| 25 | 47,525,504 | 539,650,800 | † | – | 8,595.37 | 6,027.33 |
| 30 | 114,516,604 | 1,312,110,960 | † | – | 26,085.95 | 17,314.66 |

– *exceeds the physical memory;* † *time for initialisation exceeds 24 hours;*

a state which satisfies $RafP = 2$. To represent the Embedded Markov Chain PRISM creates an MTBDD which comprises $48,149,682$ nodes, among which are $217,974$ terminal nodes. The model checking takes 201.09 seconds including state space construction and initialization. IDD-CSL requires five seconds.

The figures speak for themselves. The gap between PRISM and IDD-CSL gets larger with increasing amount of levels (tokens, boundedness degree). Our data structure is less sensitive to increasing the amount of levels, and does not care about the amount of different matrix entries in the rate matrix.

## 5   Technicalities

The tool is implemented in C++, re-using our IDD-based CTL model checking implementation IDD-CTL [HST09] and the GNU MB Bignum Library (GMP). The parsing of CSL formulae has been generated by the lexical analyser and parser generator *flex* and *bison*. For parallelisation we use the POSIX pthread library. The tool comes as an all-inclusive binary (statically linked libraries) for our development and reference test system Linux. Versions for Windows and Mac/OS are in preparation.

The Petri net models have been constructed using Snoopy [Sno08], [HRS08], a tool to design and animate or simulate hierarchical graphs, among them stochastic Petri nets as used in this paper. Snoopy provides export to various analysis tools, recently complemented by PRISM, as well as import and export of the Systems Biology Markup Language (SBML).

The tools are available at www-dssz.informatik.tu-cottbus.de, free of charge for scientific purposes. At the same web site you find also the Petri net examples (in Snoopy, APNN, and PRISM syntax), which we used as benchmarks in the preceding section. Thus, all reported computational experiments can be easily repeated.

## 6   Conclusions

We have presented a new tool for symbolic CSL model checking of ordinary stochastic Petri nets. We combine approved heuristics with an innovative approach to represent symbolically a CTMC's rate matrix by Interval Decision Diagrams. We accept potentially higher computational costs in favour of smaller data structures. The models have to be bounded, however, no a priori knowledge of the precise boundedness degree is required. Likewise, we do not depend on a suitable partitioning as Kronecker-based approaches do. A crucial point for the tool's performance are the algorithms exploiting knowledge of the network structure. The implementation benefits in particular from the chosen static variable order which also increases PRISM's performance significantly.

In total we gave the results of more than 100 computational experiments. The presented benchmarks show that our data structure used for the symbolic state space representation is relatively insensitive to increasing token numbers. The IDD hight is completely defined by the number of variables. The IDD breadth may increase with increasing token numbers, but this depends on the IDD compression effect. Our approach is not sensitive at all to an increasing amount of different entries in the rate matrix. In summary this means that we are able to do transient analysis for any SPN for which we can construct the state space, provided we have enough memory to keep the vectors $\pi$. Using our IDD-CSL prototype we are now able to compute CSL properties, which were formerly not amenable to analytic model checking, for examples see [GHL07], [HGD08].

We are working on improvements of the sketched optimisation techniques, in particular the parallelisation. Furthermore we are going to support non-ordinary stochastic Petri nets and full CSL model checking. Thus, iterative solving of homogenous linear equation systems will be realised. For the time being we model capacities of places by introducing complementary places, which does its job, but blows up the models (e.g., our circadian clock model) and prohibits the use of predefined functions as $BioMassAction$. To avoid such restrictions and eventually improve performance and memory consumption, we are going to support extended arc types, including the inhibitor arcs.

There are other stochastic Petri net tools, offering numeric analysis techniques as transient analysis which is the key to CSL model checking, e.g. SMART

[CJMS06] and Möbius [PCS07]. We will also compare our tool with them, including technical networks in a representative benchmark suite as well.

**Acknowledgements.** We appreciate the support by the PRISM tool, which we use in our back-to-back testing process as golden prototype.

Snoopy's export to PRISM has been implemented by Fei Liu, who is funded by the FMER (BMBF), funding number 0315449H. The export offers various variable ordering options for comparison and teaching purposes.

# References

[ASSB00]  Aziz, A., Sanwal, K., Singhal, V., Brayton, R.: Model checking continuous time Markov chains. ACM Trans. on Computational Logic 1(1) (2000)

[BHHK00]  Baier, C., Haverkort, B., Hermanns, H., Katoen, J.-P.: Model checking Contiuous-Time Markov Chains by transient Analysis. In: Emerson, E.A., Sistla, A.P. (eds.) CAV 2000. LNCS, vol. 1855, pp. 358–372. Springer, Heidelberg (2000)

[BL00]  Barkai, N., Leibler, S.: Biological rhythms: Circadian clocks limited by noise. Nature 403, 267–268 (2000)

[CDDS06]  Cerotti, D., D'Aprile, D., Donatelli, S., Sproston, J.: Verifying stochastic well-formed nets with CSL model checking tools. In: Proc. ACSD 2006, pp. 143–152. IEEE Computer Society, Los Alamitos (2006)

[CJMS06]  Ciardo, G., Jones III, R.L., Miner, A.S., Siminiceanu, R.I.: Logic and stochastic modeling with smart. Perform. Eval. 63(6), 578–608 (2006)

[CSK+03]  Cho, K.-H., Shin, S.-Y., Kim, H.-W., Wolkenhauer, O., McFerran, B., Kolch, W.: Mathematical modeling of the influence of RKIP on the ERK signaling pathway. In: Priami, C. (ed.) CMSB 2003. LNCS, vol. 2602, pp. 127–141. Springer, Heidelberg (2003)

[CVOG05]  Calder, M., Vyshemirsky, V., Orton, R., Gilbert, D.: Analysis of Signalling Pathways using the PRISM model checker. In: Proc. CMSB 2005, pp. 179–190. LFCS, Univ. of Edinburgh (2005)

[GH06]  Gilbert, D., Heiner, M.: From Petri nets to differential equations - an integrative approach for biochemical network analysis. In: Donatelli, S., Thiagarajan, P.S. (eds.) ICATPN 2006. LNCS, vol. 4024, pp. 181–200. Springer, Heidelberg (2006)

[GHL07]  Gilbert, D., Heiner, M., Lehrack, S.: A unifying framework for modelling and analysing biochemical pathways using Petri nets. In: Calder, M., Gilmore, S. (eds.) CMSB 2007. LNCS (LNBI), vol. 4695, pp. 200–216. Springer, Heidelberg (2007)

[HDG10]  Heiner, M., Donaldson, R., Gilbert, D.: Petri Nets for Systems Biology. In: Iyengar, M.S. (ed.) Symbolic Systems Biology: Theory and Methods, Jones and Bartlett Publishers, Inc. (in press, 2010)

[HGD08]  Heiner, M., Gilbert, D., Donaldson, R.: Petri nets for systems and synthetic biology. In: Bernardo, M., Degano, P., Zavattaro, G. (eds.) SFM 2008. LNCS, vol. 5016, pp. 215–264. Springer, Heidelberg (2008)

[HRS08]  Heiner, M., Richter, R., Schwarick, M.: Snoopy - a tool to design and animate/simulate graph-based formalisms. In: Proc. PNTAP 2008, associated to SIMUTools 2008. ACM digital library, New York (2008)

[HST09]     Heiner, M., Schwarick, M., Tovchigrechko, A.: DSSZ-MC - A Tool for Symbolic Analysis of Extended Petri Nets. In: Franceschinis, G., Wolf, K. (eds.) Petri Nets 2009. LNCS, vol. 5606, pp. 323–332. Springer, Heidelberg (2009)

[JKO+08]    Jansen, D.N., Katoen, J.-P., Oldenkamp, M., Stoelinga, M., Zapreev, I.: How fast and fat is your probabilistic model checker? In: Yorav, K. (ed.) HVC 2007. LNCS, vol. 4899, pp. 69–85. Springer, Heidelberg (2008)

[KNP08]     Kwiatkowska, M., Norman, G., Parker, D.: Using probabilistic model checking in systems biology. ACM SIGMETRICS Performance Evaluation Review 35(4), 14–21 (2008)

[LBS00]     Levchenko, A., Bruck, J., Sternberg, P.W.: Scaffold proteins may biphasically affect the levels of mitogen-activated protein kinase signaling and reduce its threshold properties. Proc. Natl. Acad. Sci. USA 97(11), 5818–5823 (2000)

[MC99]      Miner, A.S., Ciardo, G.: Efficient Reachability Set Generation and Storage Using Decision Diagrams. In: Donatelli, S., Kleijn, J. (eds.) ICATPN 1999. LNCS, vol. 1639, pp. 6–25. Springer, Heidelberg (1999)

[Noa99]     Noack, A.: A ZBDD Package for Efficient Model Checking of Petri Nets (in German). Technical report, BTU Cottbus, Dep. of CS (1999)

[Par02]     Parker, D.: Implementation of Symbolic Model Checking for Probabilistic Systems. PhD thesis, University of Birmingham (2002)

[PCS07]     Peccoud, J., Courtney, T., Sanders, W.H.: Möbius: an integrated discrete-event modeling environment. Bioinformatics 23(24), 3412–3414 (2007)

[PNK06]     Parker, D., Norman, G., Kwiatkowska, M.: PRISM 3.0.beta1 Users' Guide (2006)

[Sno08]     Snoopy Website. A Tool to Design and Animate/Simulate Graphs. BTU Cottbus (2008), http://www-dssz.informatik.tu-cottbus.de/software/snoopy.html

[Ste94]     Stewart, W.J.: Introduction to the Numerical Solution of Markov Chains. Princeton Univ. Press, Princeton (1994)

[Tov08]     Tovchigrechko, A.: Model Checking Using Interval Decision Diagrams. PhD thesis, BTU Cottbus, Dep. of CS (2008)

# Qualitative Transition Systems for the Abstraction and Comparison of Transient Behavior in Parametrized Dynamic Models

Hayssam Soueidan[1], Grégoire Sutre[2], and Macha Nikolski[2]

[1] LaBRI/ENSEiRB, Université Bordeaux I, 351 crs Liberation, 33405 Talence, France
[2] CNRS/LaBRI, Université Bordeaux I, 351 crs Liberation, 33405 Talence, France

**Abstract.** Quantitative models in Systems Biology depend on a large number of free parameters, whose values completely determine behavior of models. These parameters are often estimated by fitting the system to observed experimental measurements and data. The response of a model to parameter variation defines qualitative changes of the system's behavior. The influence of a given parameter can be estimated by varying it in a certain range. Some of these ranges produce similar system dynamics, making it possible to define general trends for trajectories of the system (e.g. oscillating behavior) in such parameter ranges. Such trends can be seen as a qualitative description of the system's dynamics within a parameter range. In this work, we define an automata-based formalism to formally describe the qualitative behavior of systems' dynamics. Qualitative behaviors are represented by finite transition systems whose states contain predicate valuation and whose transitions are labeled by probabilistic delays. Biochemical system' dynamics are automatically abstracted in terms of these qualitative transition systems by a random sampling of trajectories. Furthermore, we use graph theoretic tools to compare the resulting qualitative behaviors and to estimate those parameter ranges that yield similar behaviors. We validate this approach on published biochemical models and show that it enables rapid exploration of models' behavior, that is estimation of parameter ranges with a given behavior of interest and identification of some bifurcation points.

## 1 Introduction

Dynamic models in System Biology rely on kinetic parameters to represent the range of possible behaviors when enzymatic information is incomplete. Analysis of these parametrized models aims at the identification of parameter ranges yielding similar qualitative behaviors, or of parameter values yielding a given behavior of interest. Qualitative transient behavior can be successfully analyzed by model checking algorithms applied to models admitting a computable path semantics. However, in Systems Biology state explosion and negative decidability results limit the scope of model checking to a certain subset of models. Moreover, some published and curated Systems Biology models lack explicit semantics. Little can be assumed for these "black box" models, except the possibility of simulation.

P. Degano and R. Gorrieri (Eds.): CMSB 2009, LNBI 5688, pp. 313–327, 2009.

Mining these simulation results to identify parameter regions yielding similar behaviors is hindered by the size of the parameter space to explore, numerical artifacts and the lack of formal definition of what it means for simulation results to be similar.

In this paper, we propose the new formalism of *qualitative transition systems* for abstracting simulation results in terms of discrete objects that admit efficient similarity measures.

Indeed, simulation results for ODEs are obtained using numerical integration schemes operating on floating point numbers. The resulting approximation is problematic for the identification of precise transient properties since transient properties of interest are mathematically by equality between real numbers (e.g. $f'(x) = 0$ is necessary for a local maximum) which is inconsistent in floating point arithmetic[1]. Consequently, even analysis of basic properties (like the detection of the first time a deterministic system is in a previously visited state) fail in practice due to this inconsistency. Furthermore, different integration schemes ($n$-th order, implicit/explicit) yield different and incomparable numerical approximations of the same trajectory. Although using normalized sampling and fixed precision decimal numbers seem to solve this problem, the multiplicity of time scales in ODEs show that this solution is not completely satisfying.

For dynamic models admitting a computable path semantics, the impact of numerical artifacts is absent. Indeed, it is possible to compute a finite description of the set of trajectories of the model. Consequently, for these models, model checking algorithms can decide if a logical representation of a behavior holds, and if not, can provide a counter example. Recently, a probabilistic model checking approach was successfully used to solve the inverse problem: given a logical representation of a transient behavior, return a parameter space in which any trajectory satisfies the specified behavior with sufficiently high probability[2]. For dynamic models suitable for model checking, the intuitive notion of "similar behavior" is thus fully formalized and generally decidable.

**Contributions.** In the next section we introduce *Qualitative Transition Systems* (QTS) and define their probabilistic semantics. A novel abstraction operation is defined in section 3 with the goal of building QTSs from simulation results. We then show in section 4 that when constructing a QTS from an ODE, the QTS construction can be made independent of the numerical integration scheme. In section 5, we show that trajectory comparison using QTS can be made more resistant to noise by detecting points of interest (extremums and inflection) through the construction of a piecewise linear approximation (PLA). In section 6, we validate our approach on models from literature.

## 2    Qualitative Transition Systems

Given a set $\Sigma$, we denote by $\Sigma^*$ the set of all (finite) *words* $s_0 \cdots s_k$ over $\Sigma$. A (finite) *timed word* over $\Sigma$ is any word $W = (t_0, s_0) \cdots (t_k, s_k) \in (\mathbb{R}_{\geq 0} \times \Sigma)^*$ such that $t_i < t_{i+1}$ for $i \in [0, k)$. The nonnegative real numbers $t_i$ are interpreted as the absolute *observation times* and the $s_i$ are the *observed values*. We will focus in the paper on the particular case of $\Sigma = \mathbb{R}^n$, where observed values are vectors of reals. In this case, timed words are called (multivariate) *time series*, and are denoted $S = (t_0, \boldsymbol{x_0}) \cdots (t_k, \boldsymbol{x_k})$.

We define a *Qualitative Transition System* (QTS) in the following way.

**Definition 1.** *A* qualitative transition system *is a tuple* $\mathcal{A} = \langle Q, E, \mu, \sigma, w \rangle$ *where $Q$ is a finite set of set of qualitative states, $E \subseteq Q \times Q$ is a finite set of transitions, $\mu, \sigma : E \to \mathbb{R}$ are* mean *and* standard deviation *labelings, and* $w : E \to \mathbb{N}$ *is a* weight *labeling.*

For any transition $e \in E$, $\mu(e)$ and $\sigma(e)$ are respectively interpreted as being the mean and the standard deviation of a normal distribution that is followed by a random variable called *sojourn time*. The weight labeling $w$ induces probabilities for transitions. Formally, the *transition probability* labeling $p : E \to [0, 1]$ induced by $w$ is defined by

$$p(q, q') = \frac{w(q, q')}{\sum_{(q, q'') \in E} w(q, q'')}.$$

A QTS is thus a transition system where each transition is labeled with the amount of time the system is idle before moving to another state. The delay between two state changes follows a parametrized normal distribution. This has to be contrasted with continuous Markov chains, where the sojourn time in a state must be exponentially distributed (see e.g. [3] for a complete definition).

Suppose that a QTS is in the state $q$, and that there exists an outgoing transition $e = (q, q')$. The probability of moving from state $q$ to state $q'$ is $p(e)$, the transition probability of $e$. Suppose that the transition $e$ is selected in favor of other outgoing transitions; the system will stay in the state $q$ for a delay that is normally distributed with mean $\mu(e)$ and with standard deviation $\sigma(e)$. Let $X$ be such a normally distributed random variable that denotes the sojourn time in the state $q$, and let $F_X$ be its cumulative distribution function. The probability to move from $q$ to $q'$ between $t_1$ and $t_2$ time units is thus given by $F_X(t_2) - F_X(t_1)$. Contrary to the standard semantics of continuous time Markov chains, our semantics does not involve a race condition. That is, in a given state, the probability for the successor state is not conditioned by the delays but solely by the transition weights, similarly to a discrete time Markov Chain.

# 3 Abstraction of a Time Series in Terms of Qualitative Transition Systems

## 3.1 Abstraction of a Time Series in Terms of Timed Words

In order to represent a real-valued trajectory as an abstract-valued trajectory, each concrete observation $(t, \boldsymbol{x})$ of a time series $S$ is transformed into an abstract observation $(t, a)$ where the observation time $t$ is unchanged and $a$ is an abstract value in a finite domain $A$ called the *abstract domain*. The rationale behind abstraction is that two concrete observations that are transformed into the same abstract observation are assumed indistinguishable w.r.t. qualitative properties.

Formally, an *abstraction function* is any function $\alpha : \mathbb{R}^n \to A$ where $A$ is a finite domain. For any time series $S = (t_0, \boldsymbol{x_0}) \cdots (t_k, \boldsymbol{x_k})$ in $(\mathbb{R}_{\geq 0} \times \mathbb{R}^n)^*$, the abstraction of $S$ is the timed word $\alpha(S) = (t_0, \alpha(\boldsymbol{x_0})) \cdots (t_k, \alpha(\boldsymbol{x_k}))$ in $(\mathbb{R}_{\geq 0} \times A)^*$. Note that abstraction functions may be combined by the cartesian product. In practice, it is often desirable to use multiple-arity abstraction functions that are defined on a fixed-width "window" of observations, i.e, functions $(\mathbb{R}^n)^{d+1} \to A$ where $d \in \mathbb{N}$ is the window width.

For such a function $\alpha$, the abstraction of $S$ would be defined as the timed word $\alpha(S) = (t_d, \alpha(\boldsymbol{x_0}, \ldots, \boldsymbol{x_d})) \cdots (t_k, \alpha(\boldsymbol{x_{k-d}}, \ldots, \boldsymbol{x_k}))$. Observe that $\alpha(S) = \alpha(S')$ where $S' = (t_d, (\boldsymbol{x_0}, \ldots, \boldsymbol{x_d})) \cdots (t_k, (\boldsymbol{x_{k-d}}, \ldots, \boldsymbol{x_k}))$ is a time series over $\mathbb{R}^{nd}$. For simplicity, and without loss of generality, we only formalize our approach for unary abstraction functions (with zero width).

For example, the *sign* function can abstract a time series into a timed word over the domain domain $\{-, 0, +\}$. The *rank* function can abstract a timed word over the domain $\{1, ..., n!\}$ by mapping to each component of a vector its index in the corresponding sorted vector. For example, $sort(11, -2, 1, 2) = (-2, 1, 2, 11)$ therefore $rank(11, -2, 1, 2) = (4, 1, 2, 3)$. In the same way, the sign of the first (resp. second) derivative can distinguish between intervals where the time series is increasing (resp. rapidly increasing) or decreasing (resp. rapidly decreasing). Abstracting the value of first derivative (resp. second derivative) requires two points (resp. three points).

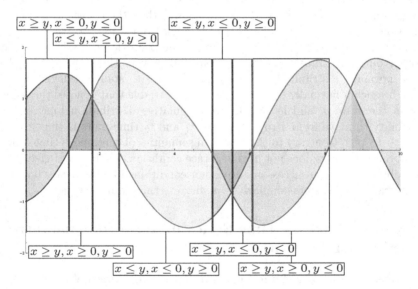

**Fig. 1.** Decomposition of a limit cycle of a two variables $(x, y)$ system. By considering the sign and rank of the variables, we map an abstract value (here denoted by a formula) to each point of the trajectory. Boxes in the figure encompass successive points that are mapped to the same abstract value. Successive points of the trajectory are collapsed whenever they have the same abstract value, that is whenever they have the same sign and rank. This trajectory can thus be abstracted as a seven state QTS.

Since the abstract domain is finite, it is often the case that the abstract time series $\alpha(S)$ has successive observations that are equal. These repeated observations are removed by collapsing them. Formally, for any timed word $W = (t_0, a_0) \cdots (t_k, a_k)$, let $collapse(W)$ be the timed word $(t_{i_0}, a_{i_0}) \cdots (t_{i_h}, a_{i_h})$ where $i_0 < \cdots < i_h$ are such that $i_0 = 0$ and $a_{i_j} = a_{i_j+1} = \cdots = a_{i_{j+1}-1} \neq a_{i_{j+1}}$ for every $0 \leq j < h$. Observe that collapsing is idempotent: for any timed word $W$, it holds that $collapse(W) = collapse(collapse(W))$. The *reduced abstraction* of any time series $S$ is then defined as the timed word $collapse(\alpha(S))$.

## 3.2   Abstraction of a Timed Word in Terms of Qualitative Transition System

The abstraction of a time series in terms of timed words abstracts the value component of the time series. In order to adequately compare two timed words, we also need to abstract the time of observations. Consider a timed word $W = (t_0, a_0) \cdots (t_k, a_k)$ over an abstract domain $A$. To qualitatively abstract this timed word, it is represented as a transition system by considering that for any $i \in [0, k)$, the pair $((t_i, a_i), (t_{i+1}, a_{i+1}))$ of successive abstract observations of $W$ is induced by a timed transition $a_i \rightarrow a_{i+1}$ between two *states* of a transition system with a delay of $t_{i+1} - t_i$. We can then consider the set of all transitions between two given states. From such a set of transitions with identical source and target, we suppose that the delays are approximately normal, and thus estimate the mean and standard deviation of the supposed underlying normal distribution. In this way, the set of concrete transitions can be abstracted by a single stochastic transition in a qualitative transition system. Formally, a timed word is abstracted in terms of QTS with the following definition. For any finite subset $X \subseteq \mathbb{R}$, we denote by $\mathbf{E}[X]$ the *mean* of $X$ and by $\mathbf{V}[X]$ its *variance*.

**Definition 2.** *The* QTS *abstraction of a timed word* $W = (t_0, a_0) \cdots (t_k, a_k)$ *over* $A$ *is the qualitative transition system* $\mathcal{A} = \langle Q, E, \mu, \sigma, w \rangle$ *with*

$$Q = \{a_i \mid 0 \leq i \leq k\} \qquad\qquad \mu(q, q') = \mathbf{E}[\Delta(q, q')]$$
$$E = \{(a_i, a_{i+1}) \mid 0 \leq i < k \wedge a_i \neq a_{i+1}\} \qquad \sigma(q, q') = \sqrt{\mathbf{V}[\Delta(q, q')]}$$
$$w(q, q') = |\Gamma(q, q')|$$

*where for any* $(q, q') \in E$, $\Gamma(q, q')$ *is the set of pairs* $(i, j)$ *with* $0 \leq i < j \leq k$ *such that* $a_i = q$, $a_j = q'$, *and* $a_{i-1} \neq a_i = a_{i+1} = \cdots = a_{j-1}$, *and* $\Delta(q, q')$ *is the multiset defined by* $\Delta(q, q') = \{t_j - t_i \mid (i, j) \in \Gamma(q, q')\}$.

Note that in the definition, the set $\Gamma(q, q')$ contains pairs of indices $(i, j)$ such that all observations between $i$ and $j$ are removed by collapsing. Therefore, any two timed words $W$ and $W'$ over $A$ satisfying $collapse(W) = collapse(W')$ have the same QTS abstraction.

# 4   Abstraction of the Transient Behavior of Deterministic Parametrized Models

Deterministic parametrized models, such as ODE systems, can exhibit different qualitative behaviors depending on the value of the parameters. When these

systems admit a simulation algorithm (e.g. numerical integration), they generate
time series. We show in this section that under assumptions concerning the
simulation algorithms, the properties of interest of a given system are preserved
by the abstraction in terms of qualitative transition systems.

**Sampling independence.** In the context of time series obtained by sampling,
the definition of QTS obviously depends on the precision of the sampling. However, we show that for convex abstraction functions if the sampling is "precise
enough" then the QTS obtained from any oversampling has the same transitions
(but with more precise delay distributions). We first introduce additional notations. For any two vectors $x, y \in \mathbb{R}^n$, we denote by $\overline{xy}$ the open line segment
between $x$ and $y$, formally $\overline{xy} = \{\lambda x + (1 - \lambda) y \mid \lambda \in \mathbb{R}, 0 < \lambda < 1\}$. A time
series $S = (t_0, x_0) \cdots (t_k, x_k)$ is a *sampling* of a (partial) function $f : \mathbb{R}_{\geq 0} \to \mathbb{R}^n$
if $x_i = f(t_i)$ for every $i \in [0, k]$. Given a time series $S = (t_0, x_0) \cdots (t_k, x_k)$, we
denote by $\Lambda_S : [t_0, t_k] \to \mathbb{R}^n$ its linear interpolation, defined by: $\Lambda_S(t_i) = x_i$
for each $0 \leq i \leq k$, and $\Lambda_S(t) = \frac{t-t_i}{t_{i+1}-t_i} x_{i+1} + \frac{t_{i+1}-t}{t_{i+1}-t_i} x_i$ for each $0 \leq i < k$
and $t_i < t < t_{i+1}$. Given an abstraction function $\alpha : \mathbb{R}^n \to A$, we say that $\alpha$ is
*convex* if $\alpha^{-1}(a)$ is a convex subset of $\mathbb{R}^n$ for every $a \in A$.

We consider for the remainder of this section a convex abstraction function
$\alpha : \mathbb{R}^n \to A$. When the sampling is precise enough, the linear interpolation
$\Lambda_S$ is often used in practice, in place of the "real trajectory". We formalize this
notion of precision with respect to the abstraction function. A time series $S =
(t_0, x_0) \cdots (t_k, x_k)$ is called $\alpha$-*adequate* if $\alpha(x) \in \{\alpha(x_i), \alpha(x_{i+1})\}$ for every $i \in
[0, k)$ and $x \in \overline{x_i x_{i+1}}$. It is called $\alpha$-*loose* otherwise. Intuitively, when $S$ is
$\alpha$-adequate, the abstraction function $\alpha$ along $\overline{x_i x_{i+1}}$ is either constant, or is
first equal to $\alpha(x_i)$ on $\overline{x_i z}$ and then equal to $\alpha(x_{i+1})$ on $\overline{z x_{i+1}}$, for some
$z \in \overline{x_i x_{i+1}}$. Indeed, if $\alpha(\overline{x_i x_{i+1}}) \subseteq \{a, b\}$ with $a = \alpha(x_i)$ and $b = \alpha(x_{i+1})$
being distinct, then, by convexity of $\alpha$, the segment $\overline{x_i x_{i+1}}$ is partitioned into
the two convex sub-segments $\overline{x_i x_{i+1}} \cap \alpha^{-1}(a)$ and $\overline{x_i x_{i+1}} \cap \alpha^{-1}(b)$, and $z$ is at
the boundary between these two sub-segments. Therefore, an $\alpha$-adequate time
series $S$ captures all changes of $\alpha$ along its linear interpolation $\Lambda_S$. However,
these changes are captured up to the precision $t_{i+1} - t_i$ of the sampling, which
leads us to the following definition. The $\alpha$-*fitting* of an $\alpha$-adequate time series
$S = (t_0, x_0) \cdots (t_k, x_k)$ is the abstract timed word $\widehat{S} = (\hat{t}_0, \hat{a}_0) \cdots (\hat{t}_k, \hat{a}_k)$ defined
by $\hat{a}_i = \alpha(x_i)$, $\hat{t}_0 = t_0$, and

$$\hat{t}_{i+1} = \begin{cases} t_{i+1} & \text{if } \alpha(x_i) = \alpha(x_{i+1}) \\ \inf \{t \mid t \geq t_i \wedge \alpha(\Lambda_S(t)) = \alpha(x_{i+1})\} & \text{otherwise.} \end{cases}$$

**Proposition 1.** *For any $\alpha$-adequate time series $S = (t_0, x_0) \cdots (t_k, x_k)$, the
respective QTS abstractions $\langle Q, E, \mu, \sigma, w \rangle$ and $\langle \widehat{Q}, \widehat{E}, \widehat{\mu}, \widehat{\sigma}, \widehat{w} \rangle$ of $\alpha(S)$ and $\widehat{S}$
satisfy $Q = \widehat{Q}$, $E = \widehat{E}$, $w = \widehat{w}$. Moreover, letting $\Delta = \max\{t_i - \hat{t}_i \mid 0 \leq i \leq k\}$,
$|\mu(e) - \widehat{\mu}(e)| \leq \Delta$ and $\sigma^2(e) \leq \widehat{\sigma}^2(e) + 4\Delta^2 + 4\Delta\widehat{\sigma}(e)\sqrt{\widehat{w}(e)}$ for every $e \in E$.*

Note that in the above proposition, we have $\Delta \leq \max\{t_{i+1} - t_i \mid 0 \leq i < n\}$.
Hence, the error on $\mu$ (w.r.t. to the $\alpha$-fitted one) is bounded by the sampling
period.

An *oversampling* of $S$ is any sampling $U = (u_0, \boldsymbol{y_0}) \cdots (u_l, \boldsymbol{y_l})$ of $\Lambda_S$ such that $t_0 \cdots t_k$ is a subsequence of $u_0 \cdots u_l$. It follows from the definitions that any oversampling of an $\alpha$-adequate time series $S$ is also $\alpha$-adequate. Note that, informally, the $\alpha$-fitting of an oversampling of $S$ is an "abstract oversampling" of the $\alpha$-fitting of $S$. According to Proposition 1, the QTS obtained by oversampling $S$ is equal to the QTS obtained from $S$, except for the imprecision on $\mu$ and $\sigma$. This shows that for qualitative analysis there is little to be gained by oversampling: $\alpha$-adequate time series are sufficient.

**Periodic orbits detection.** Oscillations are ubiquitous qualitative behaviors found in systems with a feedback loop. Although bifurcation analysis provides numerical methods to establish the presence of periodic orbits for ODEs, these methods cannot be applied to a general deterministic system such as an ODE with events. However, we show in this section that a QTS can be used efficiently to estimate the likelihood of a periodic orbit in a time series.

Under an adequate abstraction function, a QTS that abstracts the transient behavior of a system with a periodic orbit has cycles in its transition relation. Consider a QTS obtained by applying the abstraction function $\alpha$ to an $\alpha$-adequate time series $S$ obtained by sampling a continuous function $f : t \to \mathbb{R}^n$. By definition, $f$ admits an orbit if and only if there exists a time point $t$ and a period $\pi$ such that $f(t) = f(t + \pi)$. Furthermore, $f$ admits a periodic and non constant orbit if and only if there exists an intermediate time step $t' < t + \pi$ such that $f(t') \neq f(t + \pi)$. Since $S$ is adequately sampled for $\alpha$, there exist at least three successive different values in $collapse(\alpha(S))$ and consequently the resulting QTS has at least a cycle of length 1.

Since equality between the real numbers and their floating point approximation is not coherent, detection of periodic orbits for a time series must rely on estimations. To find a periodic orbit in a time series $S$ it is sufficient to find a period $\pi \in \mathbb{R}_{\geq 0}$ such that there exist two elements $(t_i, \boldsymbol{x_i}), (t_j, \boldsymbol{x_j}) \in S$ such that $(t_j, \boldsymbol{x_j}) \approx (t_i + \pi, \boldsymbol{x_i})$ for an adequate approximation relation $\approx$. However, for ODE systems integrated with an adaptive time step algorithm this scheme produces mainly false positives (successive integration steps in a quasi steady region of an ODE) and false negatives (regions with high variability).

The existence of a periodic orbit of period $\pi$ also implies that for any value $k \in \mathbb{N}, f(t) = f(t + k * \pi)$. Thus, if the system reaches a periodic orbit at point $l$, then the nearest points (according to an euclidean distance on $\mathbb{R}^n$) of $l$ contain points from all possible periods.

Therefore, we estimate the likelihood of a periodic orbit by considering a point $l = (t_l, \boldsymbol{x_l})$ of $S$ that we suppose being in the periodic orbit, and a set of sample points $P$ from $S$ such that for any point $p' = (t_{p'}, \boldsymbol{x_{p'}})$ in $S - P$, for any point $p = (t_p, \boldsymbol{x_p})$ in $P$, we have $| x_{p'} - x_l | > | x_p - x_l |$. Less formally, $P$ is a set containing the points that are the nearest to $l$ w.r.t. the euclidean distance. The likelihood $\mathcal{L}((t_l, x_l), \pi, P)$ of $\pi$ being the period of the orbit of $x_l$ given a sample $P$ of neighbors of $x_l$ is then defined by

$$\mathcal{L}((t_l, x_l), \pi, P) = \left( \sum_{\delta \in \Delta} ([\delta] - \delta)^2 \right)^{-1}$$

with $\Delta = \{(t_p - t_l)/\pi \mid t_p \in P\}$ and $[\delta]$ being the integer part of $\delta$.

Finding the period $\pi$ that maximizes $\mathcal{L}$ is difficult in practice, since this function admits local maxima that are far from the global maximum. However, the sum of the mean of the longest cycle containing the last observation in a QTS provides a good initial guess of this period. (See the case study 6.2).

## 5    Accounting for Noise by Comparing Critical Points

Qualitative transition systems can capture the dynamics of a time series, even if the time series contains numerical errors that are only local. In the case of time series admitting global noise, abstraction functions that were adequate for a smooth time series may not be resistant to noise and can generate a QTS that inadequately captures the dynamics of noise. For example, abstracting with the sign of the first derivate can adequately detect oscillations[2] but fails for time series even with little noise. Although moving average can smooth a time series and seem to circumvent this problem, the size of the window must be fixed *a priori* and this approach is thus neither general nor adaptive.

We propose here an adaptive approach to capture the most important points w.r.t. the shape of a time series. The *critical points* of continuous function $f :$ $\mathbb{R} \rightarrow \mathbb{R}$ are the set of points where $f'(x) = 0$. These are points where the function $f$ either has a peak and changes direction (local or global extremum) or presents a curvature change (inflection points). In both cases the shape of $f$ changes around the point. We generalize this definition to time series in the following way.

**Definition 3.** *The* critical point *of a time series* $S = (t_0, x_0) \cdots (t_k, x_k)$ *is the point* $(t_c, x_c) \in S$ *maximizing the function* $\Lambda(t_c, x_c) = |x_c - x_0| + (t_c - t_0) * (x_k - x_0)/(t_k - t_0)$.

The critical point of a time series is the point of maximal distance with the linear interpolation between the first and last points of the series. In a numerical context, this point is uniquely defined.

A critical point splits the time series in two time series. Since a critical point is also defined for these series, we can recursively approximate a time series by considering a piecewise function which is linear between critical points.

**Definition 4.** *The* piecewise linear approximation of order $i$ *(hereafter PLA) of a time series is the piecewise linear function on the intervals* $I_0, ..., I_k$ *where any interval* $I_j$ *has a lower bound (resp. upper bound) corresponding to the location of the* $j^{th}$ *(resp. $j + 1$) critical point.*

In order to compute the PLA of a time series, we define the piecewise linear interpolations of a set of points as the union of the linear interpolation between two successive points. The computation of the PLA of order $i$ is then performed as follows.

**PLA**$(S, i)$ returns a list of critical points of $S = (t_0, \boldsymbol{x_0}) \cdots (t_k, \boldsymbol{x_k})$ for each dimension of $S$.

1. For each dimension $d \in [0, \dim(\boldsymbol{x})]$,
   (a) Initialize the critical points with the first and the last point of $S$ projected on the dimension $d : C_d \leftarrow \{(t_0, \pi_d(x_0)), (t_k, \pi_d(x_k))\}$.
   (b) While $| C_d | < i$,
      i. Compute the linear interpolation between each successive pair of $C_d$: Let $\Lambda_j(t) = \frac{t - t_j}{t_{j+1} - t_j} \boldsymbol{x_{j+1}} + \frac{t_{j+1} - t}{t_{j+1} - t_j} \boldsymbol{x_j}$, where $(t_j, \boldsymbol{x_j})$ and $(t_{j+1}, \boldsymbol{x_{j+1}})$ are two successive points in $C_d$
      ii. Build the piecewise linear interpolation: let $\Lambda(t) = \Lambda_j(t)$ for $t \in [t_j, t_{j+1}]$,
      iii. Let $(t_c, \boldsymbol{x_c}) = \mathrm{argmax} \{\pi_d(x_c) - \Lambda(t_c) \mid (t_c, \boldsymbol{x_c}) \in S\}$,
      iv. Insert $(t_c, \pi_d(\boldsymbol{x_c}))$ in $C_d$ s.t. $C_d$ remains sorted w.r.t. the first component
2. Return $\{C_d \mid d \in [0, \dim(\boldsymbol{x})]\}$

The previous algorithm cannot append a point twice to the list of critical points. Indeed, once a point is appended to the list it becomes a bound of the piecewise linear interpolation that is used for determining the next critical point. Consequently, at this point the distance between the next piecewise linear interpolation and the time series is 0, and the distance can not be maximized. Note that this does not hold for the piecewise linear regression. Which implies that, for any unidimensional time series, the segmented linear regression of $i$ intervals

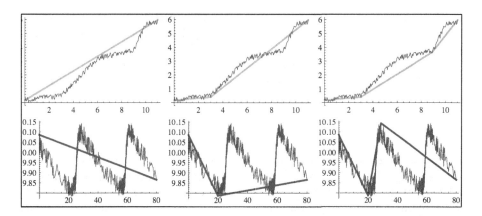

**Fig. 2.** Example of piecewise linear approximation applied to two randomly generated noisy time series. The first of the three successive plots represents the time series while the last two plots represent the result of the first two iterations of the PLA algorithm. After two iterations, the PLA algorithm selects two points (as well as the first and last point) that are considered as being representative of the global shape of the time series.

minimizing the residuals with the time series can be obtained by considering the critical points as bounds of the interval.

**Examples of critical points.** Critical points are highly related to the shape of the time series. Consider for example the sigmoid shape: a simple shape descriptor may simply specify that we start from a low plateau, follow an almost vertical increase before reaching another high plateau. Such a sigmoid shape exhibits two critical points, one at the end of the first (low) plateau, and one at the start of the second (high) plateau. Similarly, consider an oscillatory time series such as the one depicted in figure 2: its critical points contain the successive highest local maximum and the lowest local minimum.

# 6    Case Studies and Experimental Results

In this section we show how our approach can be used in practice by solving four problems related to qualitative behavior analysis. Although each problem and solution is illustrated on a specific model, the methods that are used are general-purpose. All the models used in this section were downloaded from the BioModels database[4] in the SBML V2 L1 format[5], simulated using MathSBML[6] and were used without any modification. Simulations were performed on an Intel Core2 3,2GH personal computer and each algorithm was allowed to run for at most five minute. If parameter values are not specified in the case studies, it means that those provided in the SBML file were used.

## 6.1    Searching a Trajectory with a Given Periodic Orbit

The first model we consider is a model of the cell cycle based on the interactions between the cyclin dependent kinase cdc2 and cyclin [7]. The model is comprised of six variables and ten parameters. We consider the following problem. Given the representative trajectory and its associated parameters described in the original article (left in figure 3), what kind of similar trajectories can be found in the whole parameter space ?

Abstracting the behavior of the left figure with a rank abstraction function yields the QTS depicted in the right part of figure 3. Notice the highlighted non deterministic states. These states and transitions are due to numerical errors and happen while the system reaches its periodic orbit. Consequently, the weights of the outgoing highlighted transitions are 1 while the incoming transitions are 17. All other transitions in the single cycle of the QTS have a weight of 18.

We obtained 500 random samples for the six parameters considered as being critical by the original author. For each parameter sample, we computed the trajectory, abstracted it in terms of QTS by applying the rank function and computed the Sorensen similarity index over the set of transitions to compare the sampled QTS with the representative QTS. Figure 4 depicts a subset of the results. Note that we chose parameter values exhibiting "similar" sustained oscillations, but of *different transient behaviors*. We then compared these results

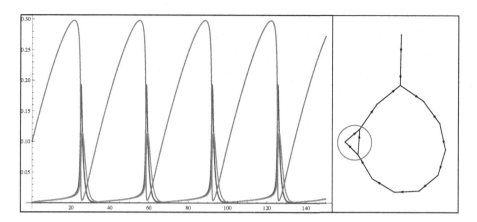

**Fig. 3.** Dynamic behavior of the cell cycle model for default parameter values. **Left**: an example of trajectory obtained by numerical integration of the Tyson cell cycle model [7]. **Right**: abstraction of this trajectory in terms of QTS by using the rank function as the abstraction function (transition labels are omitted). The total standard deviation of this QTS is 0.07. The states and transitions highlighted by the circle correspond to stochastic transitions and represent numerical integration errors.

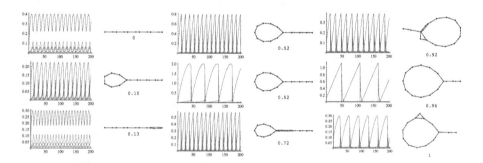

**Fig. 4.** Trajectory comparison for the Tyson cell cycle model. For 500 randomly sampled parameter spaces, we abstracted the simulated trajectory in terms of QTS by using the rank function. Trajectories were clustered in 9 bins by measuring the Sorensen similarity index between each of the 500 QTS with the QTS of the figure 3. From each of the nine clusters, a representative trajectory is depicted here together with its QTS and similarity value. These trajectories are sorted (column wise, increasing) by their similarity value.

with the one obtained with a stochastic simulation algorithm. For each simulation result, we used the PLA algorithm to reduce each noisy trajectory to its 50 most critical points, and abstracted these points in terms of QTS by applying the rank function. Trajectories similar to the one simulated with numerical integration were found for comparable parameters values.

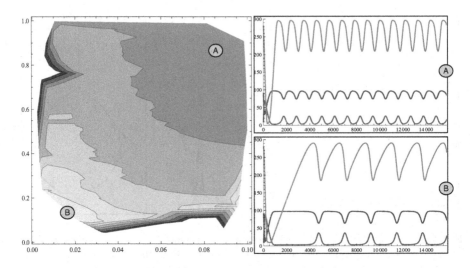

**Fig. 5.** Oscillation period for the MAPK cascade model. **Left**: contour plot representing the period of oscillations. The bottom axis (resp. left axis) represents values of the $k4$ (resp. $v5$) parameter. The contour plot was built with 500 simulations with random parameters. Regions with comparable periods are represented by a region with uniform color. **Right**: Two example trajectories exhibiting oscillations of minimal and maximal period A:1094 time units and B:2236 time units.

### 6.2   Estimating the Period of Orbits

The model of MAPK cascade from [8] describes the effect of negative feedback and ultra-sensitivity on the emergence of oscillations. We investigated the dynamics of the period of the orbits under parameter changes. To estimate these periods, we considered the parameters $\{k_4, v_5\}$ as random variables following an uniform distribution over the intervals $[0, 1]$ and $[0, 0.1]$. For parameters' sample of size 500, we abstracted the corresponding time series in terms of QTS. These QTS were then reduced by removing transitions whose probability decreased as the simulation advanced. We then approximated the orbit's period with the sum of means of the transitions of the longest cycle of the QTS. This approximation was then used in a local maximization procedure to identify the exact period value maximizing the likelihood function. In our tests, providing this initial "educated guess" of the period value to the maximization procedure yielded the global maximum in 98% of cases.

We can see from the results (figure 5) that, for this parameter subspace, oscillating behavior is very common and that the dynamics of the period does not exhibit abrupt changes.

### 6.3   Searching for Any Periodic Orbits

We consider again the MAPK cascade model [8] but with a more general objective. We consider the problem of detecting the possible oscillating behaviors and

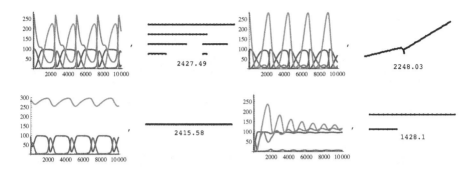

**Fig. 6.** Example trajectories of the MAPK cascade exhibiting oscillating behavior found with a random sampling of ten parameters of the MAPK cascade model [8]. On the right of each plot, the associated QTS reduced to its periodic form; under it, the period value with maximum likelihood.

of computing the probability of finding an oscillating behavior in a larger parameter subspace. The parameters of interest are $\{k_3, k_4, k_7, k_8, V_5, V_6, V_9, V_{10}\}$ and are considered as random variables following an uniform probability over the interval $[0, 1] \subset \mathbb{R}$. We built a QTS as in previous sections. In this study, only periods in the range $[200, 5000]$ with a likelihood greater than 10 were considered genuine. Although all the resulting trajectories exhibit transient or limit cycle oscillations, they follow different transient dynamics. The four example trajectories of figure 6 show a subset of the possible dynamics: each of these time series admits a specific alternation of species at their maximum concentration. Multiple instances of each of these dynamics were successfully identified by applying the method from section 6.1. The number of samples needed before finding an oscillating behavior was 57 on average. For comparison, when $k_3, k_4, k_7$ and $k_8$ were sampled in the interval $[0, 0.1]$, the average number of samples needed dropped to 10.6.

## 6.4   Searching for Given Transient Behavior in a Parameter Subspace

The extracellular signal regulated kinase (ERK) pathways plays a role in a hidden oncogenic positive feedback loop via a crosstalk with the Wnt pathway [9]. The pathological cases identified by the authors involve "an irreversible response leading to a sustained activation of both pathways". Applying our QTS construction with random samples of the $\beta$-catenin synthetic rate ($V_{12}$) yields results depicted in figure 7.

This model involves 28 species, 58 parameters, and 2 discrete events. Applying the rank abstraction yields transition systems with a state space of 600 states on average out of the possible 28! state configurations.

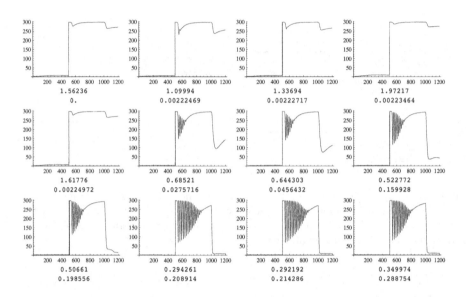

**Fig. 7.** ERK Crosstalk simulation results. From top left to bottom right, simulation results are sorted by similarity with the non pathological case (reversible activation). The sampled value of $v_{12}$ and the similarity index with the non pathological case are under each plot.

# 7   Discussion and Conclusions

In this paper, we have described the formalism of qualitative transition systems. A QTS is a transition system where each transition is labeled with the amount of time the system requires before moving to another state. The delay between two state changes follows a parametrized normal distribution. We have shown how QTS can be used to study qualitative properties of parametrized models. This is achieved by defining an appropriate abstraction function. By representing the characteristic qualitative features of a trajectory in an abstract domain that is countable, qualitative similarity can be detected by a simple equality test.

We have shown that the soundness of this approach depends on the adequacy of sampling with respect to the abstraction function. In particular, we have shown that for convex abstraction functions, if the sampling is "precise enough", then the QTS obtained from any oversampling has the same transitions. Finally, we applied this approach to some well known models. QTS were used to explore the parameter space and to detect uniform behaviors (oscillations etc.).

The limits of our approach as compared to model checking is the lack of exhaustivity. This has to be counterbalanced by the fact that our method is applicable to a large panel of formalisms, even those lacking a precise semantics. Consequently, we can avoid any model transformation. Finally, our approach can be applied independently to the data and to the model. The areas of future research for qualitative transition systems can be declined on the both technical and practical plans. As for the former, we envision a more thorough study of similarity measures of QTS and how QTS similarity relates to language equivalence. As for the latter, we plan to develop clustering techniques in order to detect the resulting behavior similarity in an experimental context.

**Funding.** This work was supported in part by the European Commission FP6 programme "Yeast Systems Biology Network" (YSBN), LSHG-CT-2005-018942.

# References

1. Chesneaux, J.: The equality relations in scientific computing. Numerical Algorithms (January 1994)
2. Rizk, A., Batt, G., Fages, F., Soliman, S.: On a continuous degree of satisfaction of temporal logic formulae with applications to systems biology. In: Heiner, M., Uhrmacher, A.M. (eds.) CMSB 2008. LNCS (LNBI), vol. 5307, pp. 251–268. Springer, Heidelberg (2008)
3. Kwiatkowska, M., Norman, G., Parker, D.: PRISM: Probabilistic symbolic model checker. In: Field, T., Harrison, P.G., Bradley, J., Harder, U. (eds.) TOOLS 2002. LNCS, vol. 2324, pp. 200–204. Springer, Heidelberg (2002)
4. Le Novere, N., Bornstein, B., Broicher, A., Courtot, M., Donizelli, M., Dharuri, H., Li, L., Sauro, H., Schilstra, M., Shapiro, B., et al.: BioModels Database: a free, centralized database of curated, published, quantitative kinetic models of biochemical and cellular systems. Nucleic Acids Research 34(Database Issue), D689 (2006)
5. Hucka, M., Finney, A., Sauro, H.M., Bolouri, H., Doyle, J.C., Kitano, H., et al.: The systems biology markup language (SBML): a medium for representation and exchange of biochemical network models. Bioinformatics 19(4), 524–531 (2003)
6. Shapiro, B., Hucka, M., Finney, A., Doyle, J.: Mathsbml: a package for manipulating sbml-based biological models. Bioinformatics 20(16), 2829–2831 (2004)
7. Tyson, J.: Modeling the cell division cycle: cdc2 and cyclin interactions. Proceedings of the National Academy of Sciences 88(16), 7328–7332 (1991)
8. Kholodenko, B.N.: Negative feedback and ultrasensitivity can bring about oscillations in the mitogen-activated protein kinase cascades. Eur. J. Biochem. 267(6), 1583–1588 (2000)
9. Kim, D., Rath, O., Kolch, W., Cho, K.: A hidden oncogenic positive feedback loop caused by crosstalk between wnt and erk pathways. Oncogene 26(31), 4571–4579 (2007)

# Author Index